I0036716

Cloud IoT Systems for Smart Agricultural Engineering

Chapman & Hall/CRC Internet of Things: Data-Centric Intelligent Computing, Informatics, and Communication

The role of adaptation, machine learning, computational Intelligence, and data analytics in the field of IoT Systems is becoming increasingly essential and intertwined. The capability of an intelligent system is growing depending upon various self-decision-making algorithms in IoT Devices. IoT based smart systems generate a large amount of data that cannot be processed by traditional data processing algorithms and applications. Hence, this book series involves different computational methods incorporated within the system with the help of Analytics Reasoning, learning methods, Artificial intelligence, and Sense-making in Big Data, which is most concerned in IoT-enabled environment.

This series focuses to attract researchers and practitioners who are working in Information Technology and Computer Science in the field of intelligent computing paradigm, Big Data, machine learning, Sensor data, Internet of Things, and data sciences. The main aim of the series is to make available a range of books on all aspects of learning, analytics and advanced intelligent systems and related technologies. This series will cover the theory, research, development, and applications of learning, computational analytics, data processing, machine learning algorithms, as embedded in the fields of engineering, computer science, and Information Technology.

Series Editors:
Souvik Pal
Global Institute of Management and Technology, India

Dac-Nhuong Le
Haiphong University, Vietnam

Security of Internet of Things Nodes: Challenges, Attacks, and Countermeasures
Chinmay Chakraborty, Sree Ranjani Rajendran and Muhammad Habib Ur Rehman

Cancer Prediction for Industrial IoT 4.0: A Machine Learning Perspective
Meenu Gupta, Rachna Jain, Arun Solanki and Fadi Al-Turjman

Cloud IoT Systems for Smart Agricultural Engineering
Saravanan Krishnan, J Bruce Ralphin Rose, N R Rajalakshmi, Narayanan Prasanth

Cloud IoT Systems for Smart Agricultural Engineering

Edited by

Saravanan Krishnan

J Bruce Ralphin Rose

N R Rajalakshmi

Narayanan Prasanth

CRC Press
Taylor & Francis Group
Boca Raton London New York

CRC Press is an imprint of the
Taylor & Francis Group, an **informa** business

A CHAPMAN & HALL BOOK

First edition published 2022
by CRC Press
6000 Broken Sound Parkway NW, Suite 300, Boca Raton, FL 33487-2742

and by CRC Press
2 Park Square, Milton Park, Abingdon, Oxon, OX14 4RN

© 2022 selection and editorial matter, Saravanan Krishnan, J. Bruce Ralphin Rose, N. R. Rajalakshmi, Narayanan Prasanth; individual chapters, the contributors

Reasonable efforts have been made to publish reliable data and information, but the author and publisher cannot assume responsibility for the validity of all materials or the consequences of their use. The authors and publishers have attempted to trace the copyright holders of all material reproduced in this publication and apologize to copyright holders if permission to publish in this form has not been obtained. If any copyright material has not been acknowledged please write and let us know so we may rectify in any future reprint.

Except as permitted under U.S. Copyright Law, no part of this book may be reprinted, reproduced, transmitted, or utilized in any form by any electronic, mechanical, or other means, now known or hereafter invented, including photocopying, microfilming, and recording, or in any information storage or retrieval system, without written permission from the publishers.

For permission to photocopy or use material electronically from this work, access www.copyright.com or contact the Copyright Clearance Center, Inc. (CCC), 222 Rosewood Drive, Danvers, MA 01923, 978-750-8400. For works that are not available on CCC please contact mpkbookspermissions@tandf.co.uk

Trademark notice: Product or corporate names may be trademarks or registered trademarks and are used only for identification and explanation without intent to infringe.

Library of Congress Cataloging-in-Publication Data

Names: Krishnan, Saravanan, 1982- editor. | Rose, J. Bruce Ralphin, editor. | Rajalakshmi, N. R., editor. | Prasanth, N. Narayanan, editor.
Title: Cloud IoT systems for smart agricultural engineering / Saravanan Krishnan, J Bruce Ralphin Rose, N R Rajalakshmi, N Narayanan Prasanth
Description: First edition | Boca Raton, FL : Chapman & Hall/CRC Press, 2022. | Includes bibliographical references and index. | Summary: "This book presents a detailed exploration of adaption and implementation of cloud IoT systems in the field of agriculture. Agro IoT bridges the gap between the conventional agricultural methods and modern technologies. Recently, cloud computing, IoT, big data, machine learning & deep learning technologies are initiated to adopt in the smart agricultural engineering. This edited book covers all the aspects of the smart agriculture with state-of-the-art Cloud IoT systems in the complete 360 degree view spectrum. This book is aimed primarily at graduates, researchers and practitioners who are engaged in agriculture engineering"-- Provided by publisher.
Identifiers: LCCN 2021040834 (print) | LCCN 2021040835 (ebook) | ISBN 9781032028279 (hardback) | ISBN 9781032028309 (paperback) | ISBN 9781003185413 (ebook)
Subjects: LCSH: Agricultural engineering. | Internet of things--Agricultural applications. | Agricultural innovations.
Classification: LCC S494.5.D3 C563 2022 (print) | LCC S494.5.D3 (ebook) | DDC 338.10285--dc23/eng/20211122
LC record available at https://lccn.loc.gov/2021040834
LC ebook record available at https://lccn.loc.gov/2021040835

ISBN: 978-1-032-02827-9 (hbk)
ISBN: 978-1-032-02830-9 (pbk)
ISBN: 978-1-003-18541-3 (ebk)

DOI: 10.1201/9781003185413

Typeset in Palatino LT Std
by KnowledgeWorks Global Ltd.

Contents

Preface.. vii

About the Editors ix

List of Contributors..xi

1. Cloud IoT Applications in Agricultural Engineering.................................1
   M. Raj Kumar, P. B. Ahir, D. Mrinmoy, and K. Utkarsh

2. Urban and Vertical Farming Using Agro-IoT Systems: The Ingredient
   Revolution – A Sustainable Production System for Urban Population 17
   K. R. Gokul Anand, S. Boopathy, T. Poornima, A. Sharmila, and E. L. Dhivya Priya

3. Sustainable Smart Crop Management for Indian Farms Using Artificial
   Intelligence 43
   S. Murugan, M. G. Sumithra, and V. Chandran

4. Smart Livestock Management Using Cloud IoT 55
   T. Vigneswari and N. Vijaya

5. GIS Systems for Precision Agriculture and Site-Specific Farming 75
   M. Kavitha, R. Srinivasan, and R. Kavitha

6. Machine Learning Approaches for Agro-IoT Systems 89
   S. A. Jadhav and A. Lal

7. A Survey on Internet of Things (IoT)-Based Precision Agriculture: Aspects
   and Technologies 109
   R. Srinivasan, M. Kavitha, R. Kavitha, and K. Saravanan

8. Novel Semantic Agro-Intelligent IoT System Using Machine Learning................ 121
   C. S. Saravana Kumar and S. Vinoth Kumar

9. Yield Prediction Based on Soil Content Analysis through Intelligent IoT
   System for Precision Agriculture.. 133
   K. Selvakumar, P. J. A. Alphonse, and L. SaiRamesh

10. Fuzzy-Based Intelligent Crop Prediction over Climate Fluctuation
    Using IoT 147
    S. Subramani, C. Chandru Vignesh, J. Alfred Daniel,
    C. B. Sivaparthipan, Balaanand Muthu, and N. Suganthi

11. Application of Drones with Variable Area Nozzles for Effective Smart
    Farming Activities... 163
    J. Bruce Ralphin Rose, V. Saravana Kumar, and V. T. Gopinathan

12. **Standards and Protocols for Agro-IoT** 187
    S. Mythili, K. Nithya, M. Krishnamoorthi, and M. Kalamani

13. **Research Issues and Solutions in Agro-IoT** 209
    N. Vijaya and T. Vigneswari

14. **Renewable Energy Sources for Modern Agricultural Trends** 223
    S. Arulvel, T. Joshva Devadas, D. Dsilva Winfred Rufuss, and M. Amutha Prabakar

15. **SPLARE: A Smart Plant Healthcare System** 243
    S. Lodha, H. Malani, N. Prasanth, and K. Saravanan

Index 263

Preface

Despite people's perception of the agricultural process, the fact is that today's agricultural industry is more data centric, reliable, and smarter than ever before. The purpose of this book is to cover the need for smart agriculture and aid in the vetting and selection of research in smart agriculture. Recently, cloud computing, IoT, big data, machine learning, and deep learning technologies have been initiated to be adopted in smart agricultural engineering. This book deals with all aspects of smart agriculture with state-of-the-art cloud IoT systems in a complete 360-degree view spectrum. It presents the rapid advancement of the technologies in the existing agri-model by applying IoT techniques. Novel architectural solutions in smart agricultural engineering are the core aspects of this book. Several use cases with IoT and smart agriculture will also be incorporated. This book can be used as a textbook and a reference book. Readers will walk away with a deep understanding of complementary features of the cloud and the IoT paradigm in smart agriculture applications.

Automated farming is an engineering problem to be solved with agro-technology. This book deals with the automation of the data collection process using agricultural drones and robots, extending remotely obtained parameters of crops, and accessing real-time data from any device at any time. Precision agriculture is the application of information technologies. This edited book comprises the concept of precision agriculture to make the practice of farming more accurate and control inter- and intrafield variability of crops. The book also presents agro-IoT tools, open-source platforms, protocols, and techniques that will help researchers and students who are interested in developing precision farming applications.

The book includes the concepts of urban and vertical farming using agro-IoT systems and renewable energy sources for modern agriculture trends. Real-world challenges, complexities in agro-IoT and its advantages will also be discussed. The wide variety of topics presented herein make this book ideal for students, academicians, and researchers who are studying and implementing cloud IoT systems for smart agricultural engineering.

About the Editors

Dr Saravanan Krishnan is a Senior Assistant Professor at the Department of Computer Science and Engineering at Anna University Regional Campus, Tirunelveli, Tamil Nadu. He has completed his ME in Software Engineering and PhD in Computer Science Engineering. His research interests include cloud computing, software engineering, the Internet of Things, and smart cities. He has published papers in 14 international conferences and 27 international journals. He has also written 14 book chapters and edited seven books with international publishers. He has done consultancy work for the Municipal Corporation and Smart City schemes. He is an active researcher and academician. Also, he is a reviewer for many reputed journals in Elsevier, IEEE, etc. He is a member of ISTE, IEI, ISCA, ACM, etc.

He has also trained and interviewed engineering college students for placement training and counseling. He has conducted many ISTE workshops in association with IIT Bombay and IIT Kharagpur. Also, he has conducted NPTEL workshops for faculty and students. He is also a coordinator for the Indian Institute of Remote Sensing (IIRS) Outreach Programme. He has delivered more than 50 guest lectures in many seminars/conferences in reputed engineering colleges.

Dr J. Bruce Ralphin Rose is faculty at the Aeronautical Engineering Department at Anna University Regional Campus, Tirunelveli, Tamil Nadu. He received his BE and ME degrees in the discipline of Aeronautical Engineering with distinction from Anna University, Chennai, India. He was also awarded a PhD degree in the field of Aerodynamics by Anna University in 2014. He has published more than 50 leading journal articles in the Aerospace Engineering archival journals (Q1 & Q2) published by Elsevier, Springer, SAGE-UK, World Scientific-Singapore, Inderscience, and Emerald UK. He is the editor of the journal, *Journal of Aircraft and Spacecraft Technology*. He has also published several book chapters on the bioinspired aerodynamic applications. He has completed two major funded projects sponsored by the Department of Biotechnology (DBT) and Tamil Nadu State Council for Science and Technology (TNSCST), India. He is a technical consultant for Buteos Aerobotics Corporation, New Delhi, which is an OEM company for drones, and Suzlon Wind Energy Ltd.

Dr Bruce has chaired many international and national conferences organized by premier institutions across the globe for the past 11 years. He is an Anna University-recognized research supervisor and has guided six PhD scholars and 101 PG scholars to date in the Aerodynamics and Aeroelasticity domains. He received the "Outstanding Reviewer" award by Elsevier Publications in 2015 and has delivered more than 100 expert talks across the country to raise the nation's aerospace potential to the next level. Dr Bruce's bioinspired aerodynamics-centric research has been continuously supported by the Government of Tamil Nadu and DBT, India.

Dr N. R. Rajalakshmi is a Professor in the Department of Computer Science and Engineering at Vel Tech Rangarajan Dr Sagunthala R&D Institute of Science and Technology, Chennai, Tamil Nadu. She has been involved in the research area of the cloud, IoT, machine learning, big data, and blockchain. She has published more than 20 leading journal articles and several book chapters in the field of the cloud, IoT, and machine learning. She is an active researcher and academician. Also, she is a reviewer for many reputed journals in Inderscience, Springer, etc.

Dr Narayanan Prasanth is an Associate Professor at the School of Computer Science and Engineering, Vellore Institute of Technology, Vellore, Tamil Nadu. He received his BTech (IT) from Pondicherry University, ME (CSE) from Anna University, India, and PhD (CSE) from MS University, Tirunelveli, Tamil Nadu. His research interest includes network switching and routing, distributed computing, and Internet of Things. He has published papers in various reputed conferences and journals. He is a life member of ISTE and a member of CSTA.

Contributors

P. B. Ahir
Agricultural and Food Engineering
    Department
IIT Kharagpur
Kharagpur, India

P. J. A. Alphonse
Department of Computer Applications
NIT Trichy
Tiruchirappalli, India

K. R. Gokul Anand
Department of Electronics and
    Communication Engineering
Dr. Mahalingam College of Engineering
    and Technology
Pollachi, India

S. Arulvel
School of Mechanical Engineering
Vellore Institute of Technology
Vellore, India

S. Boopathy
Department of Electronics and
    Communication Engineering
Kumaraguru College of Technology
Coimbatore, India

V. Chandran
Department of Electronics and
    Communication Engineering
KPR Institute of Engineering &
    Technology
Coimbatore, India

J. Alfred Daniel
Department of Computer Science and
    Engineering
SNS College of Technology
Coimbatore, India

T. Joshva Devadas
School of Computer Science and
    Engineering
Vellore Institute of Technology
Vellore, India

V. T. Gopinathan
Department of Aeronautical Engineering
Hindusthan College of Engineering and
    Technology
Coimbatore, India

Shriya A. Jadhav
School of Computer Science and
    Engineering
Vellore Institute of Technology
Vellore, India

M. Kalamani
Department of Electronics and
    Communication Engineering
KPR Institute of Engineering and
    Technology
Coimbatore, India

M. Kavitha
Department of Computer Science and
    Engineering
Vel Tech Rangarajan Dr. Sagunthala R&D
    Institute of Science and Technology
Chennai, India

R. Kavitha
Department of Computer Science and
    Engineering
Vel Tech Rangarajan Dr. Sagunthala R&D
    Institute of Science and Technology
Chennai, India

M. Krishnamoorthi
Department of Computer Science and
    Engineering
Dr. N.G.P. Institute of Technology
Coimbatore, India

M. Raj Kumar
Rajendra Mishra School of Engineering
    Entrepreneurship
IIT Kharagpur
Kharagpur, India

C. S. Saravana Kumar
Software Architect
Robert Bosch Engineering and Business
    Solutions
Coimbatore, India

V. Saravana Kumar
Department of Aeronautical Engineering
Hindusthan College of Engineering and
    Technology
Coimbatore, India

Anisha Lal
School of Computer Science and
    Engineering
Vellore Institute of Technology
Vellore, India

Srishti Lodha
Department of CSE
Vellore Institute of Technology
Vellore, India

Harsh Malani
Department of CSE
Vellore Institute of Technology
Vellore, India

D. Mrinmoy
Agricultural and Food Engineering
    Department
IIT Kharagpur
Kharagpur, India

Suriya Murugan
Department of Computer Science and
    Engineering
Vel Tech Rangarajan Dr. Sagunthala R&D
    Institute of Science and Technology
Chennai, India

Balaanand Muthu
Department of Computer Science and
    Engineering
Adhiyamaan College of Engineering
Hosur, India

S. Mythili
Department of Electronics and
    Communication Engineering
Bannari Institute of Technology
Erode, India

K. Nithya
Department of Computer Science and
    Engineering
Kongu Engineering College
Erode, India

M. G. Sumithra
Centre for Research and Development
KPR Institute of Engineering and
    Technology
Coimbatore, India

T. Poornima
Department of Electronics and
    Communication Engineering
Amrita Vishwa Vidyapeetham
Coimbatore, India

M. Amutha Prabakar
School of Computer Science and
    Engineering
Vellore Institute of Technology
Vellore, India

Narayanan Prasanth
Department of CSE
Vellore Institute of Technology
Vellore, India

E. L. Dhivya Priya
Department of Electronics and
    Communication Engineering
Sri Krishna College of Technology
Coimbatore, India

N. R. Rajalakshmi
Department of CSE
Vel Tech Rangarajan Dr. Sagunthala R&D
    Institute of Science and Technology
Tamil Nadu, India.

J. Bruce Ralphin Rose
Department of Aeronautical Engineering
Anna University Regional Campus
Tirunelveli, India

D. Dsilva Winfred Rufuss
School of Engineering
University of Birmingham
Birmingham, UK
and
School of Mechanical Engineering
Vellore Institute of Technology
Vellore, India

L. SaiRamesh
Department of IST
Anna University
Chennai, India

K. Selvakumar
Department of Computer Applications
NIT Trichy
Tiruchirappalli, India

Krishnan Saravanan
Department of Computer Science and
    Engineering
Anna University Regional Campus
Tirunelveli, India

A. Sharmila
Department of Electronics and
    Communication Engineering
Bannari Amman Institute of Technology
Erode, India

C. B. Sivaparthipan
Department of Computer Science and
    Engineering
Adhiyamaan College of Engineering
Hosur, India

R. Srinivasan
Department of Computer Science and
    Engineering
Vel Tech Rangarajan Dr. Sagunthala R&D
    Institute of Science and Technology
Chennai, India

Sangeetha Subramani
Department of Computer Science and
    Engineering
SNS College of Technology
Coimbatore, India

N. Suganthi
Department of Computer Science and
    Engineering
Kumaraguru College of Technology
Coimbatore, India

K. Utkarsh
ICAR-Vivekanand Parvatiya Krishi
    Anusandhan Sansthan
Almora, India

C. Chandru Vignesh
Department of Computer Science and
    Engineering
Vel Tech Rangarajan Dr. Sagunthala R&D
    Institute of Science & Technology
Chennai, India

T. Vigneswari
Department of Information Technology
Sri Manakula Vinayagar Engineering
    College
Pondicherry, India

N. Vijaya
Department of Computer Science and
    Engineering
K. Ramakrishnan College of Technology
Tiruchirappalli, India

S. Vinoth Kumar
Department of Computer Science and
    Engineering
Vel Tech Rangarajan Dr. Sagunthala R&D
    Institute of Science and Technology
Chennai, India

1

Cloud IoT Applications in Agricultural Engineering

M. Raj Kumar, P. B. Ahir, and D. Mrinmoy
IIT Kharagpur
Kharagpur, India

K. Utkarsh
ICAR-Vivekanand Parvatiya Krishi Anusandhan Sansthan
Almora, India

CONTENTS

1.1 Introduction..1
    1.1.1 Basic Architecture of IoT-Based Automated Irrigation Systems....................3
1.2 Micro-Irrigation System with Internet Connectivity.....................................4
1.3 ZigBee Wireless System 7
1.4 Bluetooth Wireless System 7
1.5 Wi-Fi Wireless System...8
1.6 GPRS/3G/4G Technologies...8
1.7 LoRa Technology 9
1.8 WiMAX Technology...10
1.9 Ham Radio..10
1.10 Discussion and Conclusion 11
References..13

1.1 Introduction

Agriculture is the world's primary source of income, and it also contributes to the size of the nation's gross domestic product (GDP). Because the world's water resources are dwindling, it is critical to use proper irrigation system. Drip irrigation, also known as micro-irrigation or localized irrigation, allows water to drip slowly to plant roots via a network of pumps, pipes, or emitters, which prevents soil erosion and saves water and fertilizer. Micro-irrigation is a method of artificially supplying water to a plant's root system [1]. In dry regions and during periods of inadequate precipitation, irrigation has been used to help crop growth, maintenance, and re-vegetation of upset soils. Irrigation aids crop production by protecting plants from frost, suppressing weed growth in grain fields, and delaying soil solidification. Irrigation systems are also used for dust suppression, waste removal, and mining. Watering jars, water channels that had to be manually opened and closed, or rucksack sprinklers were all used in the past for irrigation [2]. In this procedure, a lot of water is wasted. There is

DOI: 10.1201/9781003185413-1

TABLE 1.1

Comparison of Traditional and Automated systems

Agriculture System	Traditional Irrigation	Automated Irrigation
Workers	Yes	No
Water use	High	Low
Field surveillance	High	Low
Data collection	Low	High
Time	More	Less
Cost	High	Low
Yield	Low	High

a need to improve the existing or older irrigation systems. To improve crop water usage, an automated irrigation framework should be developed. An intelligent automatic irrigation system must include all the components necessary to monitor and regulate the amount of water available to the plants without requiring human intervention. In today's world, the majority of countries lack sufficient human resources in agricultural areas, which has a negative impact on the growth of non-industrial countries [3]. Thus, the present situation is an ideal opportunity to automate the area in order to overcome this problem. This system is primarily based on limiting labor and hardware costs, which are both fair for all users. The dirt level, dampness, and other elements must be physically estimated by an experienced person in the previous conventional system (Table 1.1).

The person should inspect the condition of the farm and manually turn on/off the engine to water the field. A person will check the water level in the tank, and they will have to start the engine each time to fill the tank. The methods of connecting people and the climate were devised by micro-irrigation systems using automation with or without Internet connectivity. Farmers will be able to control water system siphons and valves with the tap of a button in their phones, allowing them to monitor moisture levels and the status of water system valves from anywhere on the planet. Android phones are available and many people enjoy them due to their simplicity of use and low price. Only authorized users can monitor water system instruments and segments through a web-based remote water system robotization platform. Almost every Android phone has Wi-Fi, which will help farmers access the water system framework from their homes or from anywhere on the planet. The Wi-Fi protocol 802.11 has a low energy use, which is essential for battery life, as well as a safe convention to ensure privacy and the use of a flexible Android application with biometric and hidden password protection for controlling water device siphons and valves through the web [4]. The framework can be programmed to activate events under specific conditions, allowing the user to more flexibly monitor watering to the plants. Over the last decade, hardware cost reductions have aided the emergence of the Internet of Things (IoT) and its numerous applications. The framework proposed in this chapter employs data and correspondence advancements to allow users to examine and consider data collected by various sensors. Stickiness, temperature, dampness, and light are all measured using sensors. A mircoregulator receives feedback from sensors [5]. The mircoregulator sends the information to the user in a sequential manner. Sensor value icons will be displayed on the PC/smartphone side, and users can use the icon to turn on/off irrigation devices [4]. The data is sent to and managed by the user, who stores the sensor data in a database, allowing for simple and adaptable data transmission. Better farming practices, fewer water storages, and the development of a modern agricultural framework for the country are all possible outcomes of the anticipated framework [6]. Industrialists and analysts are working to create efficient and cost-effective automated mechanisms for monitoring various devices such as lights, fans, and climate control systems based on demand [7]. Automation saves energy, water, and a significant amount of waste [8]. Both water and manure are used in the trickle irrigation system. Water is steadily irrigated to the plants' underlying roots via thin tubes and valves, which is an excellent technique for watering plants. To avoid water logging, there should be no waste in the fields or pot plants, which could affect profitability [4]. Currently, there are frameworks for automatic trickle irrigation that water plants based on soil moisture, pH estimation, temperature, and light [9]. In large agricultural fields where crop efficiency is critical, these boundaries are crucial. Our proposed irrigation framework would be useful in small zones such as office buildings, house gardens, and so on, where watering plants at average stretch causes problems. This chapter demonstrates how to build a smart trickle irrigation system for watering plants using the Raspberry Pi and Arduino microcontrollers. The system is remotely managed via XBee, and automation is accomplished using the Python programming language. The framework requires no further maintenance after installation and is simple to use. Using intelligent IoT-enabled inlet gates capable of communicating with surrounding regulators, only the necessary amount of water is supplied to farms based on the sensor-based parameters, which is a wireless sensor network (WSN) device based on a microcontroller that allows communication between farm- and channel-level components [4].

1.1.1 Basic Architecture of IoT-Based Automated Irrigation Systems

Sensors for measuring soil moisture, humidity, and temperature are used to monitor the field in real time and send data to the "NRF24LO1" transmitter and receiver, as well as the webserver, through an "Ethernet connection" on the receiving end. A web application is used to evaluate and analyze humidity, temperature, and moisture data and threshold values. The server is responsible for deciding whether or not to water the plants [10]. The motor is turned on when the value falls below the threshold, and it is turned off when the value increases above the limit. The system architecture is shown in Figure 1.1.

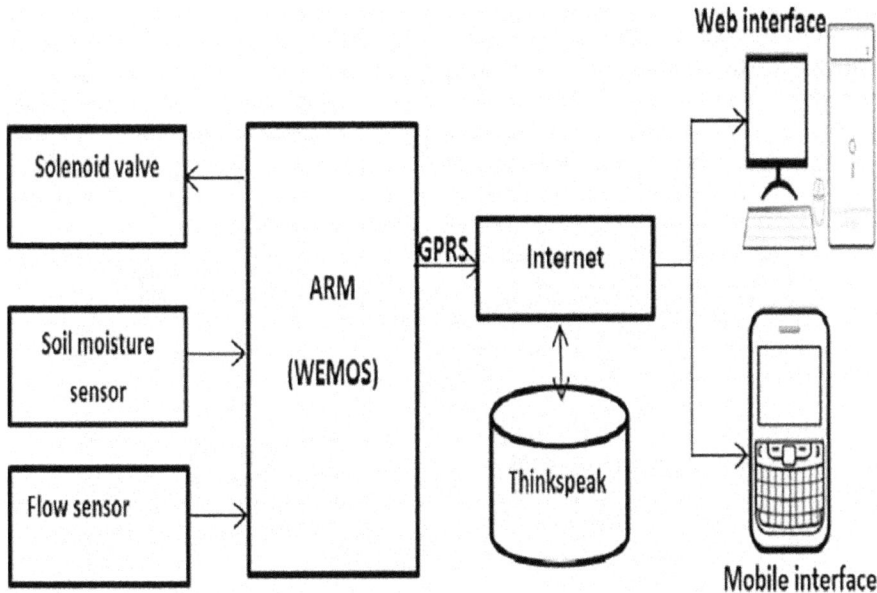

FIGURE 1.1
System architecture IoT-based automatic irrigation system [10].

Before being deployed in the field and put in a box, each sensor is connected to an "Arduino microcontroller" and programmed [11]. Two probes of the soil moisture sensor are used to pass current through the soil when the sensor is inserted into the soil. A moist soil has less resistance to current flow, while dry soil resists current flow and allows less current to pass. This resistance value is used to detect moisture. For humidity and temperature, a "DHT11" sensor is also used. Temperature and relative humidity are proportional; as the temperature increases, the relative humidity rises, and vice versa, and their readings are communicated to the user. For light detection, a "Light Dependent Resistor, LDR" is used, with resistance decreasing as light intensity decreases, and vice versa. To measure this resistance, a "voltage divider circuit" is used. The "NRF24L01 module," which has a 2.4 GHz transceiver with a "Nordic semi-conductor" and a data transmission rate of 256 Kbps/1 Mbps/2 Mbps and a voltage requirement of 1.9–3.6 V, is used for wireless data transmission [12]. The receiver and transmitter modules are both attached to "Arduino" boards, with the transmitter in the field and the receiver in the unit. An ID is specified for configuration, which is the destination address, and this is the same for all field transmitters. The webserver and the device's receiver are linked via Ethernet, an IEEE-802.11 standard for computer local area networks. The receiver receives data from the transmitter and transmits a request to the webserver through an Ethernet cable connected to the Arduino microcontroller, which assigns a specific IP address in the network range and transmits the request to the webserver. The data is saved in the database because the PHP script configures the webserver to insert values as required. The obtained sensor data is compared to threshold values that differ depending on the crop during processing [13]. They can also adjust due to climatic changes, so both of these variables are considered when calculating threshold values. The motor turns on automatically if the value of soil moisture falls below the threshold limit, or the farmer may turn it on using a phone or web application via relays, in which case control is moved from the web app to the electric switch.

1.2 Micro-Irrigation System with Internet Connectivity

Automation arrangement depends on two types of mircoregulators: the Raspberry Pi 3 and Arduino Mega. These are chosen as mircoregulators due to their computational capacity, cost, and ease of use. The variable boundaries are checked on a regular basis with various sensors, and depending on the crop type, suitable and specific irrigation is performed. Figure 1.2 outlines the conceptual work's significant empowering factors. The first

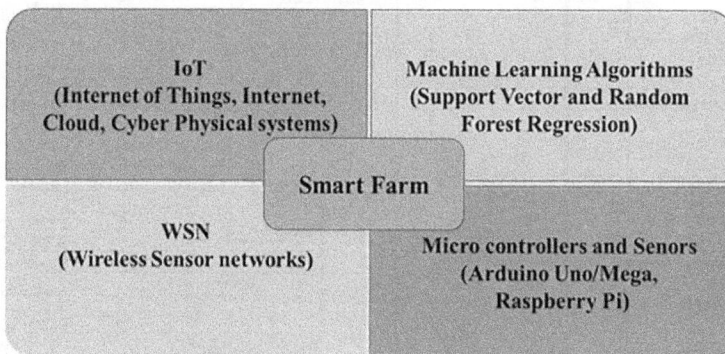

FIGURE 1.2
Conceptual parameters of a smart farm [14].

step in implementing a smart irrigation system for farm automation is to set up a remote sensor network field, in which each hub is encased in a Wi-Fi module and relays information over a typical worker, from which an automated Python programming can continue surveying the data and then send an alert/start signal for the required operation [14]. The general globally of various sensor hubs is depicted in Figure 1.3. However, the true global setup is determined by the state's socioeconomics.

Collecting data from various sensors placed in the field or nursery is the first step in building an automatic irrigation system (Table 1.2). All other sensor nodes will communicate via the Raspberry Pi 3 B+, which will act as the communication hub. Each Arduino system includes an advanced soil dampness sensor, Wi-Fi module, GSM module, Bluetooth (BT) module, temperature and humidity sensor, MQ2 gas sensor, water level pointer, alarm, clock module, battery, and relay module [15].

Any hub can send data to the Raspberry Pi 3 passage hub/base station. The Python content on the Raspberry Pi will be able to store data in the worker/cloud, from which it will be sent to the end-user via an application layer. Since Internet is not required for information transmission from the Raspberry Pi to the worker, an intranet can be used instead. Intranets are often used in areas with no or restricted Internet connectivity [14]. To connect different sensors to the mircoregulator, the basic concepts of beneficiary, transmitter, field, and positive are used; the majority of connections are made in the same way, and the Arduino Mega has a sufficient number of RXD, TXD, optical, and simple pins, as well as various VCC and GND pins. The primary period of the suggested application is completed after determining the setup's globally and collecting details. The processing of data from

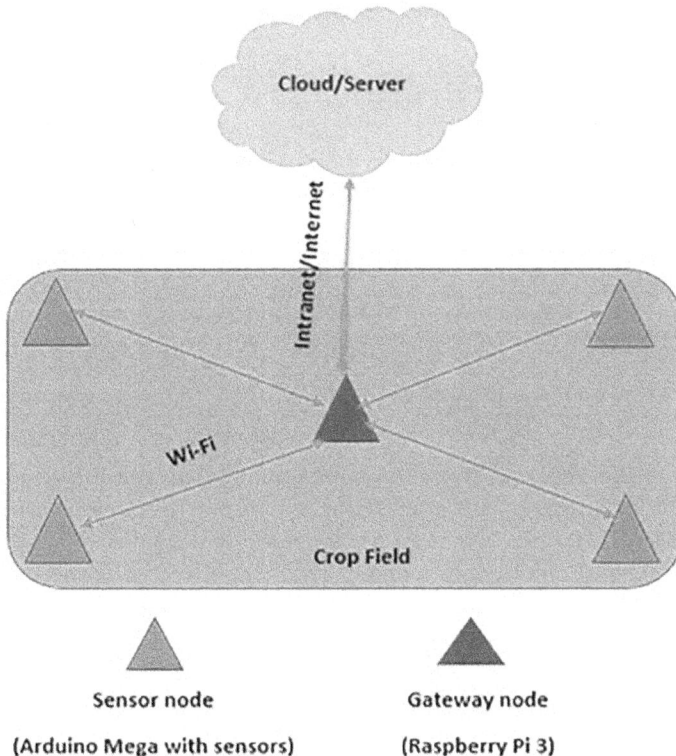

FIGURE 1.3
Wireless sensor network (WSN) [16].

TABLE 1.2

Communication Technologies Used in Automatic Irrigation Systems

Reference	Wireless Protocol/ Technology	Frequency	Data Rate	Communication Range	About
[17]	ZigBee (IEEE 802.15.4)	868/915 MHz and 2.4 GHz	20, 40, and 250 Kbps	100 meters	ZigBee is a low-cost, low-rate, and low-power-consuming device that aids in the secure and efficient transmission of data from source to destination. When a ZigBee-based wireless sensor network is applied to an agriculture field, data about the field's conditions is properly transmitted to the user at a high data transmission rate.
[18]	Bluetooth (IEEE 802.15.1)	2.4 GHz	1–3 Mbps	10–50 meters	Bluetooth is a low-power, short-range wireless sensor network device that allows data to be transmitted from one user to another.
[10]	Wi-Fi (IEEE 802.11)	2.4 GHz	11–54 and 150 Mbps	100 meters	Wi-Fi is the most successful wireless sensor network technology. Wi-Fi was previously only available on laptops, but it is now widely available on cellphones.
[19]	GPRS/3G/4G	900–1800 MHz	Up to 170 Kbps	1–10 kilometers	GPRS provides high speed of data transmission to the users.
[20]	LoRa (IEEE 802.15.4g)	868/915 MHz	50 Kbps	5 kilometers	LoRa is a long-range communication technology that consists of LoRa end devices, LoRa gateways, and LoRa network servers.
[21]	SigFox (IEEE 802.15.4g)	868/915 MHz	100 bps	10 kilometers	SigFox is an ultranarrow band wireless cellular network with low data rate applications.

various sensors is the first and the most important step in the information preparation process. Every hub requires a mircoregulator, which in this review is recommended to be an Arduino Mega because it better suits the scope of the setup, but other options such as Arduino Uno R3 and Node MCU can also be considered depending on the size of the execution. Water is supplied to the drip irrigation system from this tank through pressure and gravity, as required, and this need is determined by microcontrollers controlling valves [16]. The main drip-line is constructed as piping and is connected to a water tank over a four-meter length. Three evenly spaced drip sub-lines for a total of three crop ridges are also installed horizontally to the mainline, each with ten evenly spaced drippers for direct water distribution to the roots of ten crops, all operated by an "Arduino" micro-controller. Along each drip sub-line, a sensor box is mounted, which contains an Arduino

Uno microcontroller that reads analogous measured data about soil humidity from the sensors [3]. Five sensors are placed at an equal distance apart for each subline. The microcontroller reads the values and converts them to a percentage along with the range of values from 0 to 100. In this way, the average value for soil humidity is measured, which is then sent to the central master-control Arduino Mega via a radio transmitter – NRF24L01 – after each Arduino is given a unique identification code to distinguish it from the others. After receiving the value, the master Arduino runs an algorithm specific to the crops and opens the subsequent valve, allowing water to flow into sub-lines for field irrigation for a set period of time before closing it automatically. For large-scale agriculture, as well as for business and non-business stages, several frameworks for remote and automated irrigation monitoring and control, as well as irrigation motorization, have been planned and developed [4]. The IoT's automation and distance control have benefitted irrigation automation, security frameworks, and in-home automation. There has been a considerable amount of research into automated and remote irrigation monitoring and control all over the world, using various levels, some of which have already been listed. The most significant feature of inaccessible control and computerization is the ability of users to track devices and hardware from anywhere and at any time, as well as it results in minimizing human exertion and mediation by the use of devices and sensors.

1.3 ZigBee Wireless System

With its low-duty period, the ZigBee wireless protocol is ideal for water conservation, irrigation tracking, pesticide and fertilizer control, and is one of the most effective technologies in agricultural production [22]. The sensor nodes on the XBee Series-2 will communicate with the router over long distances of up to 100 meters in farms, while the communication range is increased to 30 meters in greenhouses or indoor fields. Numerous studies on various crops and farms were conducted in order to determine its range of communication [13].

The experiments include that on palms and orchids, for example, to investigate the signal propagation model using received signal strength indicator (RSSI), the ZigBee-wireless protocol's signal strength indicator. As a result, it was suggested [17] that a wireless channel propagation model be investigated before deploying sensor nodes in the field in order to achieve strong signals. The method of cattle localization was also investigated using this technology [18]. The link quality indicator (LQI) relation is used by ZigBee for distance calculation [23]. The weather conditions in the greenhouse were monitored and managed using ZigBee technology and GSM/GPRS [24]. When used in a WSN-based greenhouse, the ZigBee star topology is combined with artificial intelligence to save energy by switching between active and sleep modes, lowering energy consumption, and extending the battery life of sensor motes.

1.4 Bluetooth Wireless System

The BT standard was designed to enable mobile and compact devices, such as PCs, to communicate over short distances of up to 10 meters [18]. Due to its inevitability and versatility in most cell phones, BT has been used to meet multi-level agricultural requirements. Climate data, soil moisture, mechanical device location, and temperature are all verified remotely using the Global Positioning System (GPS) and BT technologies. The proposed framework [25] was created in order to use a water system to improve field profitability and control water. Using the BT remote correspondence convention, the water system framework is designed to collect field data on a continuous basis. Many technologies and devices have been developed to track overall humidity and temperature in nurseries with BT connections [4]. The BT module was used in an integrated management manner that linked soil and climate data to monitor the water system structure in nurseries, and this innovation improved [26] the leaf range, tallness, dry weight, and new weight of red and cos lettuce in nurseries. The outcome of the assessment for water use and power was also upgraded into an integrated management technique using BT technology, in contrast to the conventional technique [27] i.e., clock management process. Due to its low energy consumption, broad accessibility, and convenience for farmers, the cellphone-based BT technology has been used in a variety of agriculture applications, including dominant water system frameworks, perceptive soil and environment conditions, and dominant pesticide and fertilizer use [28].

1.5 Wi-Fi Wireless System

In the modern era, the most commonly used remote technology is the wireless local area network, which is available in portable devices such as laptops, cell phones, PCs, and work areas. A communication distance of 20 to 100 meters is deemed acceptable in both indoor and outdoor situations. A wireless local area network in agriculture broadens multiple models by interfacing various types of computers, suggesting an impromptu entity [10]. Using wireless local area network and 3G remote developments, the agricultural applications of cell phones were investigated [2]. Remote access and fast text administrations are frequently used for dominant and observant protected crops. Soil temperature, soil dampness, ambient temperature and humidity, daylight energy, and CO_2 were all stored in a very door before being transmitted to a device via an agricultural data wireless local area network setup. A Wi-Fi-based (IEEE 802.11g) sensible wireless computer network is planned for breeding check-in [H] [29]. The system includes three hubs: server, switch, and worker. In the nursery or the farming field's atmosphere, stickiness, temperature, vaporized tension, light, water level, and soil dampness are all measured. The same work focuses on lowering costs, reducing wiring links, and increasing the movability and usability of investigative work in wireless computer networks.

1.6 GPRS/3G/4G Technologies

The General Packet Radio Service (GPRS) is a GSM-based mobile phone data management system. The number of customers who share similar communication channels and properties is based on the number of customers who experience variable delays and throughputs while using GPRS. In Ref. [19], a programmed crop water system framework based on data collected by temperature and soil moisture sensors mounted at the root zone of plants using a GPRS module and a wireless computer network was built, and this framework could be viewed as a smart and realistic response to rising water quality in precision agriculture [30]. Estimating soil moisture was used to assess the feasibility of a trickle water scheme. The Arduino board was captivated by associated data, resulting in the development of a model system and a WSN-GPRS approach. The WSN-GPRS door connects the WSN and GPRS networks, allowing WSN data to be exchanged with the user. Ref. [2] presents GPRS-enabled remote hubs for measuring and transmitting soil, plant, and environmental data. The remote hubs have unrestricted independence due to their autonomy and use of solar-powered energy. A number of sensors could send data to a remote location through a GPRS network for further analysis on tablets, cellphones, or PCs. All farming devices are attached to the sensor board in order to collect data. Such information is transmitted to a distant employee for further investigation via the GPRS board, which is dependent on a GSM/GPRS transportable entity.

MICRO-IRRIGATION SYSTEM WITHOUT INTERNET CONNECTIVITY

1.7 LoRa Technology

The irrigation framework's automation also faces the challenge of data transmission over long distances [31]. The information transmitted through Wi-Fi and other radio communication devices is intended for a target individual (Figure 1.4).

The SX1278 handsets come with a LoRa long-range modem that offers super-long reach spread range correspondence and high impedance resistance while limiting current use. LoRa uses spread-spectrum ingenuity to achieve synchronization [3]. This method will transmit data up to a distance of 20 kilometers. LoRa is a low-power, wide-area radio tweaking method with a small trill spread range [32]. It is a WAN configuration that uses permit-free, sub-GHz radio-recurrence classes like 196, 433, and 868 MHz in Europe and 915 MHz in North America to enable long-range, low-piece rate communication between "things" (i.e., connected objects like battery-powered sensors). A low-power wide-area network (LPWAN) may be used to build a private remote sensor network or a third-party help or device to enable the screening and tracking of the running and turning on and off of irrigation siphons for farmers who live far away from their lands. The gadget is made

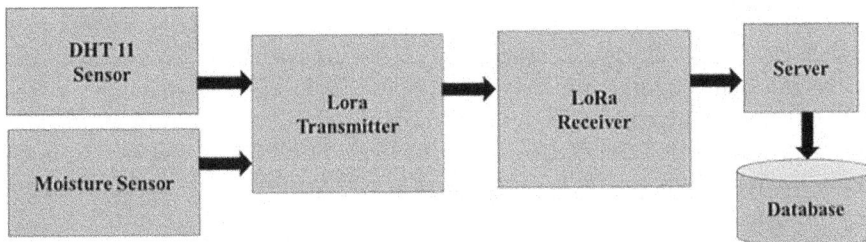

FIGURE 1.4
LoRa implementation [3].

with the most recent developments in current interchanges in mind (LoRa technology). The small size, low cost, and ability to track devices from a distance of up to 10 kilometers without the need for pinnacles for communication or other extra costs are all advantages of this invention [33]. This system is powered by a battery that lasts more than five years and burns at a low intensity level. A LoRa ESP32 for sending and receiving data, an exceptional circuit to track the electrical voltage in the setup, and eventually, an exceptional circuit to regulate and kill the electric siphon make up the primary circuit (control circuit).

1.8 WiMAX Technology

Worldwide Interoperability for Microwave Access (WiMAX) is a form of radio communication that links the Local Control Unit to the internet. It uses the IEEE 802.16 standard to provide internet connectivity to areas where conventional digital subscriber line (DSL) administration is still difficult and costly to set up or where 3G signals are unavailable. The reason for selecting WiMAX is that in certain areas of the region, such as far from midtown, there is often a lack of infiltration of media transmission lines or radio correspondence inclusion [19]. In the best-case scenario, the radio connection would span up to 70 kilometers with a clear line of sight (LOS) [11]. Different advancements, such as 3G, satellite correspondence, Wi-Fi, or asymmetric DSL (ADSL), may be used to understand distant correspondence, but none of them can match WiMAX's capabilities. Satellite correspondences provide global coverage at a considerably higher cost than WiMAX or 3G, with the exception of ADSL, which needs a link that may be prohibitively costly in these regions, and Wi-Fi, which has a very narrow range of coverage (under 100 meters). The latter is not really dynamic in these areas because building a 3G network is not lucrative for the operator [34]. WiMAX is a viable choice for this application because of these factors. This will also encourage the SC worldview to spread in places that are not technically classified as "city." In our system, researchers distributed WiMAX single transporter time division duplex (TDD) administration using a base station (BS) at 3.4845 GHz population recurrence with a transmission capacity of 10 MHz. The BS communicating reception apparatus is an Argus tilt board radio wire model SPPX310M.

1.9 Ham Radio

A WSNs' radio frequency (RF) parts distribute power more equally than information-processing devices like microcontrollers and microprocessors. Various researchers have used various radio improvement plans or strategies, such as (i) transmission power control (TPC), (ii) modification strategies, and (iii) psychological radio, to minimize the power utilization of the RF parts of agriculture sensor hubs. In the TPC scheme, sensor hubs adjust the send ability to save energy, allow interference avoidance, and set up a correspondence interface. TPC can be used in the agricultural sector, where the RF-communicated strength of sensor hubs can be adjusted to minimize power consumption by separating the sink hub and the sensor hub [8]. The use of TPC to reduce the power used by sensor hubs in precision agriculture based WSNs is being investigated. A few power levels and various collector affectability levels are thought to play a role in the said cycle. The organization layers and MAC convention in a WSN (due to CC1110) for the agriculture area are obtained to further minimize the sensor center's power utilization. The findings show that power sparing using various methods of transmitted power can be increased by around 10% as compared to the conventional model. On the other hand, increasing the send speed of the CC2420 specialized gadget improved the sensor hub's lifetime by over 8.5 [3]. An intelligent radio is a sharp remote correspondence network that can effectively pick up remote correspondence direct in the range band. An intelligent radio needs more energy than other gadgets because it uses advanced and complex battery capacity. An energy-efficient intellectual radio organization presents challenges in this way, especially in terms of battery energy consumption [8]. In remote correspondence, the electromagnetic spectrum is almost entirely used, with only a few groups remaining underutilized or unused. The use of a cognitive radio, which allows for the use of range openings, is recommended in this situation. Intertwined fluffy justification control (FLC) and a ZigBee remote sensor entertainer organization (wireless sensor actuating network [WSAN]) can detect environmental factors and respond as quickly as possible while not being obstructed by crops. In terms of adaptability, capability, profitability, and job obstruction, FLC assists masters in managing complex systems that are more in contrast to conventional control techniques. Two critical nursery environment factors (stickiness and temperature) are measured during the day and evening. An FLC focused on smart procedures would substantially increase the WSAN's life expectancy, according to the findings. As a result, the WSAN will run for a long time on a 210 mAh battery, thanks to the FLC combined with WSAN. The control boundaries can be changed to match the ideal qualities of radio modules in order to reduce their usage by the base power [8]. To limit the sensor hub's all-out power use needed to submit given information packets, the balance techniques of frequency shift keying (FSK) and minimum shift keying (MSK) are investigated. Based on their thorough testing, the developers concluded that MSK outperforms FSK in terms of forcing use. A few tweak plans cannot be included in the ZigBee 2.4 GHz remote convention because it only supports a single modification storey. In particular, the ZigBee remote convention employs offset quadrature phase-shift keying (OQPSK). The method is focused on radio streamlining plans that were previously used in agricultural applications.

1.10 Discussion and Conclusion

A survey of WSN-based agrarian applications was presented. Wi-Fi, BT, ZigBee, GPRS/3G/4G, LoRa, and SigFox were used to establish a connection between various remote innovations or conventions. The ZigBee and remote conventions were found to be more advantageous for rural applications than the others due to their low power usage and equal correspondence range for ZigBee and long for LoRa. There was also a description of energy-efficient methods or measurements, as well as energy-gathering techniques. As a result of the scientific categorization, we show that a wide variety of energy-efficient and energy-gathering methods can be used in the farming space. In order to find the best solutions for keeping the framework going, previous research was compared and contrasted with exploring the momentum problems in WSN-based farming applications. Difficulties and constraints for plan reflection were posed later in the LoRa module. To analyze cutting-edge IoT approaches in farming applications, different sensors, actuators, gadgets, IoT levels, and system layers were investigated and compared. A variety of benefits of an automatic irrigation system include low cost of operation and substantial reduction in water consumption. Furthermore, maintenance requirements are limited. An automated Internet-based irrigation system can also save you a lot of money on electricity. It can be used in greenhouses as well as areas where water shortage is a significant issue. As a result of this process, crop production increases while useless crop production decreases. The computer mechanically measures the water requirement in this system threshold. This estimate is dependent on the evaporation pan's water level dropping. Drought stress and fungal and bacterial infections are not a concern with this approach. The farmer may also use the mobile application to check the condition of the field from afar without having to physically visit it. Furthermore, human- or labor-related errors are reduced, and the optimum water sum is regarded as the best on the planet for crop quality and water conservation. Another Internet-based system for automatic irrigation is WSN, while protection is provided by a Raspberry Pi-based security system. Data from sensors at each node in the built system is used to track field parameters, especially soil moisture, and water flow can be controlled by the user from any location using any Internet-connected device. This device is very successful even at a low cost, since Internet connectivity in rural areas is now available at low prices. As a consequence, the most useful application is a web-based irrigation system based on the "Arduino Uno" and Raspberry Pi. Various field parameters such as humidity, temperature, and soil moisture can be monitored from a webpage using "HTML and PHP." The required action is taken based on the displayed state of these parameters on the web page. The device has proven to be useful in another Internet-based agriculture application because it is attentive to any animal intervention in the field, which is a major cause of crop yield reduction. At all times, you must have access to the Internet. An intelligent water-saving irrigation system is created by combining WSN and a fuzzy control system. This unit effectively monitors and regulates the ground from afar. The board's nodes are powered by solar energy and lithium battery devices, which solve the problem of inadequate power by allowing nodes to self-organize with low energy consumption and consistency by employing several hops. There is a network protocol for communication. Using a fuzzy-logic toolbox for the development and evaluation of the system, which consists of one fuzzy output for two inputs, showing a relationship between the system's input and output, the control method is made more scientific. By automatically opening valves when a threshold limit is reached, this system achieves the purpose of a water-saving irrigation system. As a consequence, it is a water-saving, productive, and reliable system. The intelligent IoT-based automated irrigation system uses Arduino and Raspberry Pi 3 as microcontrollers and preparing devices. Furthermore, soil dampness and temperature data are transmitted from sensors to detect temperature and soil dampness levels. These sensors are connected to an Arduino microcontroller, which receives data from the sensors and processes it. The water siphon actuator is also connected to Arduino for water siphoning in addition to these sensors. The detected data is then sent sequentially via sequential correspondence to the Raspberry Pi 3 control unit, where machine learning K-NN (K-nearest neighbor) calculations are performed [35]. The AI measurements are saved in the Pi 3 control units, which are used to prepare data on dampness and temperature for a variety of soil conditions, including dry, minimal dry, wet, minimal wet, and forecasted needs. The expected yield is sent to Arduino as a control signal, causing the siphon to trigger and the field to be watered as required. Finally, farmers can access the data from the flooded field via a cloud webpage. This entire model is based on the Raspberry Pi 3 climate arrangement, which is an edge-level processor where astute examination is completed by using K-NN machine learning calculations for foreseeing the dirt condition based on captured soil dampness and temperature information and on prepared informational collections relating to soil dampness and temperature for various types of soil that are dry and mounds of soil. As technology progresses, more efficient and successful work is needed, such as using "data mining algorithms," to predict crop water requirements automatically. More research is required in this area to develop a cost-effective method, as prediction is useful in deciding the correct amount of water to supply. Furthermore, rather than using mobile apps that rely on the Internet for updates, the need for constant Internet access can be avoided by creating a system to alert users directly via SMS on a cellphone using the "GSM module." A statistical model is required for a cloud-server-based irrigation system since all system information is stored in the cloud server, including values recorded by sensors, previous irrigation date and duration, and total evaporation between successive irrigations. Since the entire system is interconnected, if the power supply is cut off or a shortage occurs, the entire system, including recorded sensor values and Internet connectivity, is disrupted. As a result, future research would necessitate the use of solar-powered batteries as well as the installation of additional instruments such as an anemometer and more efficient soil moisture sensors. Furthermore, the data transfer rate and area coverage of wireless communication networks can be enhanced for better system establishment.

References

1. Madramootoo, C. A., & Morrison, J. (2013). Advances and challenges with micro-irrigation. *Irrigation and Drainage*, 62(3), 255–261.
2. Kumar, S., Sethuraman, C., & Srinivas, K. (2017). Solar-Powered Automatic Drip Irrigation System (SPADES) using wireless sensor network technology. *International Research Journal of Engineering and Technology*, 4(07), 722–731.
3. Wavhal, D., & Giri, M. (2013). Automated drip irrigation improvements using wireless application. *International Journal of Advanced Research*, 1(3), 65–73.
4. Mandlik, R. (2012). Wireless sensor network for water table monitoring in agriculture. *International Journal of Agriculture Innovations and Research*, 1(2), 51–54.
5. Singh, I., & Bhangoo, K. S. (2013). Irrigation System in Indian Punjab. MRPA. https://mpra. ub.uni-muenchen.de/50270/1/MPRA_paper_50270.pdf
6. Llaria, A., Terrasson, G. Arregui, H., & Hacala, A. (2015). Geolocation and monitoring platform for extensive farming in mountain pastures. *Proceedings of the IEEE International Conference on Industrial Technology (ICIT)*, Seville, Spain, 17–19 March 2015; pp. 2420–2425.
7. Pitì, A., Verticale, G., Rottondi, C., Capone, A., & Lo Schiavo, L. (2017). The role of smart meters in enabling real-time energy services for households: The Italian case. *Energies*, 10, 199.
8. Blume, H. R. (1979). Radio controls for gated pipe irrigation systems. (*Dissertation – Masters, Kansas State University, Manhattan, Kansas*).
9. Ali, A. H., Chisab, R. F., & Mnati, M. J. (2019). Smart monitoring and controlling for agricultural pumps using LoRa IoT technology. *Indonesian Journal of Electrical Engineering and Computer Science*, 13(1), 286–292.
10. Gujarathi, P. J., & Giri, M. J. (2012). Advanced greenhouse control and monitoring system on CAN-BUS. *International Journal of Advanced Information Science and Technology*, 22(22), 111–117.
11. Kodali, R. K., Kuthada, M. S., & Borra, Y. K. Y. (2018). Lora based smart irrigation System. *Paper presented at the 2018 4th International Conference on Computing Communication and Automation (ICCCA)*, Greater Noida, India.
12. Nithya, G. (2012). Maintenance-free solar energy drip irrigation using battery-less RF powered wireless sensors over AD-HOC network. *Paper presented at the 2012 International Conference on Emerging Trends in Electrical Engineering and Energy Management (ICETEEEM)*, Chennai, India.

13. Kumar, M. S. (2018). Research and development of virtualization in wireless sensor networks. *International Journal on Informatics Visualization*, 2(2), 96–103.
14. Giri, M., & Wavhal, D. N. (2013). Automated intelligent wireless drip irrigation using linear programming. *International Journal of Advanced Research in Computer Engineering & Technology (IJARCET)*, 2, 1–5.
15. Jiaxing, X., Peng, G., Weixing, W., Xin, X., & Guosheng, H. (2018). Design of wireless sensor network bidirectional nodes for intelligent monitoring system of micro-irrigation in litchi orchards. *IFAC-PapersOnLine*, 51(17), 449–454.
16. Van Bavel, C. H., van Bavel, M. G., Lascano, R. J., & Camp, C. (1996). Automatic irrigation based on monitoring plant transpiration. *Paper presented at the Proceedings of the ASAE International Conference.*
17. Cancela, J., Fandiño, M., Rey, B., & Martínez, E. (2015). Automatic irrigation system based on dual crop coefficient, soil and plant water status for Vitis vinifera (cv Godello and cv Mencía). *Agricultural Water Management*, 151, 52–63.
18. Rani, M. U., & Kamalesh, S. (2014). Energy efficient fault tolerant topology scheme for precision agriculture using wireless sensor network. *Proceedings of the International Conference on Advanced Communication Control and Computing Technologies (ICACCCT)*, Ramanathapuram, India, 8–10 May 2014; pp. 1208–1211.
19. Krishnamoorthy, D. (2017). A novel real-world application using wireless sensor networks. *International Journal of Current Engineering and Scientific Research (IJCESR)*, 4(10), 44–49.
20. Gungor, V. C., & Lambert, F. C. (2006). A survey on communication networks for electric system automation. *Computer Networks*, 50(7), 877–897.
21. Gu, Q. H., Lu, C. W., Li, F. B., & Wan, C. Y. (2008). Monitoring dispatch information system of trucks and shovels in an open pit based on GIS/GPS/GPRS. *Journal of China University of Mining and Technology*, 18(2), 288–292.
22. Zhang, J., Li, W., Han, N., & Kan, J. (2008). Forest fire detection system based on a ZigBee wireless sensor network. *Frontiers of Forestry in China*, 3(3), 369–374.
23. Rao, Y., Jiang, Z. H., & Lazarovitch, N. (2016). Investigating signal propagation and strength distribution characteristics of wireless sensor networks in date palm orchards. *Computers and Electronics in Agriculture*, 124, 107–120.
24. Raheemah, A., Sabri, N., Salim, M., Ehkan, P., & Ahmad, R. B. (2016). New empirical path loss model for wireless sensor networks in mango greenhouses. *Computers and Electronics in Agriculture*, 127, 553–560.
25. Gang, L. L. L. (2006). Design of greenhouse environment monitoring and controlling system based on Bluetooth technology. *Transactions of the Chinese Society of Agricultural Machinery*, 10, 97–100.
26. Hong, G. Z., & Hsieh, C. L. (2016). Application of integrated control strategy and Bluetooth for irrigating romaine lettuce in greenhouse. *IFAC-PapersOnLine*, 49, 381–386.
27. Kim, Y., Evans, R. G., & Iversen, W. M. (2008). Remote sensing and control of an irrigation system using a distributed wireless sensor network. *IEEE Transactions on Instrumentation and Measurement*, 57, 1379–1387.
28. Versichele, M., Neutens, T., Delafontaine, M., & Van de Weghe, N. (2012). The use of Bluetooth for analysing spatiotemporal dynamics of human movement at mass events: A case study of the Ghent Festivities. *Applied Globally*, 32(2), 208–220.
29. Leroy, D., Detal, G., Cathalo, J., Manulis, M., Koeune, F., & Bonaventure, O. (2011). SWISH: Secure WiFi sharing. *Computer Networks*, 55(7), 1614–1630.
30. Terrasson, G., Llaria, A., Marra, A., & Voaden, S. (2016). Accelerometer based solution for precision livestock farming: Geolocation enhancement and animal activity identification. *IOP Conference Series: Materials Science and Engineering*; IOP Publishing: Bristol.
31. Usmonov, M., & Gregoretti, F. (2017). Design and implementation of a LoRa based wireless control for drip irrigation systems. *Paper presented at the 2017 2nd International Conference on Robotics and Automation Engineering (ICRAE)*, Shanghai, China.

32. Kodali, R. K. (2017). Radio data infrastructure for remote monitoring system using LoRa technology. *Paper presented at the 2017 International Conference on Advances in Computing, Communications and Informatics (ICACCI)*, Udupi, India.
33. Nisa, N. A. B., Priyadharshini, K., Priyadharshini, K., & Devi, R. N. (2019). Agriculture irrigation water demand forecasting using LoRa technology. *International Research Journal of Engineering and Technology*, 6(3), 3050–3052.
34. Ojha, T., Misra, S., & Raghuwanshi, N. S. (2015). Wireless sensor networks for agriculture: The state-of-the-art in practice and future challenges. *Computers and Electronics in Agriculture*, 118, 66–84.
35. Mohapatra, A. G., & Lenka, S. K. (2016). Neural network pattern classification and weather dependent fuzzy logic model for irrigation control in WSN based precision agriculture. *Procedia Computer Science*, 78, 499–506.

2

Urban and Vertical Farming Using Agro-IoT Systems

The Ingredient Revolution – A Sustainable Production System for Urban Population

K. R. Gokul Anand

Dr. Mahalingam College of Engineering and Technology
Pollachi, India

S. Boopathy

Kumaraguru College of Technology
Coimbatore, India

T. Poornima

Amrita Vishwa Vidyapeetham
Coimbatore, India

A. Sharmila

Bannari Amman Institute of Technology
Sathyamangalam, India

E. L. Dhivya Priya

Sri Krishna College of Technology
Kovaipudur, India

CONTENTS

2.1 Introduction 18
2.2 Literature Survey 21
2.3 Existing System vs. the Traditional System of Cultivation 24
    2.3.1 Different Types of Vertical Farming 24
        2.3.1.1 Skyscraper Buildings...24
        2.3.1.2 Mixed Skyscraper Buildings 25
        2.3.1.3 Stacked Shipping Containers 25
    2.3.2 Three Types of Processes in Vertical Farming 25
        2.3.2.1 Hydroponics 25
        2.3.2.2 Aeroponics 25
        2.3.2.3 Aquaponics 27

DOI: 10.1201/9781003185413-2

2.4   Proposed System of Vertical Farming 27
        2.4.1   Category of Models..28
                2.4.1.1   Fixed Solution System 28
                2.4.1.2   Continual Flow System 28
        2.4.2   Method of Selection..29
                2.4.2.1   Water Culture System...29
                2.4.2.2   Ebb and Flow System 30
                2.4.2.3   Drip Systems..31
                2.4.2.4   Nutrient Film Technique Systems 32
        2.4.3   Building Structure Based on Plants 32
        2.4.4   Light – A Factor in Cultivation 32
        2.4.5   Recycling Unit for Sustainability...33
        2.4.6   Technological Role of IoT 34
        2.4.7   Future of Urban Farming...35
        2.4.8   Disruption of Farming through the IoT...37
2.5   Advantages of Proposed Model...38
2.6   Conclusion 39
References...39

2.1 Introduction

Vertical farming is the practice of growing crops at an intermediate level [1]. This often includes eco-agriculture to promote plant growth and non-essential solarium techniques such as hydroponics [2–5], aquaponics [6, 7], and aeroponics [8]. Other system options for vertical farm landing processes include buildings, containers, pits, and dumped mines. By 2020, there were approximately 30 hectares (74 acres) of agricultural land standing in the world. It was Dixon who introduced the modern concept of vertical farming in 1999. Despommier, a professor of public hygiene at Columbia University, and his students developed a sophisticated high-rise building that could feed 50,000 people. The design has not been built yet, but it has been able to spread vertical farming ideas. The vertical agricultural application, together with other modern technologies, such as today's LED lighting, has produced the following results: 10 times the traditional yield [9].

The main advantage of the vertical agricultural sector is that the required area is small for a crop to grow. Improve your ability to grow different varieties at the same time. Farming is one of the most sought-after benefits because the harvest does not share a single one. In addition, plants are resistant to time risk, leading to the demands of the domestic department to use natural resources, increase, or energy that cannot make additional tapas.

With the population density of Metro Manila rising to 20,785 people per square km and the population increasing to 12.88 million by 2015, there may be an impending call for area and resources, in particular meals and water. Mindanao drops about 90% of its local produce to deliver the needs of Luzon and Visayas, causing charges to boom because of expenses of the shipping and maintenance of those sources. Local farmers and investors are concerned about transportation because of first-class supply to the market and an excessive fee of transportation from the location to important city markets, and approximately 11–12% of lettuce coming from Mindanao are misplaced because of spoilage, which has a great effect on the profitability of wholesalers and wholesale retailers.

The idea of imparting meals in towns already exists, but the concept of a whole building packed with plants is new. The concept of vertical farming is largely city farming of various agricultural plants in an enclosed area that is designed extensively to house plants for the usage of hydroponics [2–5]. The layout is supposed to develop the plant's interior in a vertical setup while integrating aeroponics [8]. Also, for the setup, minimum area and water are needed without sacrificing the yield.

In recent years, the Internet of Things (IoT) [10, 11] has been one of the subjects that may be observed in more than one research work today. As such, a group of gadgets communicating with different gadgets in the long run ledding to smarter decisions. Through the usage of IoT, gadgets may be managed remotely through current communication technologies such as Bluetooth, Wi-Fi, and RFID. Also, this technology of the rising era has been implemented in a diversity of industries including transportation, domestic automation, forecasting, health, and security. The data from the gadgets can also be saved and processed on a data server, which also can be used while interfacing with exclusive gadgets.

Agriculture is likewise observed alongside the traces of the aforementioned industries. Plants are very dependent on how successfully factors such as light, nutrient levels, and temperature are monitored and maximized [8]. Hence, every issue has to be monitored intently through IoT well to implement automation inside an agricultural enterprise and to allow people living in city areas to have grown domestic food.

Vertical farming [12, 13] is usually designed to grow crops in modern technology homes, usually housed in city center buildings. This is a modern agricultural system with a perfect climate that eliminates traditional external environmental factors. Vertical farming can be cultivated in any building that enhances the climate of the earth, regardless of current conditions such as rain or cold season, but it should be noted that it does not rely on the use of large amounts of land.

Vertical farming [10, 14] also helps reduce the risk of burnout during soil preparation by causing burnout during this process and removing all fruitless farmland and trees. The fog and dust generated by this work cause delays. Plants that kill them for sunlight also create adverse conditions for the local or global environment.

Vertical farming [15, 16] can produce crops grown in a processing environment that can be described as mechanical engineering in agriculture. Agriculture is very important to the population, and the benefits of agriculture are clear from a human point of view. Vertical farming allows barren lands to return to their original habitat while reducing pests. Vertical farming may utilize excess space that is inactive or unused in developed cities. The presence of standing crops enables food production throughout the year. In addition, it can create an environment that supports the sustainable and healthy urban lives of those who choose to live in urban areas. In the future, agriculture will focus on urban agriculture as well as that in rural areas.

When designing a vertical garden system [11] for indoor planting, the focus is not only on planting in vertical positions or platforms but also on taking into account irrigation systems [17], room conditions, and soil conditions. Vertical farming produces minerals and organic enzymes to ensure plant growth and promote the growth of minerals essential to the nutritional value and taste of the final product produced during breeding. Therefore, direct local integration and IoT applications are important for effective plant health monitoring. Due to the manual control of the fishing industry, it is difficult to extract frozen fish that consume oxygen and produce too much ammonia nitrogen, and contaminated water makes the fish sick. When people eat this fish, it causes health problems. Smart aquaculture [7] has been developed, designed, and implemented to address key fishing industry challenges such as environmental pollution, water scarcity, and healthy aquaculture [7] conditions, consumer prices, and labor costs. This fish farm has great advantages: cleaning to improve water quality [4, 18].

The world is changing rapidly in many ways, many of which are positive, but some of these changes are challenging. Climate change can lead to disaster. By a large margin, we are bound by the original technology of the Revolution. Improving the legacy system and creating new strategies and tricks to make them sustainable is the driving force behind the cyber-physical system (CPS)/IoT revolution [19]. With the increase in global population, 90% of our diet will be expected to come from 10% of agricultural region in the world. Applying an expansion model will be one of the pillars of sustainable food production.

Vertical agriculture is defined as a multifaceted representation of the "biological" nature of agriculture, where production and integration are integrated into one system. There are four new technologies needed in the ecosystem to make this possible. The use of renewable energy is the reduction of fossil fuels, the use of green technology such as LED lighting, and the use of waste as fertilizer to reduce environmental impact, efficiency, and simple implementation. Cultivation, processing, and distribution will reduce waste samples with damage, intrusion, cooling, and transportation. This is enough to maintain a self-control system that can control the power and waste water [20].

In addition, a third backup power system is also working independently with its renewable energy source, battery storage, and backup line. While the system can automate the switching to the mains power line during a shortage of power, excess energy from the other vertical farm may not be used. Load sharing can help maximize excess energy and make sure power is provided to the vertical farm. Excessive energy consumption can reduce not only carbon dioxide emissions but also costs. A careful and balanced system requires the efficient use of energy. Renewable energy mobile microwave networks and battery storage providers have shown that sharing can reduce costs. Thus, the study of this method of urban farming will soon be essential to mankind. Furthermore, it will possibly lead to researches that will support food production outside of the earth. Further study of this applied to urban farming becomes significant as space studies on alternative human habitats are booming, implying that farming can be done anywhere. However, vertical farming cannot be optimized as without automation some plants are not easy to harvest because wall gardens are usually tall and inaccessible. In addition, producing a crop without planning the process leads to a long time taken before harvest. The application of these tasks is a cost-effective solution for crop yields [21].

The rapid development of urban housing results in overcrowding, degrades housing quality, and raises questions about the impact of temperature on urban islands. Another unusual phenomenon in Taiwan is that temperatures in the north are higher those in the south. Many meteorologists say the rising capital city of Taipei has less green space, which could cause less climate change than global warming. One of the challenges of a highly developed city is finding ways to expand vertical wall and roof space to create green space. One of the best ways is to combine green walls with roofs to create an urbanized agricultural setup and increase the influence of green vegetation [5]. This method works well with environmental management, food supplies, gardens, recycling, and food waste, in addition to life skills.

In the early stages of rooftop greening, the development process dominates. Next time, the roof will dominate, providing room for ventilation in areas with very poor urban conditions, and many appreciate its value. [22].

New roof-to-roof construction focuses on roof urban agriculture, which has become one of the most effective and efficient public health strategies for public development. The size of areas with urban agriculture ranges from small urban crops to several hectares of produce. Researchers focus on improving the quality of the greenhouse cultivation method through a series of experiments. A system developed by the researchers allows users to control micro-climate signals that activate electrical and electronic equipment when settings such as temperature, water level, and humidity are entered. The amount of data that can be managed by farmers makes it easier for farmers to assess the condition of their plants. In addition, automation can perform the fieldwork.

Urban farming, or vegetable farming, is the way to grow, process, and distribute food in suburban areas. Urban agriculture is also a term used for animal husbandry, fish farming, urban beekeeping, and vegetable farming in the cities. Suburban agriculture may have different characteristics. Urban agriculture [1, 18] can reflect different economic and social development levels. It can be a social movement for community sustainability where food processors, "foodie" and "Rocabourg", build social networks based on the common spirit of nature and society. These networks can be upgraded with government support and will be integrated with local urban systems as a "transitional city" for sustainable urban development. For some, an abundance of food, nutrition, and income generation are the main motivations for implementation. In all cases, direct access to fresh vegetables, fruits, and meat through urban agriculture can improve food security.

With the growing population of cities in different countries, it is important to design cities in a modern way that allows more space to be used. At the same time, it is important to design cities with an environmental approach so that ecosystems are not affected by the introduction of new technologies. The role of information and communications technology (ICT) is very important in designing smart cities [10].

As urbanization comes and citizens often migrate from villages to cities, government planning and building blocks of different countries need to develop cities with new modern technologies to serve them well. IoT must be implemented in cities so that past data can be collected and distributed to residents [23].

ICTs are applied in vertical agriculture, but catastrophic growth, food waste, and public inclusion have yet to be considered. Food waste educators include food prices and ignorance of the origin of food due to globalization, which consider them to be fundamental factors and separate from the food culture environment. Therefore, the growth of non-autumn crops and families causes investment in the food environment, and since food is considered a product and not an essential part of survival, various things create emotions with the charm and distraction of food [24].

2.2 Literature Survey

The ingredient revolution of this 21st century has made the agriculture sector equipped with technologies to sow and reap in a precise manner. The chapter "IoT Applications in Agriculture: A Systematic Literature Review" states the automation involved in agriculture and its vital utility by agriculturalists. The IoT technologies will enhance the throughput of the plant by assisting through technology for nurturing and monitoring its ecosystem [20]. Intensive monitoring of various systems includes remote monitoring, threshold level maintenance, smart irrigation systems, prevention from extreme weather conditions, and plant multiplicity system. The proposed work assists researchers and agriculturalists in gaining knowledge regarding the IoT. Local agriculture in the countryside and non-rural areas employ the usage of hardware and software resources and immense parameters relevant to agriculture. Crop observation assists in figuring out crop disease. Smart irrigation extremely encourages the supervision of specifications that promote huge yield for farms like humidity, temperature, and moisture-incorporated sensors [12]. IoT immensely assists farmers in monitoring the crops at all their stages, from anywhere on the globe. Inclusion involves wireless sensor networks like a wireless camcorder to monitor the status of the crops in real time and actual drones to enhance particularity agriculture and smartphones to renovate the particulars to the farmers. This ensures the 24/7 status of the farm as the entire approach is automated.

India ranks second in the world in terms of population. To proffer proper nourishment facilities, it is necessary to increase food productivity. The chapter "Adoption of the Internet of Things (IoT) in Agriculture and Smart Farming toward Urban Greening: A Review" coins emerging techniques like the IoT that come into existence to meet the demands in agriculture through effective utilization of resources. The proposed work incorporates sensors to collect gestures and offer appropriate decisions based on the particulars from the sensors and inclusive of farming techniques that can offer resources when required. The sensor monitors the pH range, temperature, and humidity level in the farm. The system promotes IoT to provide automation which depends on various gestures from the sensors and it does not depend on manpower and assists in the reduction of cost. IoT techniques offer tremendous strength to farmers from a technology perspective. The proposed system in inclusion focuses on soil fertility, water requirement, and herbicides.

The suburban farming method is used on the terrace of the building and indoor cultivation of the building is performed. The farmers in the suburban space are not much aware of traditional methods of farming and they have marginal participation in professional agricultural practices. The chapter "Smart Urban Farming Service Model with IoT Based Open Platform" explains the deployment of information technology inclusive of cloud assistance. A lot of agriculture requirement tasks are automated and requirements of human interference for monitoring and supervising sensors are incredibly reduced. The proposed suburban farming method accommodates cloud assistance with IoT-enabled sensors. The proposed work encompasses sensors like CO_2 sensor, soil knowledge sensor, nutrient sensor, sun sensor, rain sensor, and wind sensor. Gadgets other than sensors include heater, flow fan, CO_2 controller, web camcorder, heater, ceiling, weather station, recording device, and uninterruptible power supply (UPS) device. Once the particulars are gathered from the gadgets the smart greenhouse management system acquires the cloud assistance and transfers the particulars to the cloud and the mobile phone through GSM.

In 2050, the global population is predicted to be around 10 billion. Almost 60% of the population stays in cities, owing to this sustained standard of living in areas with good air and healthy food being quite an arduous task. In conclusion, there is not enough area to cultivate in suburban areas. This idea is briefly explored in the chapter "Smart Indoor Vertical Farming Monitoring Using IoT" [9, 25]. Incorporating human resources for monitoring and cultivating is an exhausting practice. Terrace or balcony gardening is the practice of expanding cultivation in the human habitat. The proposed system adopts artificial photosynthesis, which involves the assistance of flourishing light comprising red and blue LEDs as it suits indoor planting. Agriculturists can monitor the crops through mobile phones with GSM capability. The proposed system can also be employed in suburban areas where there is not enough acreage for horticulture [5, 17].

The IoT has deployment in all fields. Dominant employment of IoT requires gigantic framework coordination in terms of hardware, software, and communication systems to assure conversation between smart gadgets. The chapter "CPS/IoT Ecosystem: Indoor Vertical Farming System" [9, 19, 25] promotes IoT framework credentials for leading operations comprising smart parking, intelligent agriculture, and smart architecture. The proposed work focuses on indoor cultivation systems. Four major categories are in indoor cultivation systems: The hydroponics system [2–5] is growing crops without soil. Aeroponics [8] involves the flora being suspended in air and the water drizzled in the roots of the plants. Aquaponics [6, 7] employs water for both florae and inducing fisheries.

Indoor vertical farming [25] is an action of growing plants in an artificial habitat. The traditional routine of lengthening flora is in exigency of topology locale, meteorological conditions and the field implanting plants is out of the way from the equator to experience lessening frosty climes, sunshine, and to acquire the superior consequences. The proposed system proffers adequate sunlight, nutrients, water-assisting sensors, and actuators for providing the hygienic habitat to grow plants and to obtain finer yields from the crops.

Hydroponics is a tactic of flourishing plants wanting soil [2–5]. The system involves manipulated habitat specifications inclusive of appropriate limits of temperature and moisture extent for the finer provision of crops. The chapter "A Sensor-Based System for Automatic Environmental Control in Hydroponics" proposes the use of sensors and controllers to have supervision over the habitat climatic zone. The system minimizes the consumption of water with the assistance of a technique named drip irrigation. Automatic environmental control investigates the temperature and humidity level to determine if temperature and humidity level are beyond the threshold values. If the temperature is recognized to be above the threshold value and the humidity value is below the threshold value, the system turns ON the fans and investigates the temperature and humidity value in predetermined intervals. Once the temperature values are below the specified threshold and the humidity level is above the threshold, the system automatically turns OFF the fan. If the temperature and humidity compete for the attention of the controllers the temperature is given a preference.

At present agricultural provinces encounter protests due to arduous progression. The new techniques like the IoT, precision agriculture, and blockchain assisting in incorporating smart agriculture, in turn, pave the way to meet future requirements. The chapter "VegIoT Garden: A Modular IoT Management Platform for Urban Vegetable Gardens" [5, 15, 17] comprises a brief and protracted communication protocol to enhance the vegetable nursery administration, assisting in gathering, monitoring, investigating sensor reading changes in temperature and humidity, and investigating of particulars relevant to the specification of flourishing plants [13]. The framework concluded upon the internet facilitated Home Node, an iPhone Operating System device with mobile applications to monitor the stature of plants. The proposed work is verified for seven days and the system assists in detecting the causes of diseases in vegetables and hugely reinforces that we should expect, in the future, transition in climatic conditions, and essential resource reduction. The proposed system incorporates the IoT that grants to gather integration agriculture particulars from flora, fauna, and gases from the environment. Home Node, along with a Raspberry Pi furnished with low-power long-range and GPS communication systems, is planted in the farmer's house. The system also gathers weather variations for IoT platforms and assigns the particulars to the cloud.

The global population count is drastically increasing at present. It seems a hellacious responsibility to proffer nourishment to the people in the upcoming years. The chapter "Farm to Fork: IoT for Food Supply Chain" states that it is necessary to provide sufficient yield from farming methodologies [21]. The proposed system uplifts the IoT in the food chain to face the requirement. Depletion of nourishment due to transportation from farm to vendors as the food products are subject to changes in environmental conditions like temperature and rainfall has a huge influence on food insufficiency. The depletion of food resources results in the demolishment of resources, including water, land, air, and human labor, and the exchange of greenhouse gases. Research says most deaths occur because of food deficiency and malnourishment. Nourishment chains flow from the farmers to the vendors, from vendors to the consumers. In the future, there will be a huge demand for resources that would lead to water assets anxiety, the inadequacy of cultivation, variations in climatic conditions, an increase in global temperature, and an increase in sea level. Smart farming with IoT hugely assists in preventing nourishment deficiency and enables the effectual deployment of resources. The equipment employed in smart farming comprises drones, sensors that promote smart farming, intelligent irrigation facilities, remote monitoring systems, and cattle monitoring systems and reduces the requirement of human expediency. The livestock management incorporates sensors that are attached to keep on monitoring the body temperature of cattle, the disease that may occur due to weather conditions in the cattle. The system does not require any human power to take care of cattle. Drones assist in preventing the wastage of water, detect the area that is not provided with water supply, and provide intelligent irrigation to the fields with the assistance of mobile applications employed by the farmers. IoT accompanying RFID are housed on the heavy goods vehicle. GPS is incorporated in the gadget set up to locate the location of the heavily loaded goods vehicle. The system immensely assists in incorporating smart farming to provide accommodation to the farmers.

2.3 Existing System vs. the Traditional System of Cultivation

Vertical farming is a cultivation medium for sowing and reaping plants in a soilless platform supported with a mixed nutrient solution in a water medium. As the soil is replaced by various mechanical structures to support the plants, the absorption of the nutrient is a cautious one in such cases, as improper feed may result in hindrance to plant's growth. The nutrient solutions are prepared from bird, animal, fish manure or chemically generated solutions like general flora series. The grow bed in the vertical farming technique includes materials like gravel, wood, stone, metal, rock wool, concrete, and solids. The bed system must be exempt from sunlight as it may result in resistance to the growth of the algae in the system.

For sufficient growth of the plant both in soil-based cultivation and vertical farming method, we need to nurture it with a good amount of nutrients. Plants have different nutrient combinations for each stage of growth. The plant's growth cycle demands 16 supporting elements as follows: nitrogen, carbon, oxygen, potassium, hydrogen, sulfur, magnesium, phosphorous, iron, boron, calcium, manganese, molybdenum, chlorine, copper, and zinc. Plant's intakes of nutrients are classified according to their absorption into macro- and micronutrients. Macronutrients such as nitrogen, phosphorous, and potassium, known as NPK, form a crucial in a plant's growth. Micronutrients like copper, nickel, iron, zinc, manganese, boron, calcium, and molybdenum play a role in a plant's good health and fruit development. As the nutrient absorption directly reflects in the growth of the plant, vertical farming techniques such as hydroponics [2–5], aquaponics [6, 7], and aeroponics [8] prove to be better than soil-based cultivation. In these methods, the absorption rate is considerably found in direct proportionality to the number of concentric nutrients absorbed by the roots of the plants in it.

2.3.1 Different Types of Vertical Farming

Vertical farming is classified into three major types of systems based on the structure of cultivation: (1) skyscraper buildings, (2) mixed Skyscraper buildings, and (3) stacked shipping containers.

2.3.1.1 Skyscraper Buildings

In this structure, the vertical stacking of grow bed shelves results in increased production of crops within the closed structure with the usage of controlled environment techniques.

The grow bed shelves are not directly exposed to external climatic conditions like wind, sunlight, and temperature. The skyscrapers are completely shielded with agro-technical needs. One good advantage of vertical farming is mass production in a single space. So, it can be implemented in urban areas to meet the demand of the population. It can also be integrated into a renewable source of energy like solar energy and wind energy due to its structural advantage [12].

2.3.1.2 Mixed Skyscraper Buildings

This method of vertical farming is an integrated one in developing traditional farming techniques with modern farming techniques. This structure has both a controlled cultivation system and open cultivation system. The natural source of light is used to a greater extent in the top space and open space of the building. This method gives exceptional performance in attaining a biased production system.

2.3.1.3 Stacked Shipping Containers

The shipping containers serve the purpose of stacking upon one other in a row fashion for the cultivation of various veggies and exotic food crops like strawberries and blueberries. The shipping container is altered in such a way that it has LED lighting systems, ventilation systems for climatic balance, and sensor systems to surveil the changes and maintain a balance in growth level.

2.3.2 Three Types of Processes in Vertical Farming

2.3.2.1 Hydroponics

This is the predominant method of cultivation in vertical farming techniques [2–5]. This system enables to grow plants without soil but is supported by a rich nutrient solution. Here, the root structure of the system is immersed into the solution and it is refilled again by the water pump from the reservoir tank to the grow tray. This is completely automated with a timing circuitry system. The timer adjusts itself according to the nature of the plant to be cultivated. Once the nutrient solution is flooded in the grow-tray, the excess one is driven back to the reservoir tank through gravity. All micro- and macronutrients are available in this nutrient solution and it supports plant growth. The hydroponics system setup is shown in Figure 2.1.

2.3.2.2 Aeroponics

This system is a quite improved version of the hydroponics system [2–5], where the root structure of the plant is exposed to air with very little water content in it or mist without the soil medium. The challenges in constant exposure to the nutrient solution may result in excess growth or decay of roots. To overcome this drawback in a hydroponics system, the aeroponics system is used [8]. Here, the roots of the plants are nurtured with the solution in the root structure continuously with the help of a nozzle sprayer. This helps the roots to absorb the proper amount of oxygen from the air. This method is more efficient than the hydroponics method [2–5] as it supports not only nutrient solution but also oxygen to root structure. It helps in supporting the good life cycle of roots. Here, only 10% of water is used with a two-fold yield, which proves that the plants are potentially more nutritious in yield. Figure 2.2 is an aeroponics setup system [8].

Plants in Hydroponics System

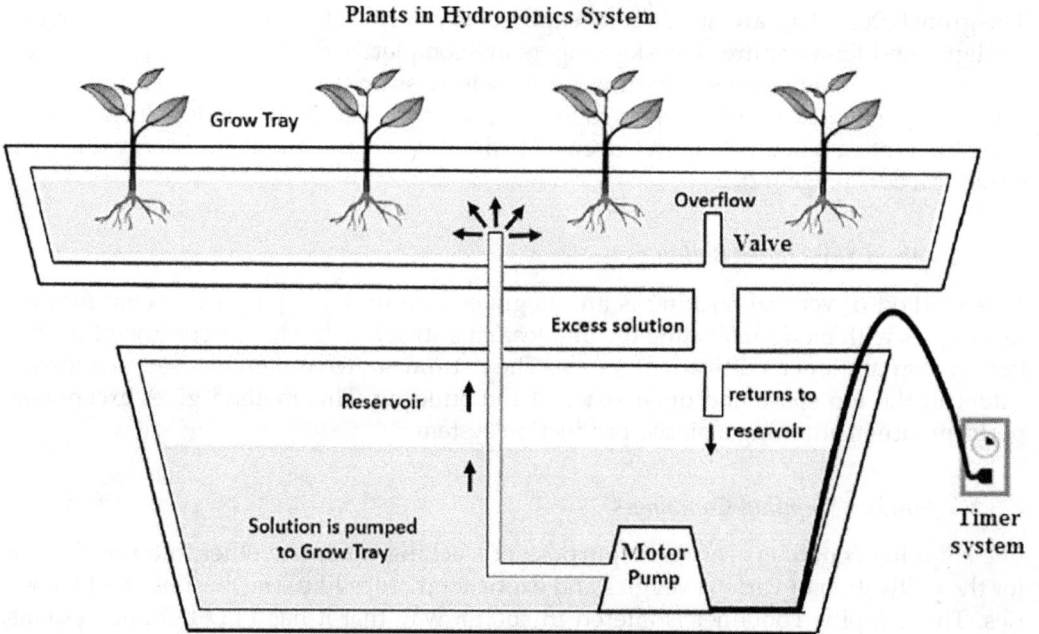

FIGURE 2.1
Diagrammatic view of hydroponics system.

Plants in grow bed

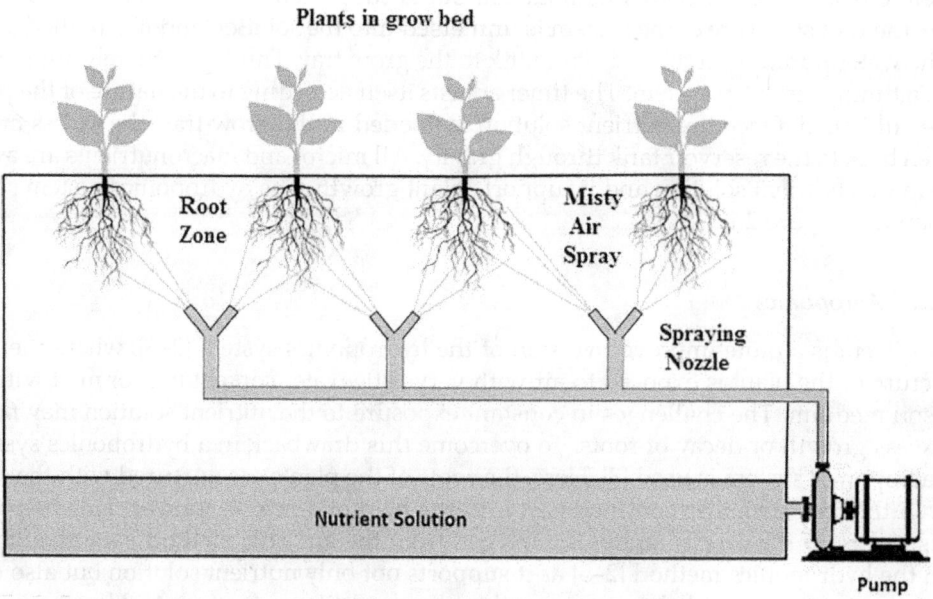

FIGURE 2.2
Diagrammatic view of aeroponics system.

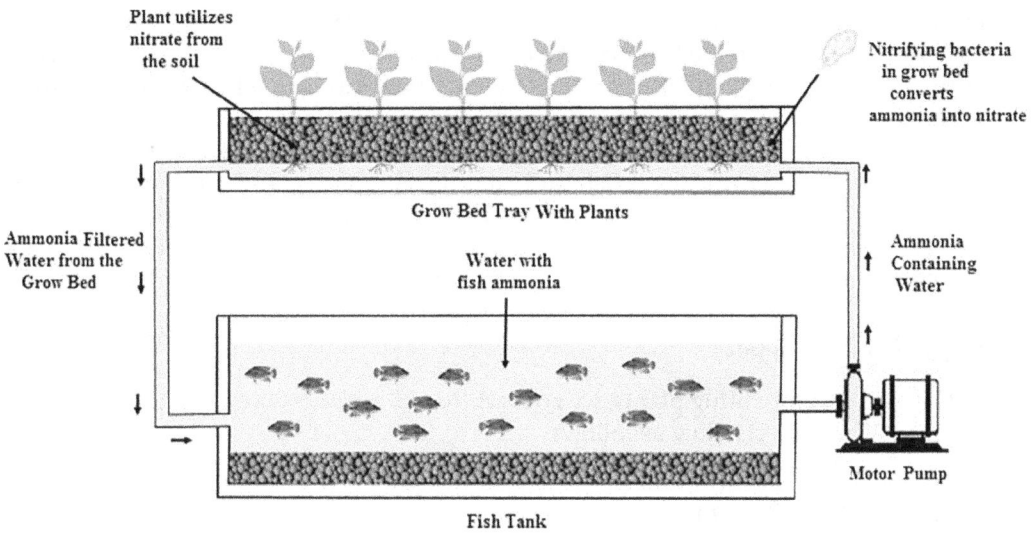

FIGURE 2.3
Diagrammatic view of an aquaponics system.

2.3.2.3 Aquaponics

Aquaculture is a combination of the fish aquarium with a hydroponics system [6, 7]. It is an ecologically balanced system between fish and grow bed. The fish in the fish tank excrete waste in the tank, which is highly nutritious to plants. So it can be used as a growth supplement for plants. The plants absorb the nitrate from the grow bed and filter the water. The process of converting ammonia into nitrite is done by nitrifying bacteria in the medium. The filtered water is driven back into the fish tank. The ammonia-rich water acts as a nutrient solution to the plants in the grow bed. The nitrifying bacterium is the key player in fixing the conversion. The ammonia solids are turned out to vermicompost, which acts as a biofertilizer to plants. Now, the ammonia-rich water is recycled back as freshwater to the fish tank. The major concern for this system is to maintain a stable pH level and ammonia level throughout the life cycle of both fishes and plants. Figure 2.3 presents a working model of aquaponics.

2.4 Proposed System of Vertical Farming

In the aquaponics [6] method of vertical farming veggies like lettuce, spinach, and mint grow well. Here, the proposed model exhibits mint plant cultivation. Mint plant belongs to the herbal variety and it is used for its flavor and aroma. Urban gardeners grow mint in their backyard and balcony through the vertical farming method. Mint plant grows successfully in a humidity level of around 70–75% and a temperature of about 22°C to 28°C and exposure to sunlight about 12 to 14 hours/day.

The proposed model chooses nutrient film technique (NFT) to grow mint plants. Here, the mint plant saplings are placed over the alluvial soil balls present in the plant holder. The NFT layering depends on the count of the production system. The NFT layering for the mint saplings is two parallel structures with a holding capacity of eight. The NFT system is directly attached to the fish tank. Here, the tank is governed by a dissolved oxygen meter to measure the oxygen level in the fish tank. Then, it is aerated with a motor pump to maintain the balance of dissolved oxygen. A timer control unit is placed, supported by a microcontroller unit to sense the data and govern the automation. By this aerated water maintains a good life balance in a fish tank and nitrated water to the mint plants helps to reach its full potential of growth. This system of mint cultivation can be increased by providing organic micro- and macronutrients to the plants through the water. Here, it is a must to closely monitor the system as it has both aquaculture and hydroponics setup [2–5].

2.4.1 Category of Models

In models used for cultivating plants, we go for different varieties according to their nature of usage. They are stated below as follows.

2.4.1.1 Fixed Solution System

In a fixed solution system, the plants are cultivated in containers with nutrient solutions. The containers can be glass jars (in-home applications), tanks, tubs, or plastic buckets. As the water is replaced only after the harvest, tender aeration must be given to the nutrient solution. Here, the water must be at a pH level around 6.5–8.5. This is because of the stagnation of water in the system. The alkaline level increases day by day, which may result in frequent replacement of water in the system. The alternative method for water replacement is aeration. And the level of the solution must be kept below the roots as this helps the roots to get adequate oxygen from the air space available between them and the solution. In this system of approach, the solution is fixed in container space. The container can be of any size and space depending on the need, such as tubs, buckets, and tanks. To maintain a healthy root life cycle, the solution must be aerated well. The dissolved oxygen helps the roots to maintain their metabolism. If the solution cannot be aerated, then it must be in a lower level with roots exposed to air. The bed system is placed above the container space. The plant's presence in the bed system is designed according to production needs. The cultivator needs to decide concerning the nature of the plant grown in it. The plant's circumference and height decide this selection system. The container space must be enclosed completely with black-colored sheets or with materials that filter the sunlight. This is done to avoid the formation of algae in the container space. Here, we need to measure the nutrient solution and its composition each week. This must be done throughout cultivation. The balancing of nutrient solution is done by increasing either the solution or water content. The floating nutrient solution should be constantly monitored with sensors to know its change in composition. The raft system is used in large-scale commercial sectors. It is floating above the solution and the root structure is completely exposed to the solution. So, we need to give an external aeration system for good maintenance. If the aeration ca not be provided frequently, then we can move toward a continual flow system (Figure 2.4) [10, 11].

2.4.1.2 Continual Flow System

In the continual flow system, the roots get immersed in the nutrient solution. Here, automation is relatively easier than the fixed method system. The temperature is keenly noted to govern the inflow and outflow of the solution. As the system is in a flow state, the concentration of the nutrient solution is stored in potential distribution space. The commercial

FIGURE 2.4
Depiction of the fixed solution method.

success of this method involves the NFT version. The streaming nutrient solution through NFT helps the plants for their continuous growth. Here, recirculation is done again and again through the NFT setup to maintain the pH at 5. Less water is consumed when compared to a fixed solution system. The roots exposed to the nutrient solution are supplied with a large amount of oxygen content through the aeration process. The NFT system can be placed in any design concerning gravity. Gravity plays a major role in this process of recirculation. The NFT system is best suited under continuous electric supply or battery power. As nutrient solution recirculation is the heart of the system, it must be taken into account. This can be managed by solar power backed up by a battery system.

Usually, the NFT system is designed with an average flow of nutrient solution at the rate of 1 L/min. During the full growth of the plant, it reaches up to 2 L/min. This flow rate must be taken care of, or else it may create a serious issue in nutrient disproportion. When we go for multi planting system, an extra subsystem must be attached with the NFT system for a nutrient supplement (Figure 2.5).

2.4.2 Method of Selection

The method of selecting a system to nurture the plants is classified based on the variety of adaptations in the design of grow beds.

2.4.2.1 Water Culture System

Here, we implant the plants in a platform and make them float on the nutrient solution in connecting with an aeration system to protect the roots with oxygen. Here, alkalinity is

FIGURE 2.5
Depiction of the continual flow solution method.

increased due to the long duration of the cultivation process. So, we need to give nutrients with acidic components to neutralize them. This must be done carefully else it may turn out to imbalance the ecosystem (Figure 2.6).

2.4.2.2 Ebb and Flow System

The working of the ebb and flow system is to make the nutrient solution get flooded below the tray. The excess solution is drained back into the reservoir. A system with a timer is imparted to monitor the pumping rate of the solution from the reservoir to the system and again back to it. Here, alkalinity and salinity are addressed with the balance of

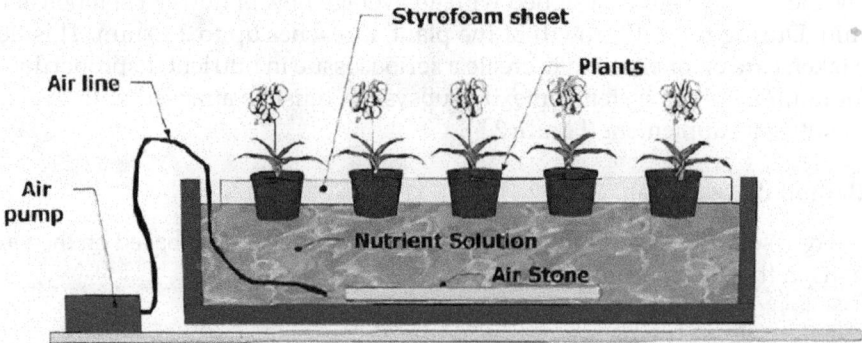

FIGURE 2.6
Depiction of the water culture system.

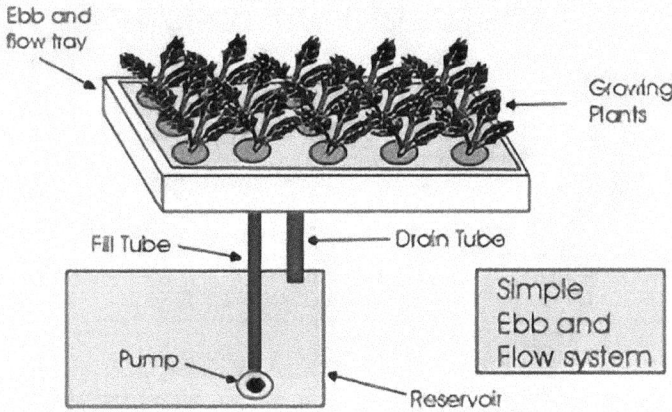

FIGURE 2.7
Depiction of ebb and flow system.

bicarbonates in the growth tray using the nutrient solution. Excess solution is drained out and a pH of 6.5–8 is maintained in the system (Figure 2.7).

2.4.2.3 Drip Systems

This is the most commonly used method of irrigation system for the plants. Here, a submersible pump is placed in the water, and as usual, a timer control unit is placed along with it. The water is dropped in the base of the plant and it is distributed to all the pipelines, which is one of the best methods comparing to previous methods. The pH balance is around 5.5–7 due to the continuous flow of water and nutrient solution (Figure 2.8).

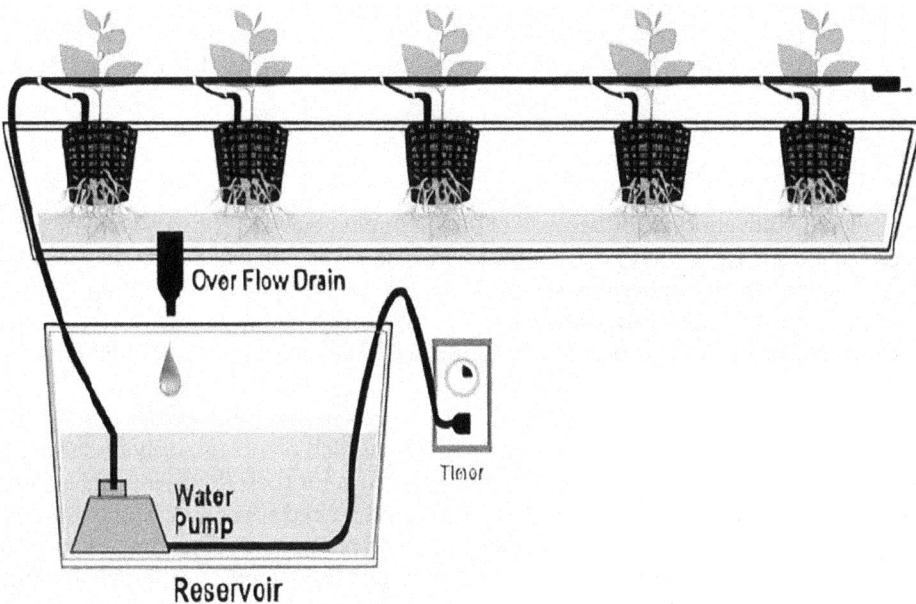

FIGURE 2.8
Depiction of the drip system.

NFT (Nutrient Film Technique)

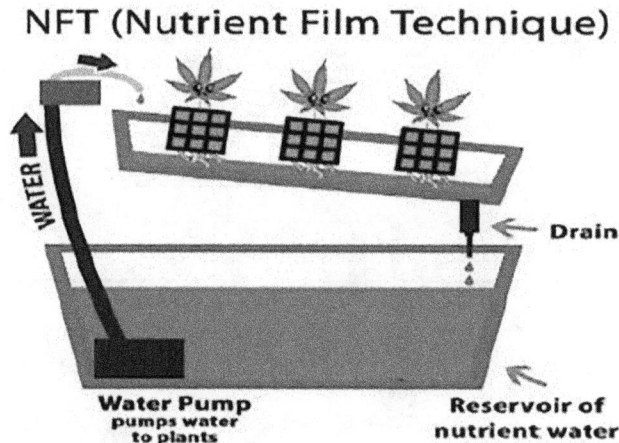

FIGURE 2.9
Depiction of the NFT system.

2.4.2.4 Nutrient Film Technique Systems

This is a well-known method in the process of hydroponics [2–5]. Here, the NFT system uses the method of inflow and outflow in a continuous loop. It is made up of plastic material, and in a few cases, the pipe is coated with black color to avoid algae formation. Here also, the pH is maintained at 5.5–7. The setup is closely similar to drip irrigation. The only difference in the system is the modeling of NFT in a cultivation fashion (Figure 2.9).

2.4.3 Building Structure Based on Plants

In building a structure for cultivating plants, the balance of the ecosystem is a necessary one. Here, the plant, fish, and bacteria are in a unique balance to establish equilibrium among them. The structure of aquaponics proposed is built through an NFT system integrated with a fish tank (Figures 2.10 and 2.11) [3].

2.4.4 Light – A Factor in Cultivation

Light is called a cultivation factor because of its life forcing energy. The exposure of the plant to sunlight makes it grow with good metabolism. The plant will express its nature of growing with abundant photosynthesis. The size of the leaves will be increased with the metabolic activity. It happens only through constant exposure to sunlight and water. The mint plants have shown a promise-able growth concerning this model [3]. Here, the nitrite is supplied from aquaculture to the mint plants. This acts as a nutrient dose to the plants (Figure 2.12).

Vertical farming has the proposal of cultivating plants in an indoor area by using an LED lighting system. Light has a variety of spectra, among which plants use only necessary spectra like red spectrum for their growth and blue spectrum for their flowering state. Here, the plant's red spectrum is further classified into a red and far-red spectrum. The plant saplings use the red spectrum to grow in their initial state. The plant, upon receiving a far-red spectrum, stops the photosynthesis process and distributes the starch to maintain its metabolism. By this process, the artificial LED lighting system provides the respective spectrum for the metabolism of the plants. This is used in indoor vertical farming [9, 19, 25] skyscrapers. The powering up of the lighting system must be backed up by a solar power system. This helps us to unleash the power of an LED lighting system for the plants in indoor cultivation.

FIGURE 2.10
NFT structure for the cultivation of the mint plant.

2.4.5 Recycling Unit for Sustainability

The recycling in the proposed model refers to the process of exchanging the values of the nutrient solution to the plants and clean water solution from the plants to the fish tank. This is the process for a sustainable production system. Various parameters such as temperature, pH levels of fish waste tanks, and dissolved oxygen are observed using sensors from the proposed model. They insinuate from the data observed that temperature must be around 26°C–28°C. When the temperature increases, the pH level starts to decrease

FIGURE 2.11
NFT structure attached with the fish tank.

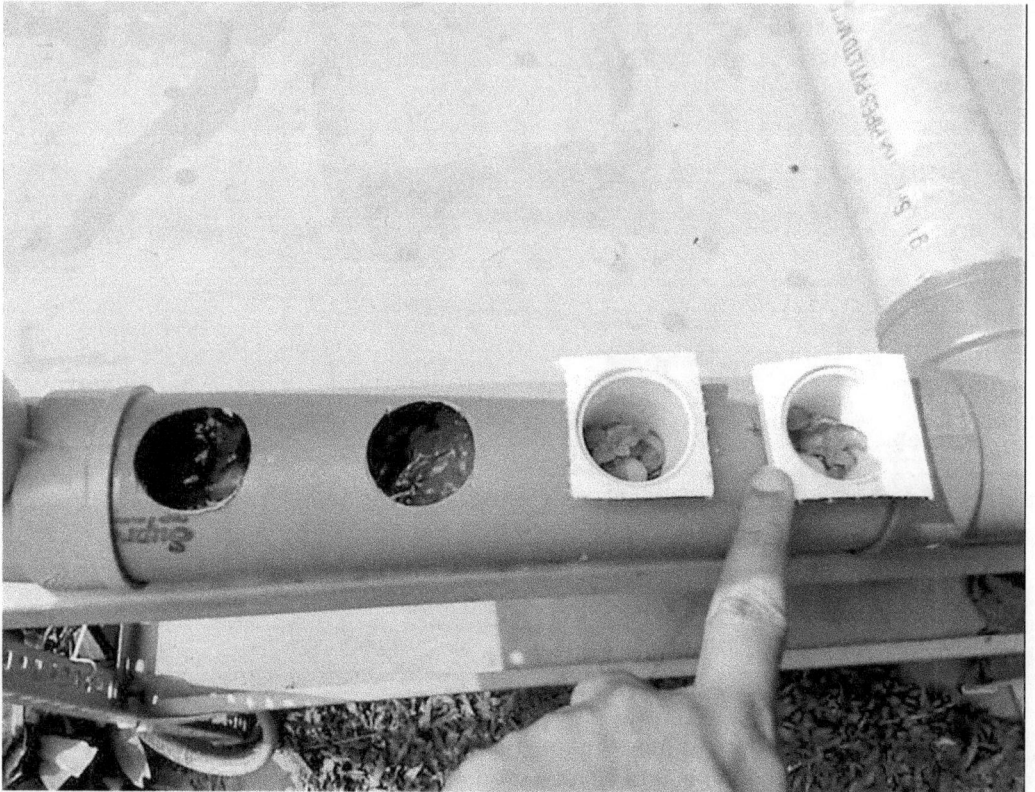

FIGURE 2.12
Mint plant growth in NFT.

gradually. As pH level is the basis for the existence of fish life cycle, it must be restored at a value around 6.8–7.6.

The second governing factor of temperature is dissolved oxygen present in the water. Oxygen is the fundamental one for life's existence [17]. Dissolved oxygen is the broken-up oxygen available in the water. Dissolved oxygen levels must be constantly maintained to have a good force of production. The size that the plants grow to is directly proportional to the size of the fish. In Figure 2.13, we could examine the mint plant growing from a small sapling to a full-size plant.

In the proposed system of cultivation, the plants must have varying pH levels when compared to the pH level in the soil. The soil is a good nutrient provider with its microorganisms, soil air, soil water, and minerals and salts in it. To compensate for the contribution of soil in vertical farming, we need to continuously monitor the pH level. This case is sensitive when aquaculture and hydroponics [2–5] are combined.

2.4.6 Technological Role of IoT

Nowadays, we use modern technology in place of traditional mechanized methods. The technology relies on the data collected from the environment through sensors and actuators. Data collected from the sensors and actuators are stored in the cloud storage space or independent hardware data spaces. By using artificial intelligence and machine learning algorithms such as deep neural networks, the complete automation of farming can be done. Here, automation gives rise to microcontroller boards like Arduino, Rasberry Pi, and

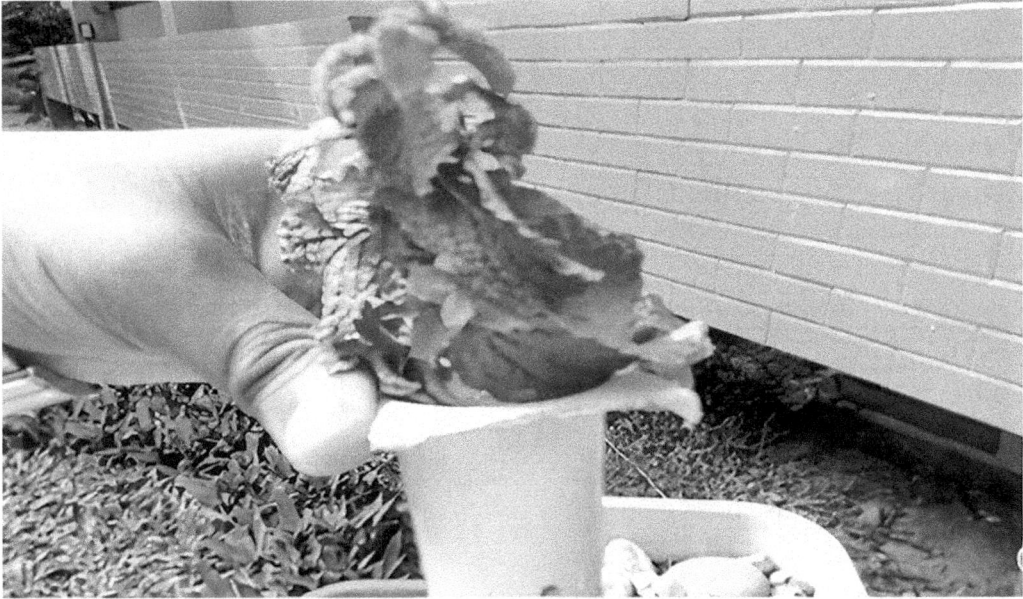

FIGURE 2.13
Fully grown mint plant in NFT system.

BeagleBone boards for the process of control and monitoring. The microcontroller boards help us to connect directly to cloud data storage through the wireless network.

Here, in this automation process, the nutrient solution tank (fish tank) must be checked with proper levels of pH and oxygen content. The system is built with nutrient conversion forms from ammonia to nitrogen. In the proposed model, mint plant cultivation is automated with an Arduino board as a controller with ThingSpeak cloud platform as a storage platform. The Arduino board is connected to the DHT11 temperature and humidity sensor along with other sensors such as pH sensor and dissolved oxygen sensor. The temperature is directly proportional to the pH level of the system. The temperature is varied along with the pH of the system and it is keenly shown in the graph. The ThingSpeak cloud platform uses algorithm status of change in pH with respect to temperature. It has been reflected in the working algorithm in the microcontroller used. The dissolved oxygen sensor in the fish tank is integrated with the pH sensor to calibrate the balance of pH levels in the water. When the pH level is stabilized between 5.5 and 7, then the growth rate is in measurable form. The plant has shown a consistent increase in growth with respect to the size of fish in the tank.

The data collected from the pH sensors in the fish tank and NFT system is shown in Figure 2.14. Here, the data show variations in levels concerning temperature and dissolved oxygen. The entire system has a balancing nature in the pH value with the aquaculture and NFT system. Hence, the continuous monitoring of pH levels in the case of soilless cultivation is achieved through techniques such as IoT, data analytics, and cloud computing [2–4, 17].

2.4.7 Future of Urban Farming

Currently, companies like General Electricals, AeroFarms, Nuvege, and Ecopia Farms have structured their vertical farms in a way to meet the demands of the world. AeroFarms has taken a good initiative of turning out a used steel factory into a vertical farming factory. It sounds to be 75% more productive than the soil-based farming technique. It consumes only 5% of water for its productivity. Here, the highly motivating thing is that AeroFarms

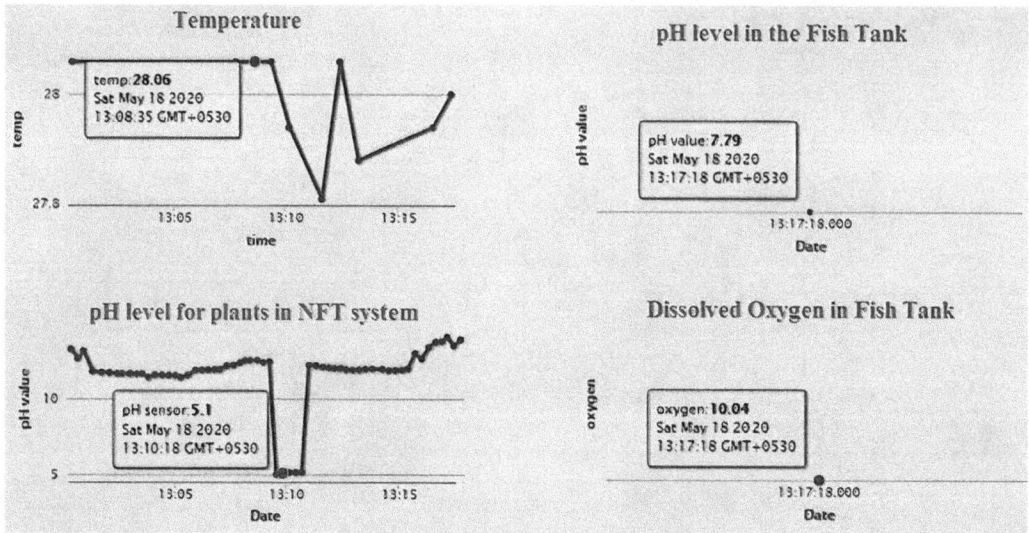

FIGURE 2.14
Data chart of the proposed system from various sensor sets.

cultivate non-genetically modified seeds with the elimination of pesticides and insecticides. This is possible because of the closed-loop system approach in vertical farming.

In Singapore, Sky Greens Farms have started the vertical farm in the year 2009 and experimented with it in a 30-feet tall tower for growing leafy vegetables. The tower can produce around 0.5 tonnes of veggies per day [18]. In Japan, during the earthquake crisis in 2011, the country faced a critical shortage in the food supply. General Electricals and Mirai jointly opened up an indoor farming system to go for a quick fix. In that farm, LEDs with the necessary spectrum for plant growth are used rather than using the full spectrum of light [23]. Philips and General Electricals combined work on the usage of the light spectrum for better growth of different plants. Mirai and General Electricals combined exposure in Hong Kong and Russia to establish vegetable factories [17].

Few local governing bodies of the different countries such as Portland and New York in the United States, United Arab Emirates, China, and South Korea have collaborated with corporate companies like General Electricals to establish vertical farming in their cities of their countries. The state universities, along with local governing authorities of the state, are developing joint venture research centers in vertical farming. One such example is the Illinois Institute of Technology in Chicago, United States. The institute established research prototypes in vertical farming. In their practice of vertical farming, methods like hydroponics [2–5], aeroponics [8], and aquaponics [6] have recorded an increase in production of up to 23 times from the traditional method and 30 times decrease in water usage [24]. These data give us the promise of being able to scope in the establishment of vertical farming for both personal and business purposes in urban areas.

In India, a Mumbai-based startup named Herbivore Farms is the country's first hyperlocal farm implementing the hydroponics method [2–5]. It can be used to grow 2500 types of plants in controlled environmental conditions and is established in a warehouse of an old industrial estate building in Andheri East in Mumbai. It grows superhealthy varieties of green leafy vegetables such as kale, lettuce, and Swiss chard.

Yet another hydroponics company in Chennai named FutureFarms has established itself as one among the agristartups in the country to use the hydroponics method [2–5] of cultivation and setting up farm structures on the rooftop to produce quality crops. It has done 32 commercially successful projects for the enterprises such as Adani Group, Dabur, Aries Agro, Parry Agro, Kalpataru Group, and many in the pipeline. FutureFarms has successfully cultivated 16 varieties of crops under English, Asian, and Indian exotics in 15 acres of land spread across 10 states of our country.

2.4.8 Disruption of Farming through the IoT

The production of crops in vertical farming involves huge data to be processed and analyzed for their effective growth. The day-to-day operations need to be taken care of based on effective usage of data from collection to analysis. This helps the system make intelligent decisions based on the data collected. It reduces human intervention in the process of indoor cultivation. Vertical farming with modern technologies appears to be a promising one, but it has its challenges. Due to its infrastructure investment, the return on investment seems to be a crucial one. With technological support such as automation of the entire process from seed to crop, we can achieve maximum yield from the crops. Bitmantis, a Bangalore-based firm, has created a solution based on the IoT termed to be TheGreenSAGE. It provides apt temperature conditions for the good growth of veggies. It provides automaticity in control of light, water, and other parameters for the growth of plants. The management of growth factors such as nutrients, irrigation, and light management are completely automated [23].

This automation is governed by technological supports such as the IoT, simulation modeling, and big data analytics. It consistently monitors, tests, and reviews the analysis. IoT forms the potential ground in vertical farming. Technologically sound companies can opt for wireless sensors and actuators to receive data from environmental factors such as light, heat, and moisture. Controlled environment automation of crucial governing factors like reducing the intensity of light, switching between the spectra, temperature control cover-structures, and pH balancer can be incorporated in vertical farms backed up with the power of technology. This helps in the proactive management of vertical farming [11, 12, 25].

Few veggies demand good aeration for their growth. In such cases, we go with monitoring the level of carbon dioxide, moisture level, and nutrient value in the solution for sustaining optimal growth, and it is achieved by integrating computerized management and operation system. This computerized mechanism uses an IoT framework for the integrated management of different layers such as sensing, delivering, and controlling. The sensing layer senses environmental conditions like temperature, pH, and soil moisture. Say, for example, a light detection sensor could show real-time variations in the intensity of light. Here, video sensors monitor the height of the plant to know its adaptability for growth. We could use more advanced analysis like spectral analysis to find out the biological stress undergone by the plants. In the delivery layer, information is acquired from the previous layer with various protocols like Zigbee, Wi-Fi, 6LoWPAN, Z-wave, Ethernet, Bluetooth low-energy near field communication, and mobile technologies (2G, 3G, 4G). The next layer comprises intelligently controlled cloud computing systems and individually operated personal digital assistants (PDA). This layer is solely responsible for rising to a better yield by using simulation modeling. It gives varieties of observations and data references for a compact analysis with various constraints like evapotranspiration, soil moisture, weather forecast, bio-stress, and crop management. These large sums of data are taken for data mining by the research groups in companies to predict the futuristic necessity of the crops. This indirectly produces employment opportunities in data science and analytics. In this way, we can use technology to fire up the performance of vertical farming for our food in the future.

The proposed model uses a simple protocol method to establish communication between the sensors and the microcontroller. The sensors are digital sensors that communicate with the microcontroller board through digital control pins in the board and it is processed through the proposed model working algorithm to channelize the output via the internet to the cloud platform. In a cloud platform, the performance analyses of sensors are taken with their data. Thus, the proposed model turned out to be a successful prototype.

2.5 Advantages of Proposed Model

Aquaponics is a combination of hydroponics [2–5] and aquaculture [6]. Here, the presence of aquaculture makes the water a nutrient solution to the plants. The metabolism of fish in fish tanks produces organic waste turning out to be a nutrient solution. The water from the fish tank is passed through a mechanized filter that removes solid substances and passes through a bio-filter that consists of gravel pebbles to convert ammonia into nitrate. The nitrated water is absorbed by the plants and recycled back to the fish tank as purified water. Here in this process, the gravel pebbles act as a bio-filter in the nitrification process.

In this nitrification process, the ammonia in fish waste is broken down into various stages of nitrogen components such as ammonia to nitrate and then to nitrite. Nitrite is the absorbable form of nitrogen by plants. Now, it seems to be an organic nutrient solution to the plants. The recycled water enters the fish tank and it is aerated by a submersible pump to have a good balance of dissolved oxygen in it. Now, the process of removal of ammonia in the water prevents it from being toxic to the fish. The nitrifying bacteria, fish, and plants all combine to make a balanced healthy environment (Figure 2.15).

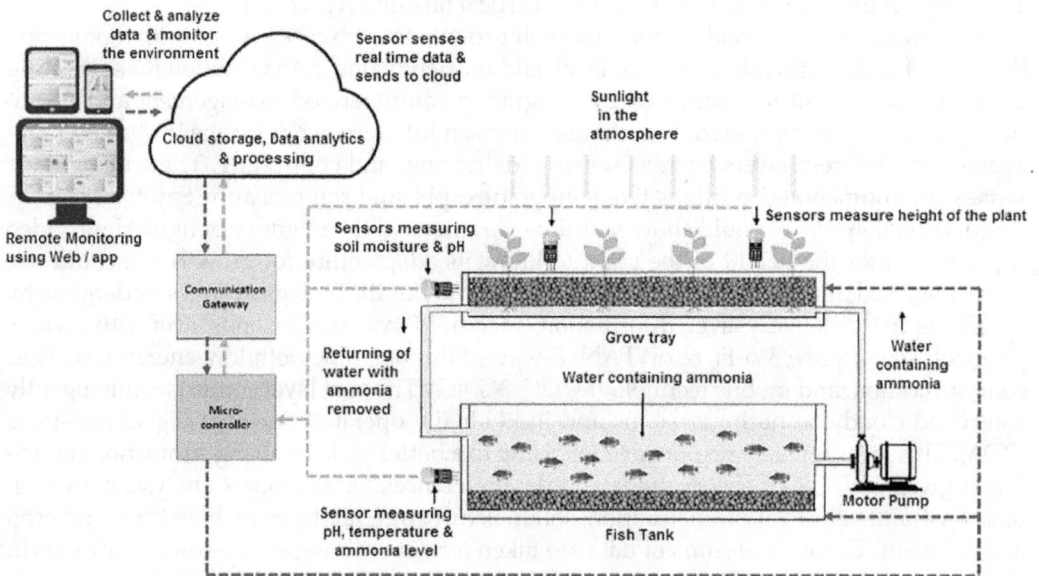

FIGURE 2.15
Diagrammatic view of the proposed model.

2.6 Conclusion

Crops, in traditional farming, are subjected to various natural challenges like global warming, climatic changes, and calamities. Here, the expectation of nature to feed the population is increasing day by day. Moreover, discrepancies in basic amenities for traditional cultivation like the shrinking of farming land and water resources and the lowering of crop yields make farming a challenging one. Here, vertical farming appears as the best alternative to sustainable production. Advanced technologies such as artificial intelligence, machine learning, and data analytics assist vertical farming in achieving technological success in managing it. The motto of technology-enabled vertical farming is to integrate the smart indoor farming system and it is governed by the controlled environment concerning customer demands. Research groups, startups, Non Government Organizations, and technologists have joined together to turn vertical farming into a cost-efficient production system with robotics, analytics IoT, and agri-simulation tools.

In vertical farming, though it has better reliability than many other techniques of farming; live monitoring for the crops is highly demanded. To overcome this demand, a smart farming technique is proposed by incorporating artificial intelligence and machine learning–based algorithms. This makes farming a smarter one as it does not require human monitoring. The idea of the proposed chapter is to make household farming possible with improved monitoring techniques. A mint plant was cultivated based on the proposed technique and tested for results. This mint plantation is made independent by the providence of a smarter environment. Parameters such as water input and growth of the plant are automated and live monitoring is provided. Higher-end challenges like pollination between plants (nature's job) need to be mechanized with automation. On the whole, it is a power-intensive system of farming. To compensate for this, intelligent artificial intelligence–powered systems must be incorporated for the sustainable existence of this farming technology.

References

1. Gupta, M. K. and Ganapuram, S., "Vertical Farming Using Information and Communication Technologies", *https://www.infosys.com/industries/agriculture/insights/documents/vertical-farming-information-communication*, December, 2019.
2. Gokul Anand, K. R., Rajalakshmi, N. R., and Karthik. S, "Hydroponics a sustainable agriculture production system". *International Journal of Innovative Technology and Exploring Engineering (IJITEE)*, ISSN: 2278-3075, (9), (pp. 1861–1867), 2020.
3. Boopathy S., Gokul Anand K. R., and Rajalakshmi N. R., "Smart irrigation system for mint cultivation through hydroponics using IOT". *TEST Engineering & Management*, ISSN: 0193-4120, (83), (pp. 13706–13714), 2020.
4. Siddiq, A., Tariq, M. O., Zehra, A. and Malik, S., "ACHPA: A sensor based system for automatic environmental control in hydroponics". *Food Science and Technology*, (40), (pp. 671–680), 2020.
5. Boopathy, S., Gokul Anand, K. R., Dhivya Priya, E. L., Sharmila A. and Dr S. A. Pasupathy, "IoT Based Hydroponics Based Natural Fertigation System for Organic Veggies Cultivation", *IEEE 3rd International Conference Intelligent Communication Technologies and Virtual Mobile Networks* (pp. 404–409), 2021.
6. Gokul Anand, K. R., Sharmila, A., Karthik, S., and Dhivya Priya, E. L., "Domestic Fertigation Management System for Improved Aquaponics", *ISMAC-CVB 2020 Proceedings of the 2nd International Conference on IoT, Social, Mobile, Analytics & Cloud in Computational Vision & Bio-Engineering*, November, 2020.

7. Gong, S., Angani, A. and Shin, K. J., "Realization of Fluid Flow Control System for Vertical Recycling Aquatic System (VRAS)", *International Symposium on Computer, Consumer and Control (IS3C)* (pp. 185–188) IEEE, December, 2018.

8. Belista, F. C. L., Go, M. P. C., Luceñara, L. L., Policarpio, C. J. G., Tan, X. J. M. and Baldovino, R. G., "A Smart Aeroponic Tailored for IoT Vertical Agriculture using Network Connected Modular Environmental Chambers", *10th International Conference on Humanoid, Nanotechnology, Information Technology, Communication and Control, Environment and Management (HNICEM)* (pp. 1–4) IEEE, November, 2018.

9. Bin Ismail, M. I. H. and Thamrin, N. M., "IoT Implementation for Indoor Vertical Farming Watering System", *International Conference on Electrical, Electronics and System Engineering (ICEESE)* (pp. 89–94) IEEE, November, 2017.

10. Labrador, C. G., Ong, A. C. L., Baldovino, R. G., Valenzuela, I. C., Culaba, A. B. and Dadios, E. P., "Optimization of Power Generation and Distribution for Vertical Farming with Wireless Sensor Network", *10th International Conference on Humanoid, Nanotechnology, Information Technology, Communication and Control, Environment and Management (HNICEM)* (pp. 1–6) IEEE, 2018.

11. Lauguico, S. C., Concepcion, R. S., Macasaet, D. D., Alejandrino, J. D., Bandala, A. A. and Dadios, E. P., "Implementation of Inverse Kinematics for Crop-Harvesting Robotic Arm in Vertical Farming", *International Conference on Cybernetics and Intelligent Systems (CIS) and IEEE Conference on Robotics, Automation and Mechatronics (RAM)* (pp. 298–303) IEEE, November, 2019.

12. Lee, J. and Chuang, I. T., "Living Green Shell: Urban Micro-Vertical Farm", *International Conference on Applied System Innovation (ICASI)* (pp. 1087–1090) IEEE, May, 2017.

13. Merin, J. V., Reyes, R. C., Sevilla, R. V., Caballero, J. M. F., Cuesta, T. R. P. D., Paredes, R. J. M., Pastor, R. M., Ramos, R. T. and Tandog, G. O., "Automated Greenhouse Vertical Farming with Wind and Solar Hybrid Power System", *11th International Conference on Humanoid, Nanotechnology, Information Technology, Communication and Control, Environment, and Management (HNICEM)* (pp. 1–6), IEEE, 2019.

14. Mishra, V. P. and Chaudhry, A., "The Role of Information and Communication Technologies in Architecture and Planning with Vertical Farming", *Amity International Conference on Artificial Intelligence (AICAI)* (pp. 1–3). IEEE, February, 2019.

15. Codeluppi, G., Cilfone, A., Davoli, L. and Ferrari, G., "VegIoT Garden: A Modular IoT Management Platform for Urban Vegetable Gardens", International Workshop on Metrology for Agriculture and Forestry (MetroAgriFor) (pp. 121–126). IEEE, October, 2019.

16. Madushanki, A. R., Halgamuge, M. N., Wirasagoda, W. S. and Syed, A., "Adoption of the Internet of Things (IoT) in agriculture and smart farming towards urban greening: A review", *International Journal of Advanced Computer Science and Applications*, (10), (pp. 11–28), 2019.

17. Boopathy, S., Ramkumar, N., Premkumar, P., and Govindaraju, P., "Implementation of Automatic Fertigation System by Measuring the Plant Parameters". *International Journal of Engineering Research & Technology (IJERT)*, (3), (pp. 583–586), 2014.

18. Dvorsky, G., "Worlds' First Commercial Vertical Farm Opens in Singapore", *http://io9. gizmodo.com/5954847/worlds-first-commercial-vertical-farm-opens-in-singapore*, 2012.

19. Haris, I., Fasching, A., Punzenberger, L. and Grosu, R., "CPS/IoT Ecosystem: Indoor Vertical Farming System", *International Symposium on Consumer Technologies (ISCT)* (pp. 47–52) IEEE, June, 2019.

20. Gómez-Chabla, R., Real-Avilés, K., Morán, C., Grijalva, P. and Recalde, T., "IoT applications in Agriculture: A systematic literature review", *2nd International Conference on ICTs in Agronomy and Environment* (pp. 68–76). Springer, Cham, January, 2019.

21. Nukala, R., Panduru, K., Shields, A., Riordan, D., Doody, P. and Walsh, J., "Internet of Things: A Review from 'Farm to Fork'", *27th Irish Signals and Systems Conference (ISSC)* (pp. 1–6) IEEE, June, 2016.

22. Perera, C., Zaslavsky, A., Christen, P. and Georgakopoulos, D., "Sensing as a service model for smart cities supported by internet of things". *Transactions on Emerging Telecommunications Technologies*, (25), (pp. 81–93), 2014.

23. Mellino, C., "World's Largest Vegetable Factory Revolutionizes Indoor Farming", *http://www.ecowatch.com*, 2015.
24. Ellingsen, E. and Despommier, D., "The Vertical Farm-the Origin of a 21st Century Architectural Typology". *Council on Tall Buildings and Urban Habitat Journal*, (3), (pp. 26–34), 2008.
25. Haris, I., Fasching, A., Punzenberger, L. and Grosu, R.,"CPS/IoT Ecosystem: Indoor Vertical Farming System", *23rd International Symposium on Consumer Technologies (ISCT)* (pp. 47–52) IEEE, June, 2019.

3

Sustainable Smart Crop Management for Indian Farms Using Artificial Intelligence

S. Murugan

Vel Tech Rangarajan Dr. Sagunthala R&D Institute of Science and Technology
Chennai, India

M. G. Sumithra and V. Chandran

KPR Institute of Engineering and Technology
Coimbatore, India

CONTENTS

3.1 Introduction...43
    3.1.1 Significance of Big Data in Agriculture..44
    3.1.2 Role of AI in Indian Agriculture 44
    3.1.3 Significance of AI and IoT in Agriculture..44
3.2 Deep Learning for Smart Farming...45
    3.2.1 Convolutional Neural Networks 46
        3.2.1.1 Pooling Layer...46
        3.2.1.2 Activation Layer 46
        3.2.1.3 Sigmoid..47
        3.2.1.4 Tanh 47
        3.2.1.5 Softmax..47
        3.2.1.6 Rectified Linear Unit 47
        3.2.1.7 Flatten Layer 47
3.3 Related Works on CNN-Based Smart Farming.......................................47
3.4 State of Art CNN Architecture 48
    3.4.1 Dataset Collection 49
3.5 Result and Discussion 49
    3.5.1 Confusion Matrix...50
3.6 Conclusion 52
References..52

3.1 Introduction

With increasing demands and the need for sustainable agriculture, farmers, and other stakeholders must invest heavily in expertise as well as more sophisticated machines and equipment. Smart farming objectives must address concerns such as how much fertilizer to apply, when to apply it, and where it should be applied, as well as the resources needed for plant protection and other related issues. Agricultural companies can better monitor the condition of their crops and receive recommendations that can help them minimize pollution and pesticide use by using the Internet of Things (IoT) to gather data on their crops and then analyzing the data with technologies like machine learning. These advances make farming not only more sustainable but also more productive and profitable. The IoT offers a potential approach to these issues. By adding tracking devices to their crops and sending the data back to a central center that the farmer can access, agricultural companies can track their crops in real time and with greater precision. Artificial intelligence (AI) is being used to diagnose pests, predict the best time to plant, and measure production costs. A novel automated quality assessment method based on convolutional neural networks (CNNs) will be used to evaluate crop quality in the Indian context in this report. The challenges of implementing CNN-based quality assessment in India are complicated by the large amount of data needed to train the models.

3.1.1 Significance of Big Data in Agriculture

Big data applications in agriculture are addressing critical issues such as sustainability, global food security, protection, and increased productivity. Traditional players such as technology and input suppliers provide forums and solutions to farmers. Farmers will either become franchisees in interconnected long supply chains or will partner with suppliers and the government to participate in short supply chains as a result of the application growth of big data in agriculture. Data from GPS-equipped tractors, soil sensors, and other external sources assists in improved crop, pesticide, and fertilizer management, as well as increased production to feed the world's rising population. Sensor data collection removes inaccuracies in manual labor and provides useful insights into yield forecasting. Data-driven farming has minimized crop failures caused by changing weather conditions. By detecting microbes and other pollutants early, data on temperature, humidity, and chemicals can help to reduce the risk of food spoilage. The agriculture industry saves a lot of money thanks to AI and data analytics–driven farming.

3.1.2 Role of AI in Indian Agriculture

AI technology has already begun to change the agriculture market. Large-scale farmers are now harnessing the power of technology, such as autonomous tractors that use GPS, satellite imagery, and AI to plant crops more effectively. Many companies are employing automation and robotics technology to assist farmers in seeking more effective ways to practice sustainable farming, such as avoiding the use of pesticides for weed control, harvesting crops, and diagnosing soil defects, among other things.

Emerging technologies play an important role in allowing farmers to continue on their path to sustainable agriculture. In the current pandemic scenario, technology can be an efficient way to keep farming ongoing by overcoming all restrictions. In addition, problems like population growth, climate change, and food security necessitate novel approaches to increasing crop yield. As a result, comprehending AI's application in agriculture becomes essential.

3.1.3 Significance of AI and IoT in Agriculture

According to a recent market research study, by 2024, the AI agriculture market will have expanded at a compound annual growth rate of 28.38%. Farmers had previously

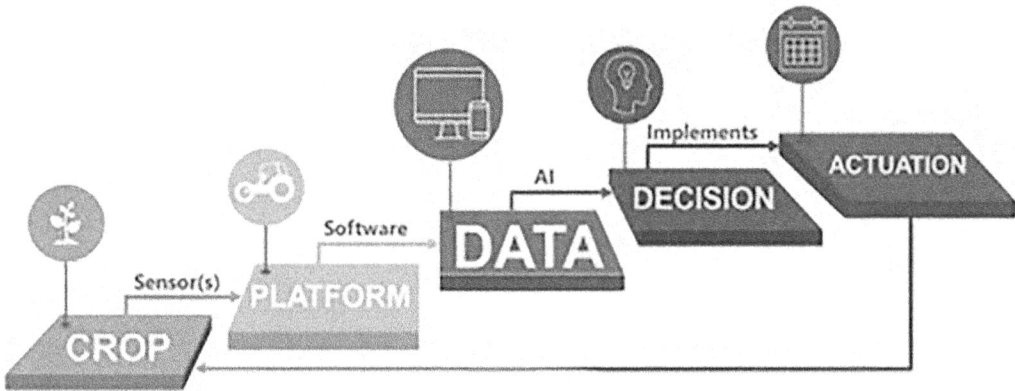

FIGURE 3.1
Overview of AI and IoT toward smart farming.

relied on manual data collection during their working hours to gather information about their crops. This resulted in two major issues. For starters, when doing something by hand, there is usually more space for error. Second, manual monitoring should only be done in intervals, should the farmer spend the whole day checking on the crops and ignoring the rest of his or her responsibilities. Agribusinesses can better track the condition of their crops and obtain recommendations that can help them minimize contamination and pesticide usage by using the IoT to collect data on their crops and then analyzing the data with technologies like machine learning. Both of these problems can be solved by the IoT [1]. However, thanks to software-as-a-service (SaaS) cloud systems, smart farming technology has the ability to go beyond individual farms. In particular, SaaS and IoT hardware can be combined to collect data and provide insights into how farm operations can be handled. To name a few examples, this includes details on crop conditions, weather cycles, harvesting, and soil quality. All of this information will be processed in the cloud, where it will be organized and available at all times, allowing field operations to be tracked from anywhere and at any time. Figure 3.1 depicts how IoT sensors are used in the field and data is stored in the cloud, providing an overview of AI and IoT for smart farming. How the data is analyzed using AI techniques for prediction and decision-making is also addressed [2].

3.2 Deep Learning for Smart Farming

Deep learning is commonly used in a variety of areas including medicine, industry, and transportation, and agriculture is no exception. Shallow learning suffers from a lack of feature speech as well as dimensionality issues. Furthermore, the features must be produced by human experts [3]. Deep learning addresses these issues by automatically extracting these features from raw data. Deep learning is a new AI research direction. Various applications like image recognition, natural language processing (NLP), text processing, facial recognition, and voice recognition, and a variety of other fields have all seen success with it. There are various deep learning architectures like CNNs, deep belief networks, and

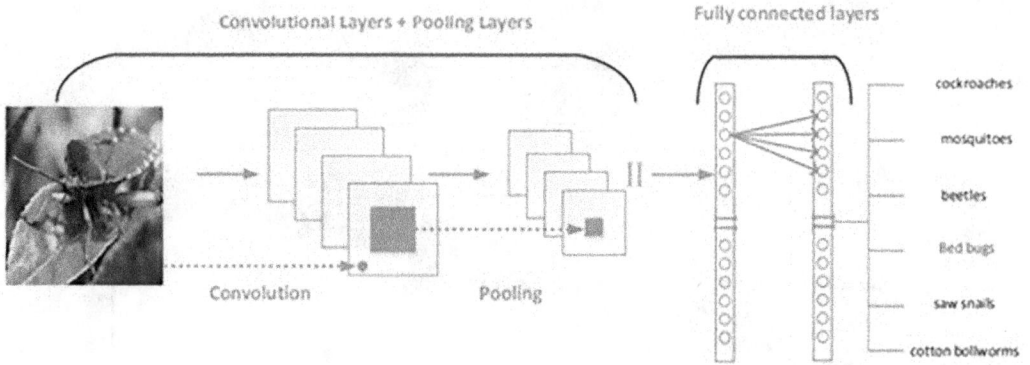

FIGURE 3.2
CNN model for pest identification.

recurrent neural networks. A CNN is a popular supervised learning model that has a lot of adaptability and is frequently used to process image data [4].

3.2.1 Convolutional Neural Networks

A CNN is a multilayer feed-forward neural network consisting of learnable parameters such as weight and bias. In each layer the neurons receive a huge number of inputs and weights and then forward the weight to the activation node. It works on the principle of convolution. Each neuron in this layer is only bound to a part of the actual input image. The mapping of weights shared by all the neurons in the output of the convolutional layer is referred to as a "feature map."

A traditional neural network receives its inputs as a one-dimensional vector, though the CNN processes the inputs as multichannel images. The CNN is built on two fundamental ideas that enable it to perform complex tasks such as image classification and recognition, object detection, object segmentation, and localization. Figure 3.2 depicts a sample CNN model for pest identification for reference [5, 6].

3.2.1.1 Pooling Layer

This layer is the CNN's most powerful and basic building block, which can minimize the number of parameters in the network by reducing the spatial scale of the input dimension. The pooling layer, also known as the downsampling layer, is extended to all layers if the computation expense needs to be reduced. There are three different pooling methodologies available: Min, Max, and Average pooling. The pooling layer is considered to be one of the primary network layers that avoids overfitting and enables the network to be used as a basic model.

3.2.1.2 Activation Layer

The purpose of this layer is to generate some amount of non-linearity between the input and output. This layer's main job is to transform a node's associated signal degree into a signal that can be used by the next layer in the stack. The signal would be nothing more than a simple rectilinear regression function if the activation function was not used.

It is a simple method to decipher, but it is limited in terms of power and complexity. Furthermore, if this activation function is not used in the neural network, then the network cannot learn and model complex data like images, videos, and audio and voice, among other things. Sigmoid, Tanh, Rectified Linear Unit (ReLu), Leaky ReLu, and Maxout are some of the activation functions used.

3.2.1.3 Sigmoid

For binary classification, the sigmoid activation function is used, with values ranging from 0 to 1.

3.2.1.4 Tanh

A shifted version of the sigmoid hyperbolic function is the tangent hyperbolic function. Since the range of the tanh function is between –1 and 1, it is not a zero-centered function, making optimization much easier. Even though it is a major improvement over the sigmoid feature, it still has diminishing gradient issues.

3.2.1.5 Softmax

When there are a lot of classes in a multiclass function, softmax is used. The softmax is simply a probability function that predicts outcomes as a specific class with a predicted value.

3.2.1.6 Rectified Linear Unit

In CNN, ReLU is commonly used as an activation mechanism to prevent vanishing gradients and increase training speed. The following is the definition of ReLU's function:

$$\text{ReLU}(x) = \begin{cases} x, \text{if } x > 0 \\ 0, \text{if } x \leq 0 \end{cases}.$$

3.2.1.7 Flatten Layer

The flatten layer transforms n-dimensional vector representations into a single-dimensional column vector to operate at dense or fully connected layers. The function map is converted into a flatten output by the flatten layer (column vector).

3.3 Related Works on CNN-Based Smart Farming

A review of potential applications using deep learning techniques, specifically CNN, are analyzed and outlined below in Table 3.1 for various smart farming applications.

TABLE 3.1

Existing Works on CNN-Based Smart Farming

Type of Problem	Article	CNN Model Used	Accuracy (%)
Plant detection and localization	Ref. [7]	VGG-CNN-VD16 and VGG-CNN-S	73.07
Plant segmentation	Ref. [8]	CNN-based segmentation approach (U-net)	84
Leaf detection, localization, and counting	Ref. [9]	CNN	88.5
Leaf segmentation	Ref. [10]	Mask R-CNN	91.5
Leaf tracking	Ref. [11]	R-CNN	91.8
Land cover identification	Ref. [12]	FCNN	–
Classification and regression of mutants and treatments	Ref. [13]	CNN	96.8
Weed management	Ref. [14]	CNN	98
Pest management	Ref. [15]	CNN	81.4
Disease diagnosis	Ref. [16]	VGG-FCN-VD16 and VGG-FCN-S	97.95 and 95.12, respectively
Precision livestock farming	Ref. [17]	YOLO-v3	98.46%
Smart irrigation	Ref. [18]	R-CNN	80

3.4 State of Art CNN Architecture

Figure 3.3 displays an overall model of the proposed CNN, where the data is divided into training and validation datasets. The images are 64 × 64 RGB images, which are given as input to the deep neural network. The process of convolution is performed for a filter size of 128 and kernel size of 3. The Leaky ReLu activation is applied to find the probability of prediction and further, Maxpooling is performed. This process is repeated with a convolution filter size of 256 and 512. An uncertainty matrix is used to measure the performance assessment metrics [19, 20].

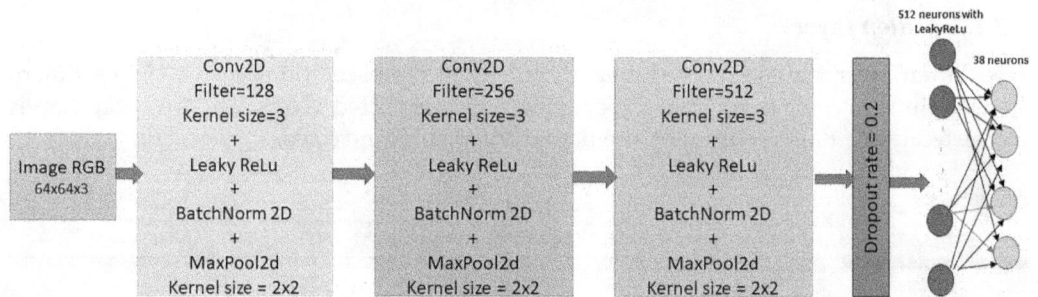

FIGURE 3.3

State of art CNN model for disease classification.

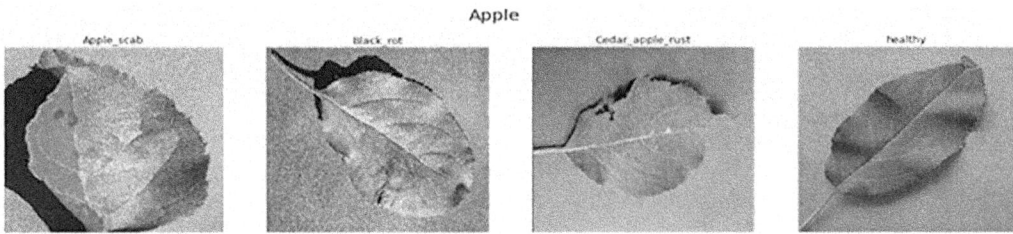

FIGURE 3.4
Sample dataset.

3.4.1 Dataset Collection

The plant village dataset is collected from Kaggle for this study. This dataset contains approximately 87.5K RGB images of stable and diseased crop leaves that are divided into 38 classes of species and diseases. The entire set of data is divided with an 80/20 split, where 80% belongs to the training dataset and 20% belongs to the validation dataset. The training set has 70,295 images and the validation set has 17,572 images. The sample image dataset is given in Figure 3.4 and the split ratio of the dataset based on training and validation is depicted in Figure 3.5.

3.5 Result and Discussion

The proposed state-of-the-art CNN was constructed and evaluated using the parameters stated in Table 3.2.

The performance metrics for the model using CNN are obtained by running the model till epoch 25 and the corresponding values are outlined in Table 3.3. The learning rate for the model is also given in the table. From the data, it can be deduced that increasing the epoch leads to an increase in accuracy as well as a decrease in loss when compared to

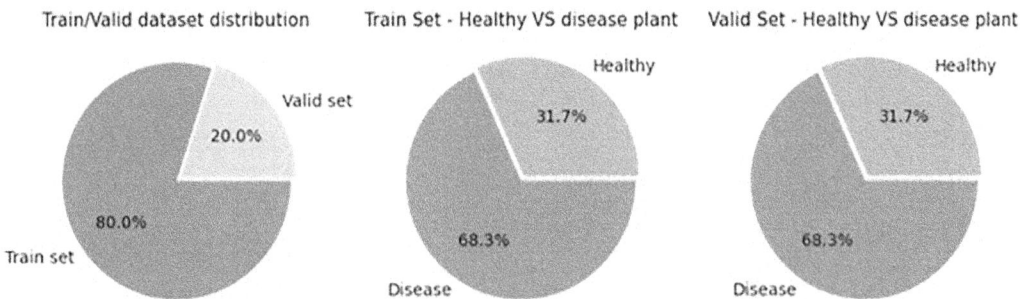

FIGURE 3.5
Split-up ratio for the dataset.

TABLE 3.2

Existing Works on CNN-Based Smart Farming

Type of Optimizer	Adam
Learning rate scheduler	Linear wise, step size = 5, gamma = 0.2
Amount of loss	Categorical cross entropy
Value of learning rate	0.001
Epochs	25

previous iterations. Table 3.3 displays the effects of running the model for epochs from 1 to 25 and the hyperparameters are adjusted to improve the efficiency with the Adam optimizer.

3.5.1 Confusion Matrix

A confusion matrix is a table that shows how many right and incorrect predictions a classifier made. It's a metric for evaluating a classification model's results. It can be used to calculate various metrics to determine the performance of a classification model. The uncertainty matrix, also known as an error matrix, is a tabular format that displays how well a classification model performs on validation data when the positive values of the validation dataset are available. The matrix is used to measure the following parameters:

- **Error rate**: The total number of incorrect predictions divided by the whole dataset.

$$\text{Error rate} = (\text{False Positive} + \text{False Negative})/(\text{Positive} + \text{Negative})$$

- **Accuracy**: A measure of total number of correct predictions made.

$$\text{Accuracy} = (\text{No. of true positive} + \text{No. of true negative})/ \\ (\text{Tot. no of positive} + \text{Tot. no of Negative})$$

- **Sensitivity**: A measure of the fraction of actual positives ranging from 0 to 1.

$$\text{Sensitivity or Recall} = \text{True Positive}/\text{Positives}$$

- **Precision**: A measure of the fraction of actual positives predicted as positive.

$$\text{Precision} = \text{No of True Positive}/(\text{Tot. no of True Positive} + \text{Tot. no of False Positive}).$$

TABLE 3.3

CNN Performance Analysis Using Adam Optimizer

No of Runs (Epoch)	Training Accuracy	Validation Accuracy	Training Loss	Validation Loss	Learning Rate
10	99.68	98.22	0.02	0.05	0.002
15	100	99.14	0.006	0.03	0.00004
20	100	99.37	0.005	0.027	0.000008
25	100	99.38	0.004	0.026	0.0000016

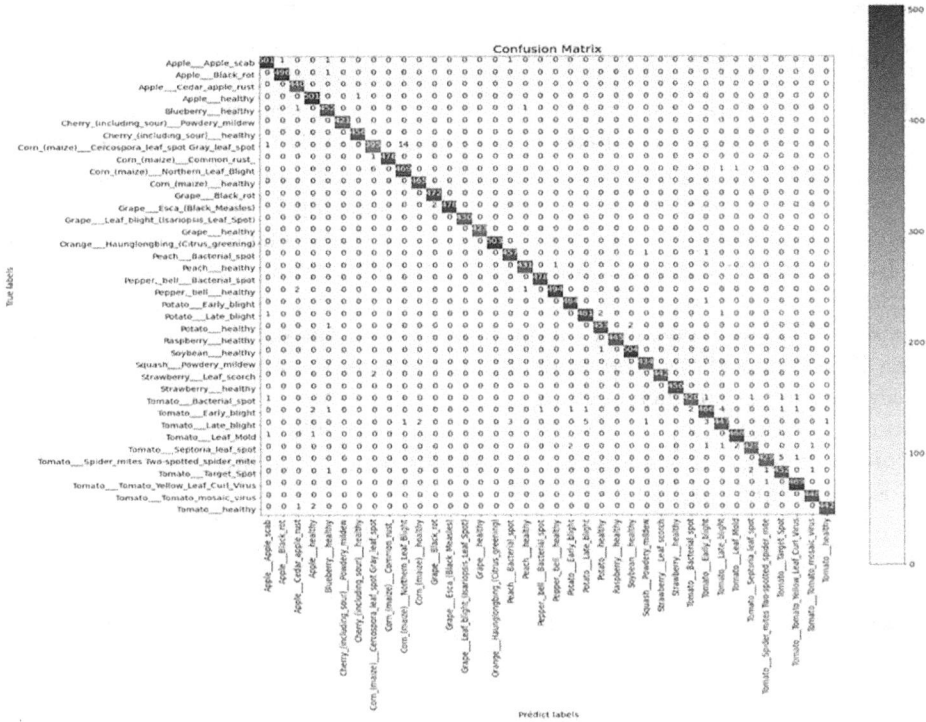

FIGURE 3.6
Confusion matrix.

The matrix known as the "confusion matrix" is considered in order to test the efficiency of the proposed procedures and it is illustrated in Figure 3.6. All the accuracy values of total positive calculations of all 38 classes of unhealthy classifications are obtained over the diagonal.

Due to hyperparameter tuning, an accuracy of 99.38% is achieved by the proposed model, Figure 3.7(a) and (b) shows the final validation loss and accuracy graph after epoch 25, respectively.

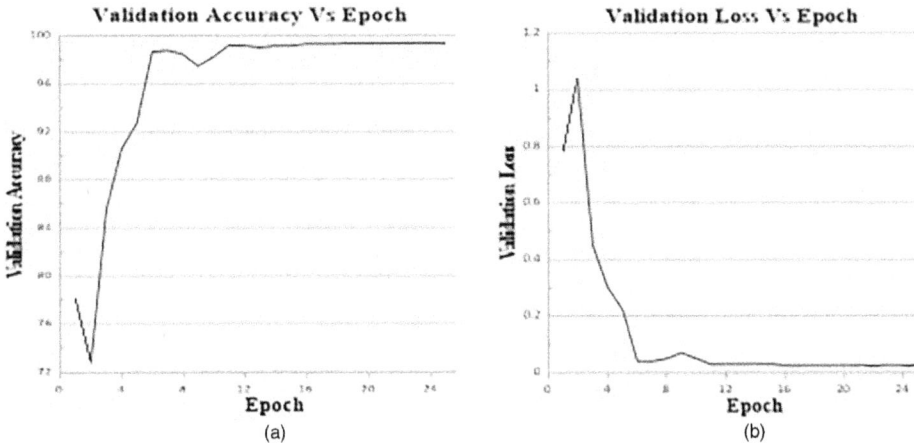

FIGURE 3.7
(a) Training loss at epoch 25. (b) Training accuracy at epoch 25.

3.6 Conclusion

Data-driven agriculture, aided by IoT technologies that incorporate AI techniques, lays the foundation for future sustainable agriculture. Currently farmers and food producers are battling the Covid-19 pandemic to explore new opportunities for smart farming. This study emphasizes the importance of using a deep learning CNN architecture for a variety of smart farming applications involving data analysis, image analysis, and computer vision. In this chapter, the significance of IoT in smart farming has been explored and an illustration of how AI plays a vital role in the prediction of healthy and unhealthy plants is done using CNN. The experiment was performed using the standard plant village dataset from the Kaggle repository. The model's performance was analyzed for over 25 epochs with necessary hyperparameter tuning and it achieved improved accuracy of 99.38% with minimum loss. Indian crops face huge sustainability problems and this can be overcome using smart farming technologies like AI and IoT. And hence this model is applicable in scenarios where smart prediction is highly required.

References

1. N. N. Misra, Y. Dixit, A. Al-Mallahi, M. S. Bhullar, R. Upadhyay and A. Martynenko, "IoT, Big Data and Artificial Intelligence in Agriculture and Food Industry," IEEE Internet of Things Journal, doi: 10.1109/JIOT.2020.2998584.
2. S. Y. Liu, "Artificial Intelligence (AI) in Agriculture," IT Professional, vol. 22, no. 3, pp. 14–15, 1 May–June 2020, doi: 10.1109/MITP.2020.2986121.
3. Z. Ünal, "Smart farming Becomes Even Smarter with Deep Learning – A Bibliographical Analysis," IEEE Access, vol. 8, pp. 105587–105609, 2020, doi: 10.1109/ACCESS.2020.3000175.
4. S. Ninomiya, F. Baret and Z. M. (M.). Cheng, "Plant phenomics: Emerging transdisciplinary science," Plant Phenomics, vol. 2019, no. 2765120, p. 3, 2019.
5. M. Ayaz, M. Ammad-Uddin, Z. Sharif, A. Mansour and E.-H.-M. Aggoune, "Internet-of-Things (IoT)-based smart agriculture: Toward making the fields talk," IEEE Access, vol. 7, pp. 129551–129583, 2019.
6. M. Suriya, V. Chandran and M. G. Sumithra, "Enhanced deep convolutional neural network for malarial parasite classification," International Journal of Computers and Applications, 2019, doi: 10.1080/1206212X.2019.1672277.
7. M. H. Saleem, S. Khanchi, J. Potgieter and K. M. Arif, "Image-based plant disease identification by deep learning meta-architectures," Plants, vol. 9, no. 11, p. 1451, 2020, doi: 10.3390/plants9111451.
8. T. Kattenborn, J. Eichel and F. E., Fassnacht, "Convolutional neural networks enable efficient, accurate and fine-grained segmentation of plant species and communities from high-resolution UAV imagery," Scientific Reports, vol. 9, p. 17656, 2019, doi: 10.1038/s41598-019-53797-9.
9. W. Li, P. Chen B. Wang et al., "Automatic localization and count of agricultural crop pests based on an improved deep learning pipeline," Scientific Reports, vol. 9, p. 7024, 2019, doi: 10.1038/s41598-019-43171-0.
10. K. Yang, W. Zhong, F. Li, "Leaf segmentation and classification with a complicated background using deep learning," Agronomy, vol. 10, p. 1721, 2020, doi: 10.3390/agronomy10111721.
11. L. Wang and W. Q. Yan, Tree Leaves Detection Based on Deep Learning. Geometry and Vision: First International Symposium, ISGV 2021, Auckland, New Zealand, January 28–29, 2021, Revised Selected Papers, 1386, 26–38, doi: 10.1007/978-3-030-72073-5_3.

12. C. Persello, V. A. Tolpekin, J. R. Bergado and R. A. de By, "Delineation of agricultural fields in smallholder farms from satellite images using fully convolutional networks and combinatorial grouping," Remote Sensing Environment, vol. 231, Sep. 2019.

13. B. Veeramani, J. W. Raymond and P. Chanda, "DeepSort: Deep convolutional networks for sorting haploid maize seeds," BMC Bioinformatics, vol. 19, no. S9, pp. 85–93, Aug. 2018.

14. S. Ferreira, D. M. Freitas, G. G. da Silva, H. Pistori and M. T. Folhes, "Weed detection in soybean crops using ConvNets," Computers and Electronics in Agriculture, vol. 143, pp. 314–324, Dec. 2017.

15. R. Li, X. Jia, M. Hu, M. Zhou, D. Li, W. Liu, et al., "An effective data augmentation strategy for CNN-based pest localization and recognition in the field," IEEE Access, vol. 7, pp. 160274–160283, 2019.

16. J. Lu, J. Hu, G. Zhao, F. Mei and C. Zhang, "An in-field automatic wheat disease diagnosis system," Computers and Electronics in Agriculture, vol. 142, pp. 369–379, Nov. 2017.

17. X. Huang, Z. Hu, X. Wang, X. Yang, J. Zhang and D. Shi, "An improved single shot multibox detector method applied in body condition score for dairy cows," Animals, vol. 9, no. 7, p. 470, Jul. 2019.

18. C. K. G. Albuquerque, S. Polimante, A. Torre-Neto and R. C. Prati, "Water spray detection for smart irrigation systems with Mask R-CNN and UAV footage," 2020 IEEE International Workshop on Metrology for Agriculture and Forestry (MetroAgriFor), Trento, Italy, 2020, pp. 236–240, doi: 10.1109/MetroAgriFor50201.2020.9277542.

19. A. Picon, A. Alvarez-Gila, M. Seitz, A. Ortiz-Barredo, J. Echazarra and A. Johannes, "Deep convolutional neural networks for mobile capture device-based crop disease classification in the wild," Computers and Electronics in Agriculture, vol. 161, pp. 280–290, Jun. 2019.

20. Y. Jiang and C. Li, "Convolutional neural networks for image-based high-throughput plant phenotyping: A review," Plant Phenomics, vol. 2020, pp. 1–22, 2020. doi: 10.34133/2020/4152816.

4

Smart Livestock Management Using Cloud IoT

T. Vigneswari

Sri Manakula Vinayagar Engineering College
Pondicherry, India

N. Vijaya

K Ramakrishnan College of Technology
Tiruchirappalli, India

CONTENTS

4.1 Introduction..55
4.2 Monitoring the Health of Livestock 57
    4.2.1 Sensing Layer 57
        4.2.1.1 Vital Parameters of the Livestock 57
        4.2.1.2 Environment Monitoring..61
        4.2.1.3 Movement and Behavior Monitoring....................................62
    4.2.2 Communication and Edge Computing Layer.....................................63
        4.2.2.1 Distributed Ledger..63
        4.2.2.2 Machine Learning Techniques for Decision Making 63
    4.2.3 Cloud Computing Layer...64
    4.2.4 Service Layer 64
4.3 Location Tracking System..65
    4.3.1 Location Tracking Using RFID...65
    4.3.2 Location Tracking Using WSN...66
    4.3.3 Location Tracking Using Ultrasonic Sensor 67
    4.3.4 Location Monitoring Using Auditory Command 68
4.4 Smart Feeding 69
4.5 Conclusion 71
References...71

4.1 Introduction

Livestock refers to domesticated animals raised in an agricultural locale to generate labor and commodities such as wool, leather, milk, egg, and meat. The livestock species plays an inevitable economic and socio-cultural role in the economic upliftment of the rural population. Often farmers face many problems due to health issues of livestock, theft, etc. They have to continuously monitor the activities, location, and health condition of

DOI: 10.1201/9781003185413-4

the livestock so that they do not face a higher sum of mortality loss. Usually, the monitoring of livestock is done manually, which is less reliable and prone to more human errors when it comes to large farms. This necessitates reliable monitoring of livestock that supports the farmers by providing an efficient system that utilizes the recent advancements in computing and communication technologies. The management and tracking of livestock have been made more expedient by deploying the Internet of Things (IoT) and cloud computing paradigm [1, 2].

Livestock management, also known as livestock monitoring or precision livestock farming (PLF) [3], involves deploying IoT devices to track and monitor the health status of livestock. PLF is defined as: "the application of process engineering principles and techniques to livestock farming to automatically monitor, model and manage animal production." The major scope of the PLF is to make livestock farming economically profitable to the farmer, which is made possible by comprehensive management of the farm by continuous monitoring and control of the livestock.

The confluence of the cloud and the IoT has paved the way to a new perspective technology termed cloud IoT, where the services offered by both the technologies are aggregated. Recent research presents many efficacious cloud IoT-based smart livestock management systems that are affordable and easy to use [4]. These technologies have enhanced livestock management by collecting both physiological and environmental parameters that affect the health of the livestock through the sensors and providing vital information regarding them to the farmers via the cloud. The desirable design characteristics of livestock monitoring systems discussed in various works of literature are:

- Low cost
- Robustness
- The design of the device not affecting the behavior of the animal
- Energy consumption of the device
- Reliable communication
- Very less packet loss

The key benefits of smart livestock management using cloud IoT are as follows:

- Creating a census of livestock by providing them with a unique identification (UID) number
- Monitoring the health levels of the livestock and quickly treating the animals and preventing the spread of many contiguous diseases and illnesses
- Automated feeding system to monitor the quantity and quality of food intake by the livestock
- Enabling remote monitoring and tracking of the location of animals and aiding in the prevention of animal theft
- Collecting and storing the data of the livestock on the cloud for future traceability
- Identifying the patterns in cattle health or trailing the transmission of infectious diseases by analyzing historical data available on the cloud
- Voice-based assistants create a user-friendly system to control and monitor the livestock

In this chapter, a comprehensive overview of the existing livestock monitoring system available in the current scenario is detailed. Most of the literature provides solutions for the following three major management components:

- Monitoring the health status of livestock in real-time
- Tracking the location of grazing animals and identification of livestock using a UID for animals
- Smart feeding system

The succeeding sections provide a detailed description of the deployment of cloud IoT for the components listed earlier.

4.2 Monitoring the Health of Livestock

There is an essential need to oversee the health status of livestock to prevent and diagnose diseases before a large number of them get affected. The conventional health monitoring system involves the manual inspection of farm animals to identify the symptoms of disease or any wounds that are prone to the occurrence of more inaccuracy. IoT-enabled livestock management provides enhanced solutions to monitor the health of livestock [5]. Normally they are wearable devices attached to the animal while the built-in sensors help data acquisition and notify farmers about several factors that may eventually affect the livestock health.

This section discusses the layered architectural model for health monitoring that implements cloud IoT for better efficiency. The architecture consists of four layers, as seen in Figure 4.1.

- Sensing Layer
- Communication and Edge Computing Layer
- Cloud Computing Layer
- Service Layer.

4.2.1 Sensing Layer

This layer consists of various sensors to monitor the health of the livestock. The sensors present in this layer monitor the following details to predict the well-being of the livestock:

- Vital parameters of the livestock
- Environment of the livestock
- Behavior monitoring of the livestock.

4.2.1.1 Vital Parameters of the Livestock

The vital parameters measured in general include temperature, blood pressure, heartbeat, and rumination. These parameters are measured by the appropriate sensors [6–8] and data is sent to the cloud through edge nodes. The section below gives a brief outline of sensors that are commonly used for monitoring.

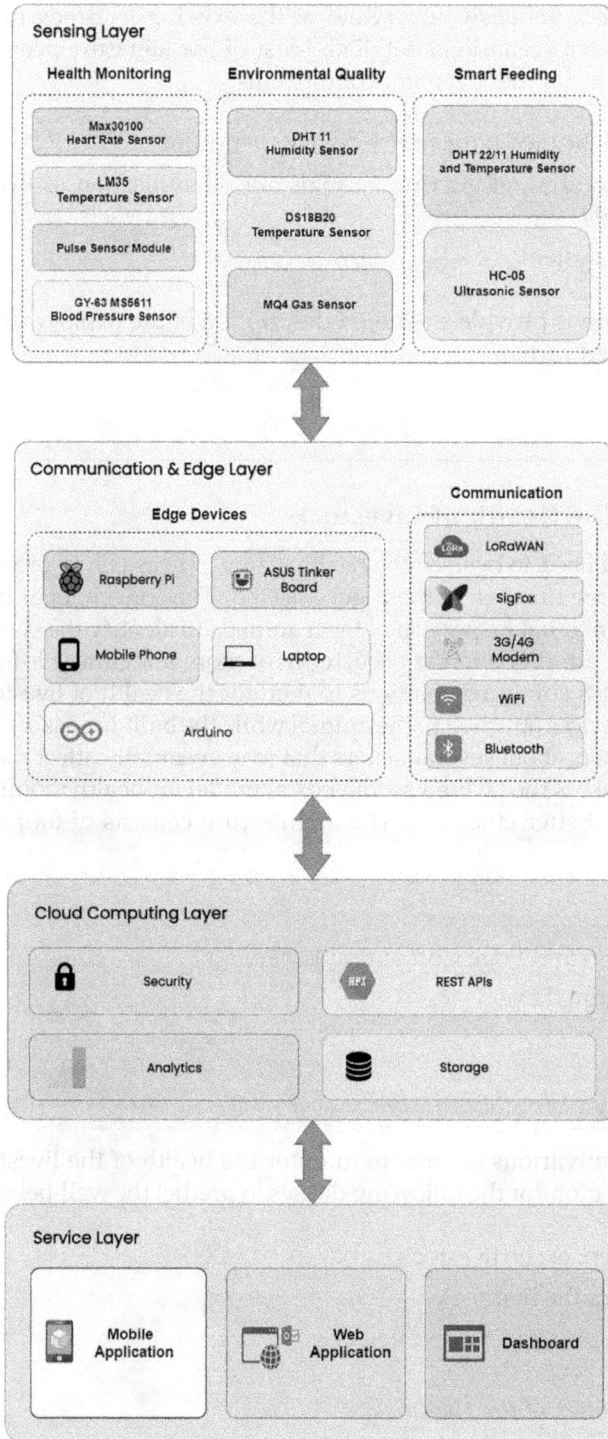

FIGURE 4.1
Architecture of health monitoring system.

4.2.2.1.1 Temperature

The health status of livestock can be predicted by monitoring the temperature as it is a primary indicator for many diseases. The normal temperature of livestock ranges between 38.5°C and 39.5°C. Fluctuations in core temperature can lead to many infections in cattle. Hence it is inevitable to monitor the temperature of the cattle. When the temperature is below the lower limit, diseases like indigestion, milk fever, and poisoning can occur. Similarly, if it exceeds 41°C, foot and mouth diseases, anthrax, influenza, etc., may occur.

Another important factor monitored among livestock is heat stress (HS). The increase in temperature is also a sign of increased heat stress which leads to less milk production that indirectly affects the farmer's profit. The increase in stress level may even lead to the death of the livestock as it causes an increase in temperature and respiration rate. Temperature also has a major impact among livestocks on issues regarding fertility. The time recommended for artificial insemination is 16 hours at a particular temperature, and hence it is essential to monitor the temperature. This ensures that optimum time is utilized for making the procedure of artificial insemination successful.

Different types of sensors that operate at different temperature ranges and varied efficiencies are enlisted below. The sensors chosen based on the requirement are connected to microcontrollers and utilized for monitoring the temperature.

- DS18B20 Body Temperature Sensor Cable – has a unique one-wire interface that provides easy communication.
- ISB-TS45D – an infrared thermometer sensor that measures the body temperature of the livestock; small size, low cost, and better accuracy are the features of this sensor that make it the most commonly used one among such health monitoring applications.
- The LM35 – an integrated circuit temperature sensor that operates between the range of −55°C and +150°C. The temperature output measured by this sensor is directly proportional to the output voltage. The different models of the LM35 sensor operate on different levels.

4.2.2.1.2 Heartbeat

The heart rate of the animal is one of the vital parameters to be constantly monitored to assess the health status of the livestock. The rate of heartbeat per minute acts as a predominant factor in evaluating health status. The normal heart rate of the adult cattle ranges between 48 and 84 beats per minute (BPM). Stress, exertion, anticipation, anxiety, increased movement, and various diseases occur when there is a significant change in the range of heartbeat. The beat per minute for different animals is given in Table 4.1.

The heartbeat of the livestock is normally measured by an electronic device known as Pulse Sensor KG011. The BPM is given as the output along with the live heartbeat waveform by this sensor. The sensor can be easily attached to the earlobes of animals and directly connected to an Arduino controller with necessary extension cables. The heart rate can be visualized as a graph in real time by using the mobile application provided along with the sensor. The pulse sensor enables faster and easier reliable pulse reading as it is enhanced with amplification and noise cancellation mechanism. A timer can be used to analyze the time difference between the digital pulses produced by the transmitter and receiver to provide a time basis for calculating the heart rate. Polar sport tester (PST)-based heart rate monitoring is also suggested as the best method to measure the heartbeat rate of cattle.

TABLE 4.1

Normal Pulse Rate of the Livestock in Beats/Minute

Sl. No	Animal	Heartbeat Per Minute (BPM)
1	Cattle	40–80
2	Young calves	100–140
3	Sheep/goat	70–130
4	Pig	90–100
5	Horse	35–40

4.2.2.1.3 *Electrocardiogram*

Electrical activity of the heart can be detected by using an electrocardiogram (ECG). It is a non-invasive test that records the rhythm, rate, and conductivity of the cattle's heart. The ECG module consists of three Ag–AgCl electrodes where a pair of electrodes are attached to the foreleg and another on the left hind limb. ECG pads and crocodile clips are used to attach the electrodes to the cattle.

AD8232 is an ECG sensor that can be integrated with Arduino, Raspberry Pi, etc., to acquire the electrical signals of the heart. This sensor can even be operated in noisy conditions that occur due to the cattle movement or remote electrode placement.

4.2.2.1.4 *Respiratory Rate Module*

When the respiration rate of cattle increases, it represents stress or pain or weakness or may be a sign of respiratory disease. If the temperature is high in the barn, the livestock may pant to compensate for the heat loss through evaporation. If a livestock pants for more than 100 breaths per minute, it is an indication of severe heat stress.

The rate of respiration can be measured by using a respiratory sensor and the respiratory rate for a different animal is given in Table 4.2. A relative pressure sensor and flexible belt fixed around the chest of livestock are the components of the respiratory rate sensor module. The inspiration and expiration of the lungs create pressure in the chest cavity, which is sensed by the pressure sensor. It calculates the pressure difference to estimate the rate of respiration.

4.2.2.1.5 *Rumination*

Rumination [9] is an essential activity of cattle through which the animal digests the food and it spends almost nine to ten hours for the same. When an abnormality is seen in the rumination process, we can easily conclude that the animal is suffering due to some illnesses like indigestion, metabolic calving, or mastitis. The direct and accurate health status of animals can be measured by monitoring the rumination.

TABLE 4.2

Normal Respiratory Rate of Livestock in Breaths/Minute

Sl. No	Animal	Breaths per Minute
1	Cattle	10–30
2	Young calves	30–60
3	Sheep/goat	16–34
4	Pig	32–58
5	Horse	10–14

Rumination can be monitored by using the following designated devices:

- Non-invasive accelerometers like ADXL 335 to record animal movement
- Built-in microphone for the acquisition of sounds
- Jaw movement monitoring by using pressure sensors.

4.2.1.2 Environment Monitoring

The environment is another important factor to consider in maintaining the health of live-stock. The temperature and humidity of the barn play an indispensable role in the health of the livestock. Also, other harmful gases in the environment can harm the health of livestock to an unexpected level.

Air in the environment are a mixture of various gases such as CO_2, O_2, O_3, etc. Sensors are used to detect the air quality index to determine the quality of air and keep the live-stock stay healthy. An efficient way is by using sensors to detect all possible harmful gases so that we can keep livestock from getting affected by infectious diseases. The architectural components of the environment monitoring module are as shown in Figure 4.2.

Some of those gases that are harmful to the livestock are:

- CH_4 (methane)
- H_2S (hydrogen sulfide)
- NH_3 (ammonia gas)
- CH_2O (formaldehyde)

To detect the above gases sensors such as MQ-4, MQ-136, MQ-137, and MQ-138, which help in detecting these gases, are used. Table 4.3 lists the sensors relative to the type of gas that they detect.

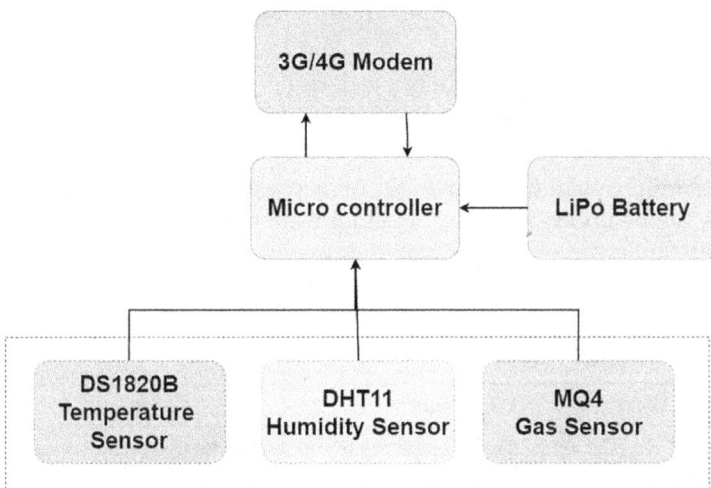

FIGURE 4.2
Environment monitoring module.

TABLE 4.3

Gas Sensors

Sensor	Detected Entity
SHT20	Barn temperature and humidity
MQ-4	Detects CH_4
MQ-136	Detects H_2S
MQ-137	Detects NH_3
MQ-138	Detects CH_2O

The barn of the livestock should be maintained at an appropriate temperature and humidity. The temperature and relative humidity values are combined to obtain the temperature–humidity index (THI), which indicates the stress that affects the livestock and directly affects milk production and reproductive efficiency:

$$THI = 0.8T + \left((RH / 100)(T - 14.3) \right) + 46.4 \tag{4.1}$$

where T is the air temperature in degrees Celsius and RH is the relative humidity expressed in a percentage between 0% and 100%.The normal THI value should be between 70 and 80. The details of variation in the THI and its indication in the health of livestock are given in Table 4.4.

4.2.1.3 Movement and Behavior Monitoring

The well-being of the livestock can be predicted by monitoring their movement and behavior. These sorts of observations, when done manually, provide less accurate results and delayed diagnoses. Hence there is a necessity to observe livestock farms using emerging technologies. Some of the monitoring systems and their scope available in the literature [10–13] are listed below.

- The abnormal movement of animals infers that they have some physical defects in their limbs.
- Deploying cameras that utilize motion detection technology at top angles in the barn to analyze the movement. Animals normally stand together closely when they face some threat. And also, they tend to move at a rapid pace at some time.
- Global Position System (GPS) and 3D collar–mounted accelerometers are used to supervise the behavior. Initially, behavior patterns are recorded and changes in patterns are analyzed to predict well-being.

TABLE 4.4

THI Stress Levels and Consequences

THI	Stress Level	Comments
<70	None	Optimum productive and reproductive performance
70–79	Mild	Increasing respiration rate, seek for shade, dilation of blood vessels
80–98	Moderate	Increasing body temperature, breaths per minute, and salival secretion
>98	Danger	Death of livestock

- The shaking motion of livestock during coughing is differentiated from other motions by using motion history image (MHI)-based technology.
- GY-25 is an accelerometer and gyroscope sensor that is used to monitor livestock movement.

4.2.2 Communication and Edge Computing Layer

The previous layer has an array of heterogeneous sensors to monitor the livestock, which leads to problems related to the management of devices, data acquisition, processing, and transmission. These constraints are solved by the edge computing paradigm. Edge nodes are capable of coordinating and managing all the resources in an IoT environment. Data collection from heterogeneous sources is also made possible through edge nodes which aid in effective data management through homogenization.

The edge layer [14, 15] is responsible for acquiring and pre-processing the information collected by the IoT devices in the sensing layer through low-cost microcomputers like Raspberry Pi, Arduino UNO, Raspberry Pi Pico, and Orange Pi. Even smartphones and laptops at the user end act as edge devices. Hence only the filtered information reaches the cloud for storage and further processing.

Edge-enabled services operate in two different ways. The first one is node-centric services, which operate independently of the cloud, and cloud-centric services, which rely on at least one service from the cloud for operation. Some IoT devices introduce delays due to intermittent connection. Based on this, another type of edge service called delay-tolerant edge service, which is also a cloud-centric service, is introduced that could withstand intermittent connection for a certain amount of time.

The addition of edge computing in between the sensor module and cloud computing offers the following advantages:

- Decreased latency
- Increased support for scalability
- Computation offloading
- Effective bandwidth utilization
- Modularity
- Reliability
- Data compression

The edge layer may also possess these add-on capabilities.

4.2.2.1 Distributed Ledger

The data collected by edge device is hashed and stored in the blockchain by using crypto IoT chips, thus ensuring traceability by maintaining the inviolability of the data. At the same instance, the data also becomes part of distributed ledgers such as Hyperledger, Ethereum, etc. Transmission costs and data traffic are reduced with the usage of distributed ledgers.

4.2.2.2 Machine Learning Techniques for Decision Making

Data analytics techniques are employed for pre-processing and filtering of the data and knowledge is generated at the edge where data is received to reduce the volume of data transferred. Machine learning frameworks such as TensorFlow Lite, Keras, scikit-learn, etc. are deployed for this purpose. Machine learning algorithms like multilayer perceptron, random forests, extreme gradient boosting, k-nearest neighbors, convolutional neural networks, etc. may be used as discussed in Refs. [16–18].

The communication between sensor nodes, edge devices, and cloud platforms is done by protocols [19–21] such as IEEE 802.11, IEEE 802.16, low-rate wireless personal area networks (LR-WPAN), Bluetooth, LoRaWAN, SigFox, and Narrowband IoT (NB-IoT). A suitable communication protocol is selected based on the range, energy constraint, and cost of the device.

4.2.3 Cloud Computing Layer

Cloud computing provides the necessary backbone infrastructure and extends a variety of services that are used by sensing devices, edge nodes, and users. The most common available cloud services are AWS IoT Platform, Google Cloud's IoT Platform, Cisco IoT Cloud Connect, Microsoft Azure IoT Suite, IBM Watson IoT Platform, Thingworx 8 IoT Platform, etc.

Each of these services is built as microservices that have specific tasks or operations to perform and are independent of each other. Communications between the microservices are through REpresentational State Transfer (REST) Application Programming Interface (APIs) [22, 23].

- Handling the pre-processed data obtained from the sensors through edge nodes and storing in the data storage provided by cloud
- Computational power is provided by the cloud to analyze the raw data and provide information about animal health, location, etc.
- Provides decision support system that makes a decision based on the data gathered from various sensors
- Alerts sent to the users to notify any perilous situation that may occur and impart suggestions to handle the same
- Provides data visualization techniques that facilitate the visualization of the data gathered in the form of a graph or relevant visuals
- Security services including authentication and authorization of users and the resources

4.2.4 Service Layer

The service layer acts as a user interface to all the services provided in this cloud IoT-based health monitoring system. All the services deployed in the health monitoring architecture are adopted by the end-users such as farmers, veterinary doctors, and animal welfare officials through this layer, which acts as the frontend. This interface is available either as a web-based application or an android application for obtaining the following services [24, 25].

- Registering livestock and the status of registered livestock can be monitored from any remote area.
- The sensor data stored in the cloud storage is analyzed and any findings are given in the dashboard. Suppose if there is a rise in temperature among the cattle that belong to the same area, a warning may be given to the authorities or farmers of that particular area for taking precautionary methods.

- Individual farmers receive messages about vaccination or health checkups for their livestock from the veterinary doctor. Queries from livestock owners are also sent to veterinary doctors through the same interface.
- Providing a visual representation of the information available and results to the end-users for better understandability. For example, the visualization of any vital parameter monitored among different groups of livestock in different locations may be represented as a graph. This will provide a vivid picture if there is an abnormality in the value of the vital signs among any herd. The concerned authorities can take proper precautionary methods and prevent the spread of infectious diseases based on the symptom.

4.3 Location Tracking System

Livestock location monitoring is a tedious process since the cattle do not stay in a fixed location. Traditionally the livestock is monitored by manual inspection and by fencing cattle. Fencing of cattle is considerably costly and it requires the physical presence of a farmer to keep an eye on the grazing cattle. The farmer needs to prevent the cattle from crossing the boundaries of the fence. The process of visual tracking and fencing is more time consuming and it is a backbreaking task for the farmers. Smart IoT-based livestock location tracking and creating virtual fences by using IoT devices, termed geofencing, can be used to monitor the livestock remotely without the physical intervention of the farmer. This kind of smart livestock monitoring system can reduce the cost of farming, prevent the cattle from infectious disease, and also enable remote tracking and monitoring. Real-time livestock location tracking and monitoring are made possible by advanced technologies [26, 27].

There are several methods to track the position of the cattle. The livestock can be extensively tracked by using navigation satellites and GPRS. The other technologies that are used for geofencing that help to keep the cattle in confined areas include radio frequency identification (RFID), wireless sensor networks (WSN), and low-power wide area network (LPWAN) [28].

It is more essential to monitor the location of livestock as it

- Is used to track the real-time location of livestock in a very large grazing area and bring back them to the barn,
- Notifies the farmers immediately in case of any emergency where an animal separates itself from its herd,
- Safeguards the cattle against theft,
- Enables remote monitoring by providing accurate data at any place, any time, and any environment too, and
- Segregates infectious cattle from healthy cattle and avoids the mass spread of disease among livestock.

4.3.1 Location Tracking Using RFID

Introduction of RFID-based techniques in precision livestock management has made data management and retrieval extremely efficient and offers some advantages like a decrease in recording errors, automation of farm implements, reduction in labor costs, overall productivity optimization, and cost-prohibitive to farmers. RFID [29] can be used to monitor and track livestock accurately and it is integrated with the cloud, which enables continuous tracking of livestock location.

The process of locating the livestock using RFID is as shown in Figure 4.3. Mainly active RFID is used to find the livestock across the farms and also the diseased livestock's location, if any. Here each animal is attached with an RFID tag and a unique ID is allotted to each animal. The RFID-based card reader is composed of a reader module, a microprocessor, and a wireless transceiver. The wireless transceiver sends the location of livestock from the RFID tag by generating a radio frequency signal to the reader. The microcontroller unit processes the information and the details are sent to the farmer's mobile device and also to the cloud storage.

4.3.2 Location Tracking Using WSN

The previous section detailed the use of RFID mechanisms to monitor livestock. Even though the RFID electronic tags are visible and they are easy to administer, one of the biggest demerits of using RFID ear tags is it fails to prevent livestock theft. The tags are easily susceptible to wear and tear. Moreover, the cattle must be within the range of its zone. There is also a chance of losing the tag when the animal is caught in fights or by tree branches. Therefore, this inaccurate history of movement of livestock results in moving to some other techniques like using WSNs and existing GSM networks for enabling real-time tracking and identification of cattle. In recent times, WSN has offered an opportunity to collect data remotely. WSN is a self-configured and infrastructure-less wireless network that has geographically distributed and dedicated sensors. It monitors and records surrounding environmental conditions and organizes the collected data at a central location.

For tracking the animal using WSNs [29] each cattle is given a UID. This UID is used to identify an animal among a group of livestock and the type of the cattle can also be identified. A WSN consists of sensor nodes that are equipped with a power supply, Zigbee module, and microcontroller unit, as seen in Figure 4.4. The master node is equipped with GPRS, which can send data via a mobile communication network to the controller. The sensors are connected to the master node through the Zigbee protocol. In addition, the

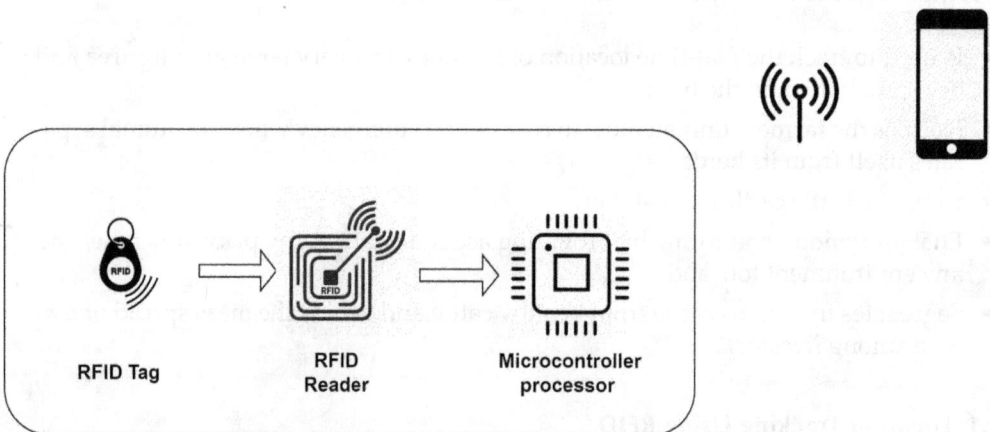

FIGURE 4.3
Location tracking using RFID.

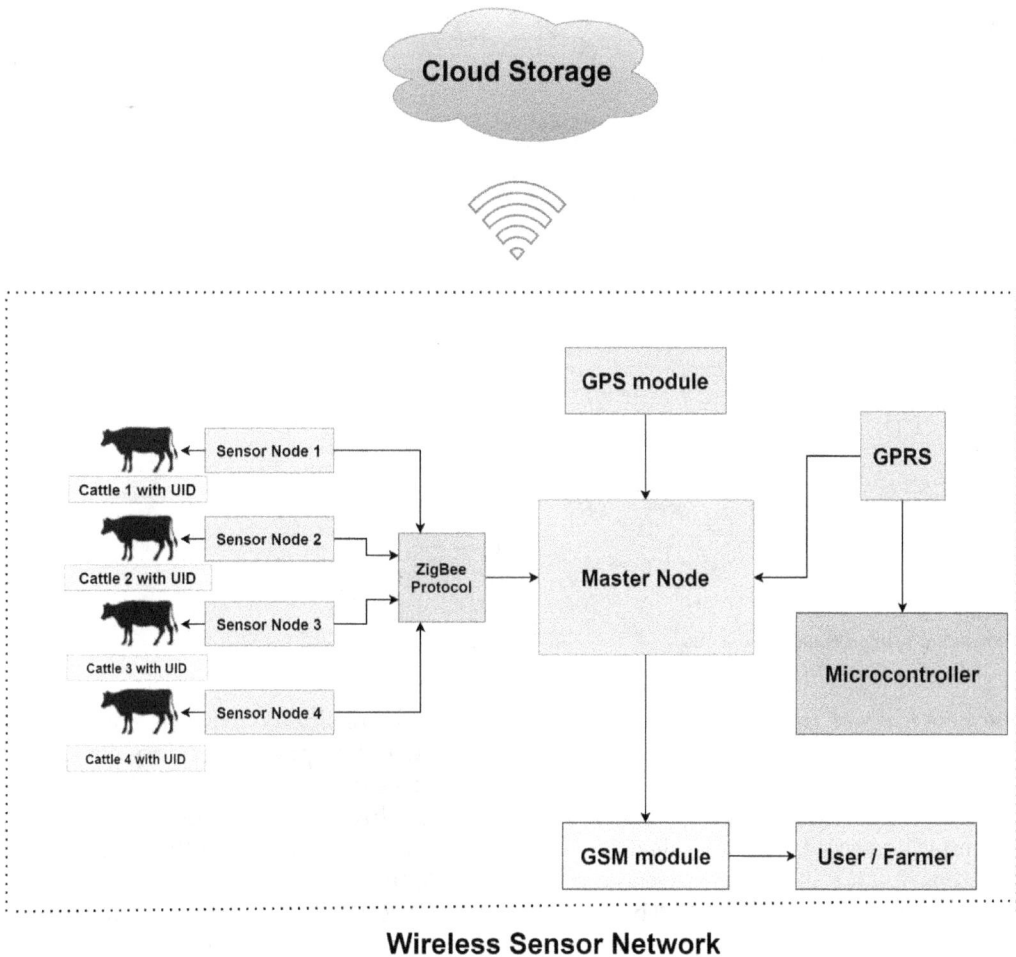

Wireless Sensor Network

FIGURE 4.4
Tracking using a wireless sensor network.

master node is equipped with a block device called GPS, and the GSM module is connected to the master tracking node to give a proper alert to the user or the farmer. Using GSM, the coordinates of latitude and longitude of the animal's location are sent to the user.

Monitoring a wide, open field where usually the grazing of cattle occurs and tracking of animal movement are made easier by using the WSN. A WSN makes it easier for the farmers to know the exact location of the cattle in case of any emergency or cattle loss due to wandering. The activity and the behavior of the cattle can be monitored over a long range of distance and time-to-time communication between remote servers and tracking systems can be achieved effectively with the help of WSN.

4.3.3 Location Tracking Using Ultrasonic Sensor

Location tracking by WSN consumes more device power since it uses sensors like GPRS and GPS. An enhanced system to monitor and safeguard the livestock by providing a geographical safe zone for the livestock by using ultrasonic sensors is suggested as an

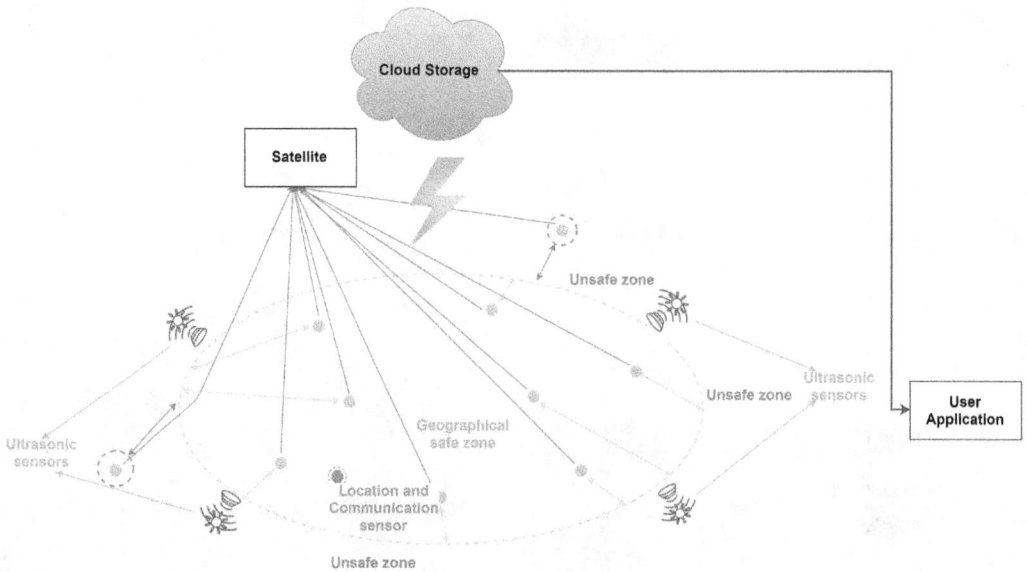

FIGURE 4.5
Geofencing using ultrasonic sensors.

alternate method. In this method, a virtual ellipse that is considered to be the safe zone for the cattle is created, which acts as a geofence. The virtual elliptical boundary consists of ultrasonic sensors which are used to propagate the ultrasonic waves to locate the existence of the livestock. To obtain the location of the livestock through navigation satellites, each animal is equipped with a navigation sensor. If an animal goes beyond the geographical safe zone and if the distance crosses the specified safe threshold value, then the communication navigator gets activated. Then the farmer receives a notification on his smartphone about the livestock crossing its boundary and its exact location.

The mechanism of creating geofencing [30] is explained in Figure 4.5. Consider a herd containing a variety of livestock. To receive the location coordinates and to connect individual livestock with communication satellites each cattle is equipped with a navigation sensor. The location of the livestock is identified by using the location coordinates sent to the satellites through a navigation sensor. By monitoring this location, the farmer can be notified if the cattle are not in the safe zone. The GPS sensor is activated to send the alert/ alarm to the farmer if the ultrasonic sensor detects any cattle crossing the safe geographical boundaries. The geographical sensor communicates with the satellite to calculate the location coordinates of the cattle. The difference between the current location coordinates and the safe zone coordinates is calculated to find how far the cattle are from the defined safe grazing zone.

4.3.4 Location Monitoring Using Auditory Command

The livestock can be monitored by using digital or virtual auditory assistants powered by artificial intelligence. Based on voice command, an intelligent virtual assistant (IVA) can perform a task or a service. In livestock location monitoring systems, the AI-based virtual assistant acts as an entity that issues auditry commands or voice commands to the livestock, as in Figure 4.6. Already in the previous sections, we have seen various

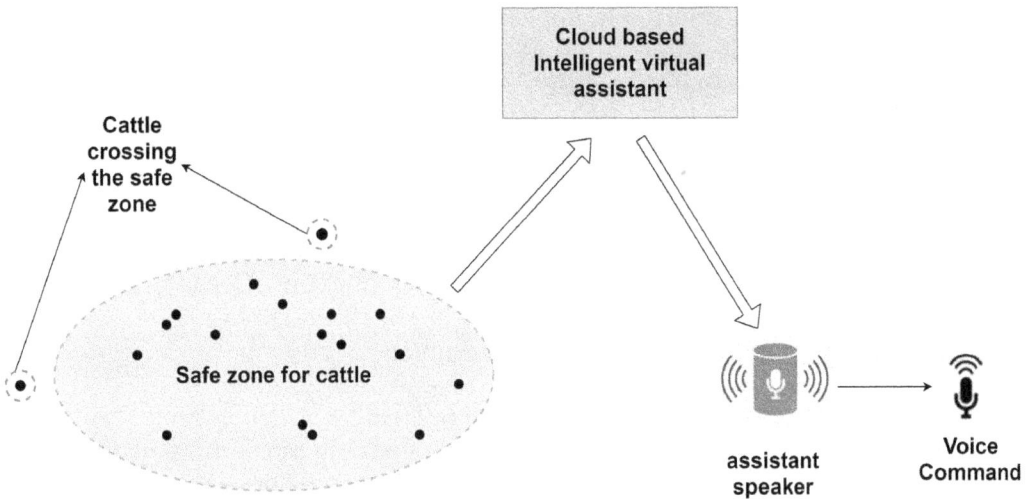

FIGURE 4.6
Monitoring through auditory command.

methods to monitor the location of livestock. The output provided by any of those techniques is fed as input for this IVA, which in turn issues voice-based commands to the livestock.

For example, if we are monitoring the livestock by using WSN, the output representation may be as given in Figure 4.6. When this image is received as input by the IVA, it identifies the livestock that is nearing the boundary of the safe zone. Then it sends a command through the speakers to the cattle that go beyond the geographical safe zone. Thus by deploying voice assistants livestock losses can be prevented, and also it does not always require the physical presence of farmers in the field. Today there are several virtual assistants available on the market. The most popular virtual Assistants are Google Assistant by Google, Cortana by Microsoft, Alexa by Amazon, and Siri by Apple.

4.4 Smart Feeding

Feeding the livestock plays a vital role in livestock management. It includes feeding at an appropriate time interval, quality of food, and the correct mixture of nutrients needed. Improper feeding of livestock affects behavior and health status. The food quality must be ensured as it may lead to many foodborne diseases [32]. Thus feeding and food quality of livestock accounts for the major part of livestock management. The significance of feed and feeding has not gained much attention and it needs more concentration to enable unrelenting livestock monitoring.

Smart feeding or precision feeding utilizes emerging information and communications technologies (ICTs) to increase productivity by associating appropriate feeding techniques to deliver feeds to livestock in a precise manner. Smart feeding enables feeding animals by integrating IoT devices, cloud platforms, and decision-making tools [33]. The smart feed system provides the following functionalities:

- Scheduled feeding of animals at proper time intervals
- The level of food in storage is informed to the farmer to ensure enough amount of food is available to feed the livestock
- The correct quantity of food supply provided to each livestock to avoid excess or less eating, which may lead to ailments like coli infection
- Monitoring the quality of the food to avoid foodborne diseases
- Remote monitoring of whether animals have acquired their food or not
- Suggesting a correct mix of nutrients to be added to the animal feed through data analysis
- Reduced human effort in feeding animals manually

An overall architecture for smart feeding is given in Figure 4.7.

The smart feeding system [34] is deployed with sensors to monitor the quantity and quality of the feed inside the silo. The smart feeding system provides the quantity of food at any instant inside the container by using the IoT device mounted at the top of the silo. It consists of a GH-311 RT ultrasonic sensor mounted on a servo motor that measures the distance between the top of the silo and grains at different angles to calculate the volume of the grain inside the silo. Apart from this, the DHT 11 sensor is also available to measure the temperature and humidity of the grain inside the silo, which implies the quality of the feed. An automatic timer triggers the filling of grains in an animal feeder at scheduled time intervals and a Wi-Fi module transfers data from these sensors to an intelligent cloud server that uses an agent-based architecture. This cloud application does the

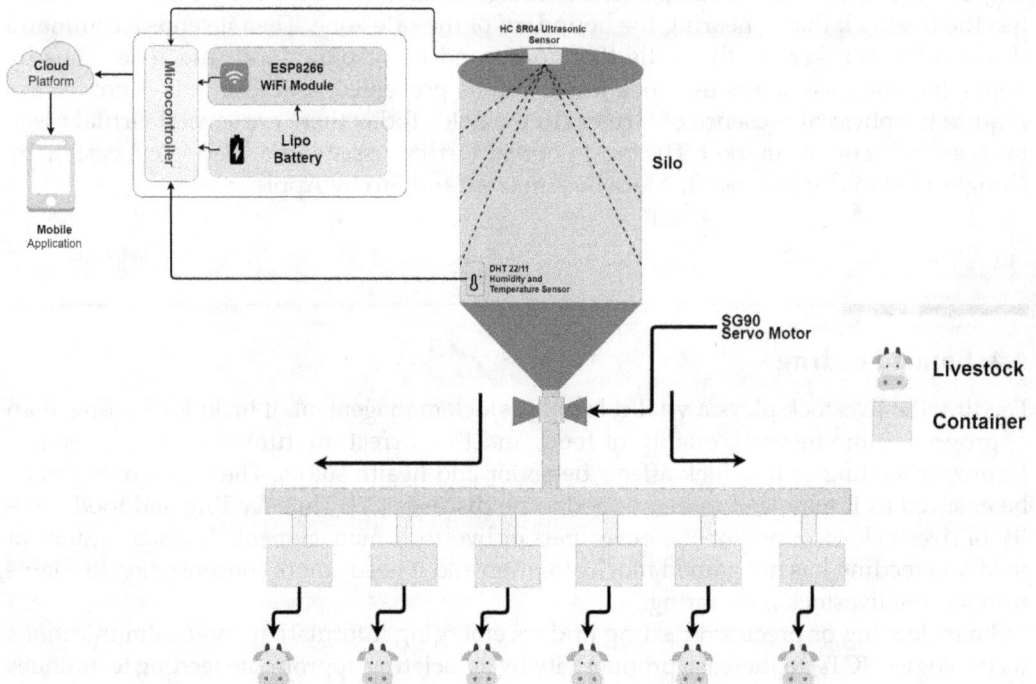

FIGURE 4.7
Architecture of a smart feeding system.

decision-making regarding refilling the silo and the quantity of food to be fed to animals so that excess or low feeding can be avoided and also ensures the quality of food inside the silo. The status of whether the livestock has consumed food or not is also monitored by knowing the volume of grains in the silo. The cloud server receives data from the sensors inside the silo at various time intervals. The data is analyzed and the inference can be viewed through an Android or web application. The farmer receives a notification whenever the volume of feed decreases in the silo. The farmer can refill the silo to prevent the scarcity of food. When the quality of feed deteriorates, the farmer is notified immediately such that feed is not supplied to the livestock. At proper time intervals, the livestock is automatically fed and status is informed to the farmer after every feed [35].

4.5 Conclusion

Precision livestock monitoring includes ensuring cattle health and well-being, prevention from infectious diseases, offering safer and nutritious food, and safeguarding from thefts by using the frontier technologies like cloud and IoT. The integration of these two provides better data storage, management, and data analysis, which aid in decision-making regarding livestock management. This helps framers effectively manage with less human effort and higher accuracy, even from remote locations. This chapter has provided a detailed study of various tools and techniques deployed through cloud IoT for efficient livestock management.

References

1. Saravanan, K., and S. Saranya. "An Integrated Animal Husbandry Livestock Management System." *Journal of Advances in Chemistry* 13, no. 6 (2017):6259–6265.
2. Akhigbe, Bernard Ijesunor, Kamran Munir, Olugbenga Akinade, Lukman Akanbi, and Lukumon O. Oyedele. "IoT Technologies for Livestock Management: A Review of Present Status, Opportunities, and Future Trends." *Big Data and Cognitive Computing* 5, no. 1 (2021): 10.
3. Thakur B.S., and J. Sheetlani. "Analyzing a Cattle Health Monitoring System Using IoT and Its Challenges in Smart Agriculture." In Satapathy S., Bhateja V., Janakiramaiah B., Chen Y.W. (eds.) *Intelligent System Design. Advances in Intelligent Systems and Computing*, Vol. 1171. Springer, Singapore (2021). https://doi.org/10.1007/978-981-15-5400-1_79
4. Berckmans, Daniel. "General introduction to precision livestock farming." *Animal Frontiers* 7, no. 1 (2017): 6–11.
5. Unold, Olgierd, Maciej Nikodem, Marek Piasecki, Kamil Szyc, Henryk Maciejewski, Marek Bawiec, Paweł Dobrowolski, and Michał Zdunek. "IoT-Based Cow Health Monitoring System." In International Conference on Computational Science. Springer, Cham (2020): 344–356.
6. Caria, Marcel, Jasmin Schudrowitz, Admela Jukan, and Nicole Kemper. "Smart Farm Computing Systems for Animal Welfare Monitoring." 2017 40th International Convention on Information and Communication Technology, Electronics and Microelectronics (MIPRO) (2017): 152–157.
7. Yadavalli, Pavan Kumar, Anil Kumar Mutyala, Vijay Kumar Palla, Abhishek Pappu, and Noble Prathipati. "Smart IOT System for Monitoring and Controlling Livestock Parameters." *International Conference of Advance Research & Innovation (ICARI) 2020*. Available at SSRN 3621946 (2020).

8. Neethirajan, S., S. K. Tuteja, S. T. Huang, and D. Kelton. "Recent Advancement in Biosensors Technology for Animal and Livestock Health Management." *Biosensors and Bioelectronics* 98, (2017): 398–407.

9. Kumari, S., and S. K. Yadav. "Development of IoT Based Smart Animal Health Monitoring System Using Raspberry Pi." *International Journal of Advanced Studies of Scientific Research* 3, no. 8 (2018): 24–31.

10. Pratama, Y. P., D. K. Basuki, S. Sukaridhoto, et al. "Designing of a Smart Collar for Dairy Cow Behavior Monitoring with Application Monitoring in Microservices and Internet of Things-Based Systems." International Electronics Symposium IEEE (IES) (2019): 527–533.

11. Bello, R. W., and S. Abubakar. "Framework for Modeling Cattle Behavior through Grazing Patterns." *Asian Journal of Mathematical Sciences* 4, no. 1 (2020): 75–79.

12. Van Nuffel, Annelies, Ingrid Zwertvaegher, Stephanie Van Weyenberg, Matti Pastell, Vivi M. Thorup, Claudia Bahr, Bart Sonck, and Wouter Saeys. "Lameness Detection in Dairy Cows: Part 2. Use of Sensors to Automatically Register Changes in Locomotion or Behavior." *Animals* 5, no. 3 (2015): 861–885.

13. Berckmans, Dries, Martijn Hemeryck, Daniel Berckmans, Erik Vranken, and Toon van Waterschoot. "Animal Sound … Talks! Real-Time Sound Analysis for Health Monitoring in Livestock." *Proceedings of Animal Environment and Welfare* (2015): 215–222.

14. Alonso, Ricardo S., Inés Sittón-Candanedo, Óscar García, Javier Prieto, and Sara Rodríguez-González. "An Intelligent Edge-IoT Platform for Monitoring Livestock and Crops in a Dairy Farming Scenario." *Ad Hoc Networks* 98, (2020): 102047.

15. O'Grady, M. J., D. Langton, and G. M. P. O'Hare. "Edge Computing: A Tractable Model for Smart Agriculture?." *Artificial Intelligence in Agriculture* 3, (2019): 42–51.

16. Riaboff, L., Sylvain Poggi, Aurélien Madouasse, S. Couvreur, S. Aubin, Nicolas Bédère, E. Goumand, Alain Chauvin, and Guy Plantier. "Development of a Methodological Framework for a Robust Prediction of the Main Behaviours of Dairy Cows Using a Combination of Machine Learning Algorithms on Accelerometer Data." *Computers and Electronics in Agriculture* 169, (2020): 105179.

17. Mansbridge, Nicola, Jurgen Mitsch, Nicola Bollard, Keith Ellis, Giuliana G. Miguel-Pacheco, Tania Dottorini, and Jasmeet Kaler. "Feature Selection and Comparison of Machine Learning Algorithms in Classification of Grazing and Rumination Behaviour in Sheep." *Sensors* 18, no. 10 (2018): 3532.

18. Meonghun Le. "IoT Livestock Estrus Monitoring System based on Machine Learning." *Asia-pacific Journal of Convergent Research Interchange* 4, no. 3 (2018): 119–128.

19. Dizdarević, Jasenka, Francisco Carpio, Admela Jukan, and Xavi Masip-Bruin. "A Survey of Communication Protocols for the Internet of Things and Related Challenges of Fog and Cloud Computing Integration" *ACM Computing Surveys (CSUR)* 51, no. 6 (2019): 1–29.

20. Mehboob, Usama, Qasim Zaib, and Chaudhry Usama. "Survey of IoT Communication Protocols Techniques, Applications, and Issues." *xFlow Research Inc., Pakistan* (2016).

21. Germani, Lorenzo, Vanni Mecarelli, Giuseppe Baruffa, Luca Rugini, and Fabrizio Frescura. "An IoT Architecture for Continuous Livestock Monitoring Using LoRa LPWAN." *Electronics* 8, no. 12 (2019): 1435.

22. Pan, Liwu, Mingzhe Xu, Lei Xi, and Yudong Hao. "Research of Livestock Farming IoT System Based on RESTful Web Services." *2016 5th International Conference on Computer Science and Network Technology (ICCSNT)*, IEEE (2016): 113–116.

23. Saravanan, K., and S. Saraniya. "Cloud IOT Based Novel Livestock Monitoring and Identification System Using UID." *Sensor Review* 38, no 1 (2018): 21–33.

24. Park, J. K. and E. Y. Park. "Animal Monitoring Scheme in Smart Farm using Cloud-Based System." *ECTI Transactions on Computer and Information Technology (ECTI-CIT)* 15, no. 1 (2021): 24–33.

25. Jeong, Seokkyun, Hoseok Jeong, Haengkon Kim, and Hyun Yoe."Cloud Computing Based Livestock Monitoring and Disease Forecasting System." *International Journal of Smart Home* 7, no. 6 (2013): 313–320.

26. Bello, R. W., A. Z. H. Talib, and A. S. A. B. Mohamed. "A Framework for Real-Time Cattle Monitoring using Multimedia Networks." *International Journal of Recent Technology and Engineering* 8, no 5 (2020): 974–979.

27. Zhu, Qiming, Jinchang Ren, David Barclay, Samuel McCormack, and Willie Thomson. "Automatic Animal Detection from Kinect Sensed Images for Livestock Monitoring and Assessment." *2015 IEEE International Conference on Computer and Information Technology; Ubiquitous Computing and Communications; Dependable, Autonomic and Secure Computing; Pervasive Intelligence and Computing* (2015): 1154–1157.

28. Casas, Roberto, Arturo Hermosa, Álvaro Marco, Teresa Blanco, and Francisco Javier Zarazaga-Soria. "Real-Time Extensive Livestock Monitoring Using LPWAN Smart Wearable and Infrastructure." *Applied Sciences* 11, no. 3 (2021): 1240.

29. Anu, V. M., M. I. Deepika, and L. M. Gladance. "Animal Identification and Data Management Using RFID Technology." *International Conference o7n Innovation Information in Computing Technologies* (2015): 1–6.

30. Risteska-Stojkoska, B., D. Capeska-Bogatinoska, G. Scheepers, and R. Malekian. "Real-time Internet of Things Architecture for Wireless Livestock Tracking." *Telfor Journal* 10, no. 2 (2018): 74–79.

31. Ilyas, Q. M., and M. Ahmad. "Smart Farming: An Enhanced Pursuit of Sustainable Remote Livestock Tracking and Geofencing Using IoT and GPRS." *Wireless Communications and Mobile Computing*, 2020 (2020): 1–12.

32. Gheorghe, Donca. "IoT Application to Sustainable Animal Production." *Annals of the University of Oradea, Fascicle. Ecotoxicology, Animal Husbandry and Food Science and Technology* 16, (2017): 103–111.

33. Makkar, Harinder P.S. "Smart Livestock Feeding Strategies for Harvesting Triple Gain– the Desired Outcomes in Planet, People and Profit Dimensions: A Developing Country Perspective." *Animal Production Science* 56, no. 3 (2016): 519–534.

34. Prieto, Javier, and Juan M. Corchado. "Intelligent Livestock Feeding System by Means of Silos with IoT Technology." *Distributed Computing and Artificial Intelligence, Special Sessions II, 15th International Conference* Springer 802 (2019):38.

35. Fakharulrazi, Alyani Nadhiya, and Fitri Yakub. "Control and Monitoring System for Livestock Feeding Time via Smartphone." *Journal of Sustainable Natural Resources* 1, no. 2 (2020): 21–26.

5

GIS Systems for Precision Agriculture and Site-Specific Farming

M. Kavitha, R. Srinivasan, and R. Kavitha
Vel Tech Rangarajan Dr Sagunthala R&D Institute of Science and Technology
Chennai, India

CONTENTS

5.1 Introduction...75
5.2 Literature Survey 76
    5.2.1 Geographic Information System Role...76
    5.2.2 Geospatial Technologies for Precision Farming............................76
    5.2.3 Remote Sensing for PA...77
        5.2.3.1 Satellite Remote Sensing 79
5.3 Proposed Framework 79
    5.3.1 Precision Agriculture Using Geospatial Information 79
    5.3.2 Precision Agriculture Using Remote Sensing 80
        5.3.2.1 Data Gathering 80
        5.3.2.2 Yield Monitoring 80
        5.3.2.3 Indicators of Vegetation 81
5.4 Result and Discussion..83
    5.4.1 Classification...83
5.5 Conclusion 85
References...86

5.1 Introduction

Precision agriculture (PA) is a way to control farms that use statistical analysis to ensure that vegetation and soil receive the right nutrients needed for maximum fitness and productivity. The goal of PA is to make sure profitability, sustainability, and environmental concerns are satisfied. Satellite TV for farming, or web-based concrete crop control, is a farming control concept based on detecting, quantifying, and responding to differences in vegetation between fields (Alahi et al., 2018).

PA is a current and sustainable technology that offers possibilities to optimize productivity and decrease natural resource strain. This generation primarily focuses on integrating the unique agricultural understanding of farmers with features of geographical information system (GIS), global positioning system (GPS), remote sensing (RS), and statistics generation (Akkaş & Sokullu, 2017). The use of PA is revolutionizing farm management and reducing agriculture's dependence on harmful chemicals. It is changing people's perspectives on how to farm. Currently, transparency is increasing to reduce harmful chemical use. The use of diversified methods and approaches in agricultural production is allowing it to be more sustainable and safer while causing less harm to the environment. PA is essential for achieving a good result. The control of variability rests at the core of Precision Farming (PF). The generation might be new in India, but the notion of precision control is not. In Indian agriculture, soil conditions, fertility, moisture levels, and so on differ tremendously between fields. Various factors within fields are responsive to unique types of inputs and cultural practices in their respective regions. It is likely that the PF generation that has been thoroughly adopted by developed nations will not be followed in Indian farming structures because of Indian socioeconomic circumstances differing from those of developed nations. India's agriculture continues to face challenges in relevance and adaptability (Bhanumathi & Kalaivanan, 2019b). An important use of geospatial technology could be in PA, according to a recent literature review.

5.2 Literature Survey

5.2.1 Geographic Information System Role

GIS is a machine-based framework for gathering, preserving, interpreting, analyzing, planning, and visualizing spatial data. GIS portrays a closer perception of information for identifying collaborations, designs, and situations that help individuals make more informed decisions in their daily lives (Adeyemi et al., 2017). A GIS software program is largely used to maintain datasets and integrate PA information via remote sensing, as well as to provide many options for analyzing geospatial information. There are numerous modern GIS programming languages combining raw materials, such as maps, base maps, imagery, spreadsheets, and features, freely standardized by many standards like GeoMedia, OpenStreetMap, ArcGIS, GRASS GIS, and QGIS. In addition, it is embedded with publicly available tools for tackling difficult situations (Barik et al., 2018). Rather than providing the user's location, the GPS displays the user's altitude, longitude, and latitude, making it ineffective for finding a place. GIS, on the other hand, uses a computerized tracking process to provide information about where you are on a map. Moreover, the topography surface can be viewed in two-dimensional and three-dimensional modes. In the process, a GIS method is combined with GPS, producing important information about data transmission locations and satellite imagery mapping to the associated farmer's enrolled cropland.

5.2.2 Geospatial Technologies for Precision Farming

The common geospatial technologies are (i) remote sensing, which is the collection of images from space or from camera and sensor systems in an aircraft (Bhanumathi & Kalaivanan, 2019a). There are several satellite communication picture suppliers that develop images with one meter or key characteristics. (ii) The GPS, a satellite network operated by the United States Department of Defense. It provides a precise unit vector to military users and civilians with the appropriate receiving devices. In a few years, a parallel European program known as Galileo will be active, and a Russian network is operating, but it is limited (Bhardwaj et al., 2016). (iii) GIS stands for Geographic Information

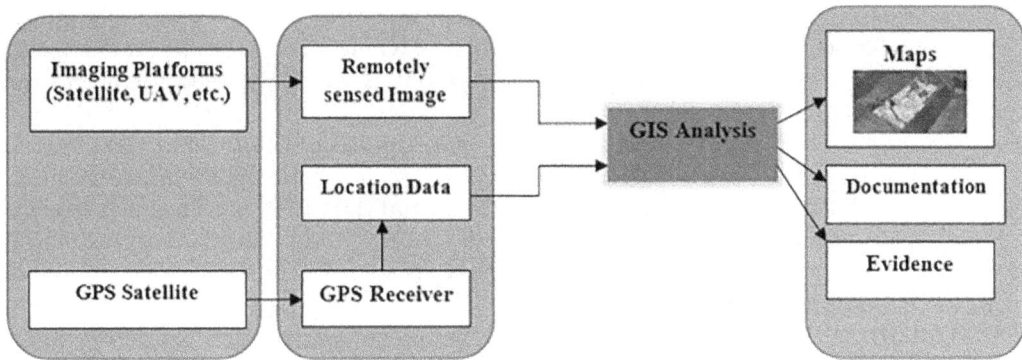

FIGURE 5.1
Geospatial technologies.

System, which is a structure for capturing, analyzing, storing, organizing, and presenting various forms of geographical and spatial data. For any other type of information, GIS uses space and time as the important variables. Figure 5.1 explains the geospatial technologies.

5.2.3 Remote Sensing for PA

Remote sensing can be defined as the process of gathering data from non-contact measurements of scattered or generated radiation from a particular substance (Campos et al., 2019). Reflection and radiation are two properties of the item that are commonly used for remote sensing. In addition to the physical and chemical characteristics of the specific object, the geographical environment, such as leaf moisture, chlorophyll content, and temperature, influences signal reflection or emission from the object. Chlorophyll, which is a chemical substance found in plants, releases radiation that is inversely proportional to the absorption of infrared radiation (Fang et al., 2018). Multispectral images are used to measure emitted signals at various levels including Green Blue Red (GBR), Near InfraRed (NIR), and Short-wave Infrared (SWIR). The infrared, green, and red indices are frequently used to determine estimation methods in agriculture. The Normalized Difference Vegetation Index (NDVI) Index is a standardized method for calculating vegetation index (Foughali et al., 2018). These indicators are used to analyze specific properties such as organic manure, LAI, crop count, moisture content, staple crops, and water level. Chlorophyll content in agriculture is very sensitive to changes in the red and green spectrum (Hammoudi et al., 2018).

For example, plants that produce high amounts of chlorophyll reflect more blue and red spectrum light than ultraviolet or green spectrum light. On the other hand, the high red frequency reflects less chlorophyll. The study of optical to infrared signal transduction can provide information on the cellular architecture of plants and can therefore be used to measure stress and production. This spectral range has a better response to stress due to changes in chlorophyll concentrations, and it also calculates the LAI more accurately than the red and green ranges (Im et al., 2016). Satellite-based platforms and ground-based platforms can both be used for satellite data.

The ability to determine spatial patterns of crop yield, N stress, salinity, soil temperature, potassium, calcium, phosphorous, carbon, moisture, soil pH, water stress, and soil organic matter (Srinivasan et al., 2019) is essential with earth remote observation, also known as local satellite data, which is when IoT devices are mounted to a tractor, sprayer, or other pieces of farm equipment. Sensors are used to control the agricultural production process, such as irrigation, fertilizer, and insecticide. Remote earth observation is less impacted by weather and is a better option than cloud satellite data because of its less weather-dependent characteristics. López-Martínez et al. (2018) reported a method for analyzing earth's spatial data that used a linear motion irrigation system with six thermal cameras and a rolling radar station to detect aerial temperature and agricultural dryness. With the data gathered, the grower was able to open the irrigation actuator.

The following are some examples of surveillance applications in agriculture: (i) analysis of the cropping (Mulla, 2013) method; (ii) agriculture dryness evaluation and tracking; (iii) soil analysis and tracking; (iv) controlling water management; (v) assessment and controlling the agricultural area; (vi) frequency prediction; and (vii) pest and diseases diagnosis The following are the constraints of remote sensing in agricultural production:

1. The sensor's simultaneous monitoring of the crop's reflection radiation generates different spectral confusion. Figure 5.2 shows the remote sensing process in PA.

2. Scaling concerns arise as a result of the inadequacy of the relationship between the real consideration and distant sensing data, and it also provides insufficient information for analyzing past data.

3. Low precision in images obtained from soil and water because of organic matter, microstructure, and wetness. Describing soil and water properties is extremely difficult.

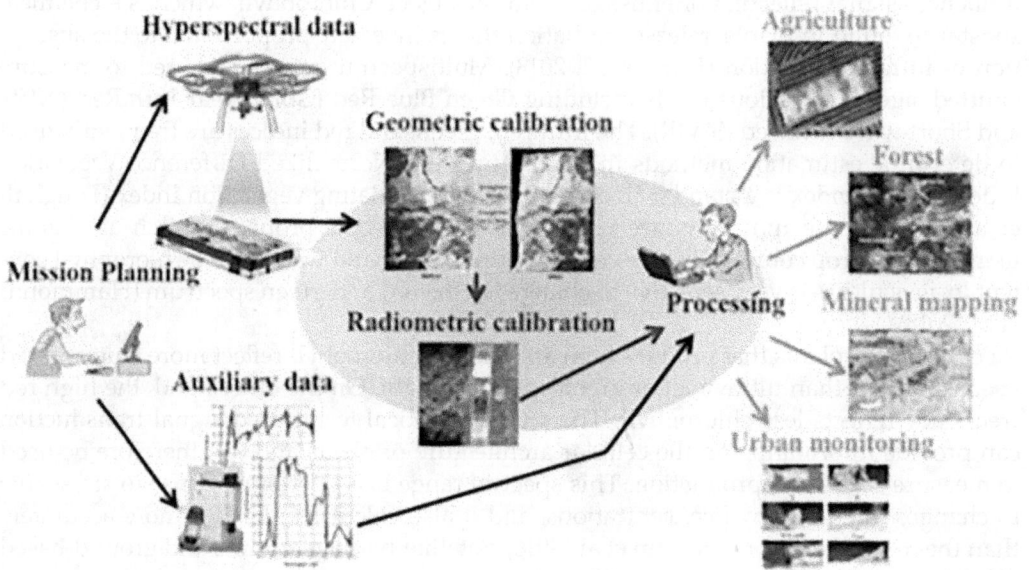

FIGURE 5.2
Architecture for the ground, remote, and aircraft sensing methods organized for PA.

4. Crop varieties are difficult to classify.

5. Weather conditions have a significant impact on the use of passive sensors in remote sensing.

5.2.3.1 Satellite Remote Sensing

Satellites have been used for remote sensing data in agriculture since the 1970s. They are specifically designed for large-scale crop classifications. Remote sensing programs in traditional agriculture quickly led to PA initiatives. Satellite imagery was used to measure physical geographical styles in soil natural organic matter, which was used in combination with floor-based total measurements (Nagarajan & Minu, 2018) to measure geographic styles in plant nutrients and flour grain yield. This was the first remote sensing application in Pennsylvania. Modern PA operations rely on GPS satellites, Landsat, and SPOT for their geographical planning (45–55 m). Eventually, satellite television began to be developed for computer imaging structures that would have better spatial decision-making and faster revisit periods. As new spectral and spatial sensors have become available, higher correlations have been discovered. Images from the IKONOS satellite have been used to detect nitrogen deficiencies in sugar beet, pesticide performance (Pajares, 2015) in wheat, and insufficient synthetic flow in wheat subject web pages. QuickBird images of olive fields in Spain were used to estimate olive crop areas, tree counts, geographic distributions of tree canopies, and olive yields. These satellites have steadily gained a large base of commercial subscribers interested in PA programs, compared to older satellites such as SPOT or Landsat.

5.3 Proposed Framework

5.3.1 Precision Agriculture Using Geospatial Information

PA was described as an agricultural control system that collects real-time data, processes and analyses it, and then provides farmers with options to minimize resources and increase crop yield throughout the decision-making process. Innovations in technology paved the ground for its implementation in a variety of industries (Pradhan et al., 2018). With the use of IoT and reliable software, the agriculture sector and automation vendors are attempting to identify and utilize prospects for boosting productivity, efficiency, and agricultural practices. Figure 5.3 indicates the typical design for integrating geospatial data with IoT to achieve PA. As can be seen in the diagram, the detectors in the farm and farm machinery give actual information and alarms over the Internet for additional processing.

After the information is gathered, it is processed with big data and analytics and saved on a cloud server for use in decision-making. Decisions are made after gathering geospatial data. Sensor data is typically exchanged to and from users in sensor-based PA via a database machine or the cloud. The stability of the transmission network and the Internet play an important role in sharing information among users (Sahani et al., 2018). Using mobile devices and consumer applications, the generated information can be accessed in a timely and effective manner. According to the research, geospatial data affects crop productivity,

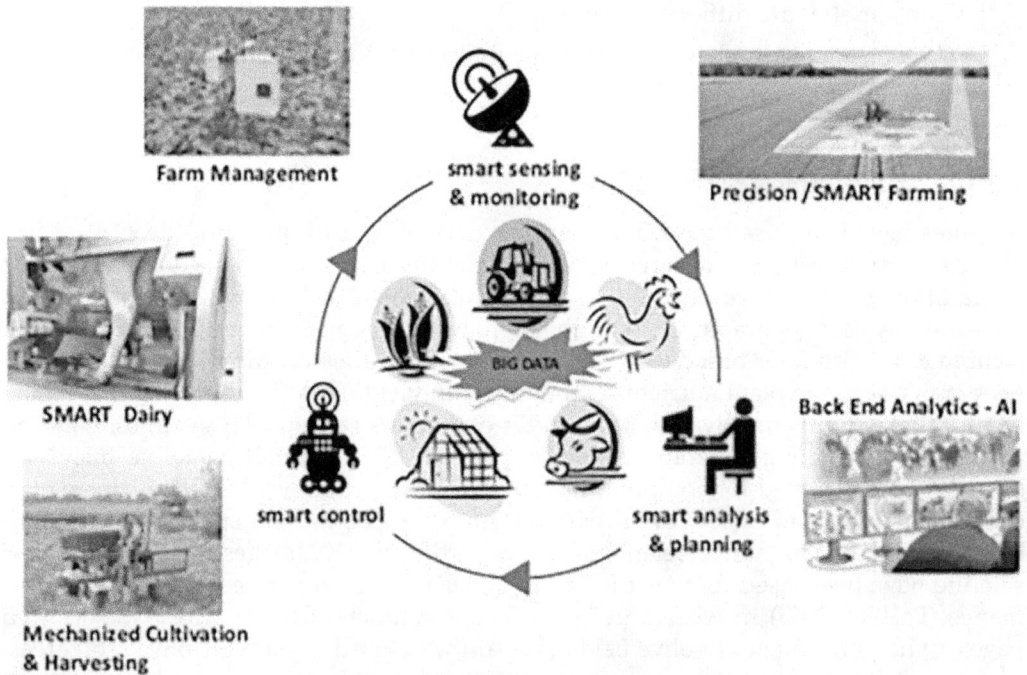

FIGURE 5.3
Concept of IoT.

watering, and soil quality. As a result, IoT-based PA aims to simplify remedial and preventive activity. A better understanding of accessible and needed water, nutrients, and pest control is possible. Due to advances in technology, PF has been gaining popularity among farmers worldwide. Farmers need this to generate maximum yield with limited resources (Sarkar et al., 2016). In this climate change era of extreme weather, it is impossible to anticipate efficient agricultural production without these smart technologies.

5.3.2 Precision Agriculture Using Remote Sensing

5.3.2.1 Data Gathering

The great degree of variability in agricultural-based conditions within fields prompted the adoption of PF technologies. Establishing an agricultural credit data system that provides information on plant status, crop field cultivation, soils, and other factors is one of the criteria for introducing PA technologies. Data from this source can be a starting point for growth and yield estimation. It is crucial to use advanced data collection and processing tools to establish such an information system. Monitoring the earth's crust and its changes was done with the most effective technology.

5.3.2.2 Yield Monitoring

Information about crop production fluctuations within a field is becoming more relevant for crop management as more PA techniques are used. Yield maps incorporate a variety of geographical variables, including soil conditions, elevation, growing conditions, nutrient

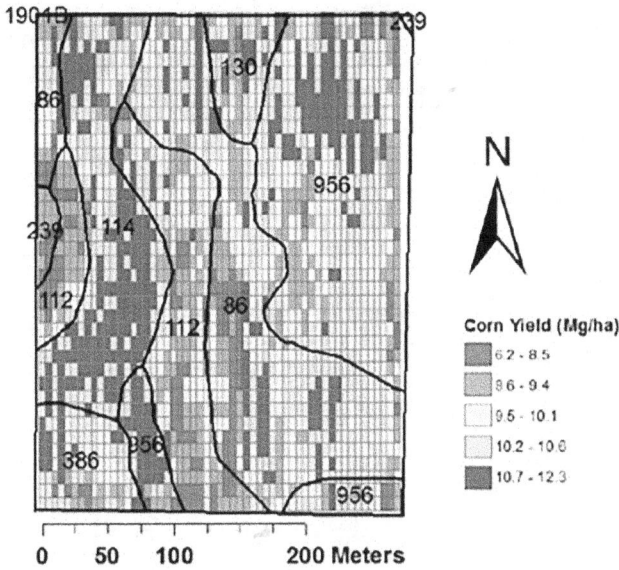

FIGURE 5.4
The effect of the direction of harvest on the quantity of crop measured.

replacement techniques, and applied agriculture technology. Thus, a production map, alone or in combination with other spatial data, can be an important input for site-specific activities. By mounting yield monitors on the harvester, yield data can be obtained at harvest time. A yield measuring device was first applied in the early 1980s, and it has now become standard equipment on most combines. In the near future, farmers will have access to detailed and reliable yield data due to technology improvements like GPS and harvester-mounted production sensors. The construction of productivity maps may be carried out immediately following the collection of data in order to identify trends in productivity across specific areas (Thorp et al., 2015). Based on the results of the analysis, after-season management can be implemented. A yield monitor's accuracy is determined by the brand and type of yield monitor, as well as harvest conditions, validation regime, and flow rate. In order to achieve the highest level of accuracy from the yield sensor, the yield monitor needs to be validated. Figure 5.4 presents the yield monitoring process for crop production measurement.

Although yield monitors are commercially available, many agricultural harvesters lack them. To analyze the map effectively, additional information is needed, such as information on plant stress during the growing season. Within a season, yield monitor data cannot be used to detect problems and generate maps. The use of remote sensing can be beneficial for estimating crop yield fluctuation and identifying stress within seasons. Spatial data may be used to produce productivity maps for both after-season and within-season crops during the growing season (Figure 5.5).

5.3.2.3 *Indicators of Vegetation*

By measuring the reflectivity of the plants at different frequencies, it is possible to gather more information about their condition. Light spectrum transmittance is influenced by tissue water content, plant type, and other intrinsic variables (Ayaz et al., 2019). Because chlorophyll absorbs light for photosynthesis, vegetation has short wavelengths in the blue

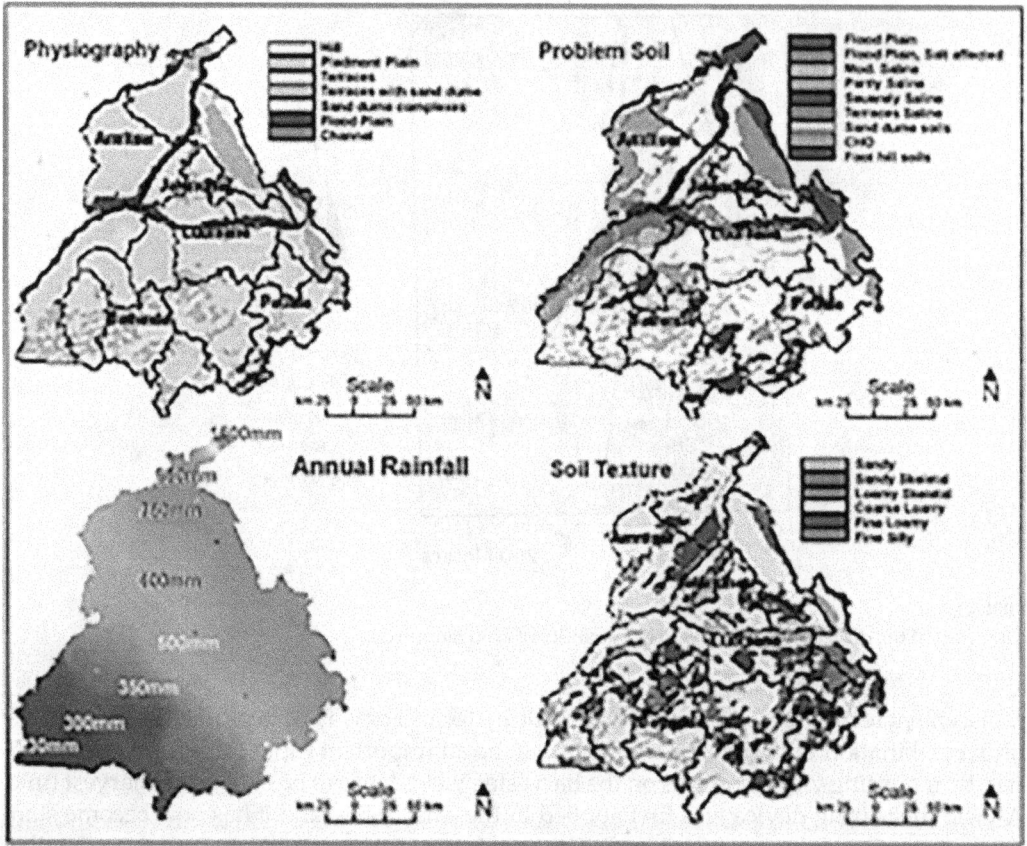

FIGURE 5.5
Cell yield of maize created from Sentinel data: Input data and output.

and red range of perceptible light. It reaches its peak in the green zone when vegetation takes on a green hue. Because of the cellular structure in the leaves, there is a much greater NIR area than there is in the wavelength spectrum. There is higher moisture content in the mid-infrared region of the spectrum (Figure 5.6).

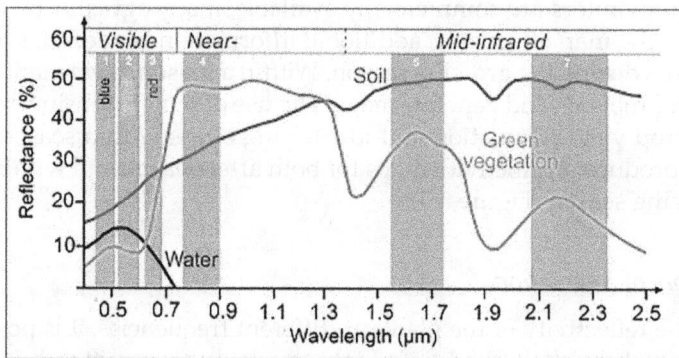

FIGURE 5.6
Spectral curve of the light reflected from the plant.

As part of the statistical analysis of spatial data obtained from vegetation, individual light spectrum bands or a collection of bands can be used to extract vegetation data for information analysis. A useful tool for determining the state of the vegetation is the construction of value iteration (VI) algorithms. The derivation of vegetation data from remotely sensed pictures is based on differences and disparities in plant vegetative leaves as well as canopy spectral features. As required, near-infrared (1–2.0 m) and red (1.1–1.3 m) or other bands are blended using a variety of methods. The amount of organically produced by an increasing plant population causes a decrease in red reflectance and an increase in total near-infrared reflection. According to earlier studies, there is a correlation between the indices using those two parts of the spectrum and the amount of vegetation. These estimates can be used to infer the data population provided by the detectors, which can also be used to predict future crop yields. We can calculate the NDVI to determine the crop's health when its plant cover is poor. A low index value indicates that there is little healthy vegetation, while a high index value indicates that there is a lot of healthy vegetation. Several indices have been developed to better represent the actual amount of vegetation on the ground. A number of studies have been conducted to determine the correlation between a crop's vegetation index assessed at a specific time and its yield. The study was conducted by Tzounis et al. (2017). Numerous indices have been developed to better design the total amount of plants on the ground. The study of land cover change often makes use of spectral indices derived from satellite data. The landscape index produces individual images by displaying vegetation quantity, or vegetation strength, using various multispectral remote sensing methods. This may reduce the amount of data needed for analysis and provide a more comprehensive picture of changes than any single band.

The number of bands acquired by satellite data was increasing as high-resolution spectroscopic equipment was developed. Numerous studies have addressed the relationship between vegetation indices and yields. The relationship between yield and index depends not only on the type of index but also on the time of data collection and the stage of the plant. Long-term yield estimation requires multiple time points during vegetation. NDVI is one of the most widely used indices. It is typically used to assess canopy vigor or development. It has been compared to the LAI, which is defined as the surface area of a leaf per square meter of ground. Figure 5.7 shows the change of vegetation index (NDVI) depending on vegetation growing.

5.4 Result and Discussion

5.4.1 Classification

Different methodologies and statistical approaches have been used to separate different types of habitats. A pixel-based and object-based approach was used in this study to map the spatial variability within a field. The researcher does not specify the natural vegetation or ecosystem types in an uncontrolled categorization (Figure 5.8). In an image analysis program, the data from the spectral analysis is separated into several groups without awareness of the image's spectral response. Spectral variance within each class and the number of data classes can both be limited by the user. An analyst must then assign labels to these groups based on their understanding of these photos and what the different

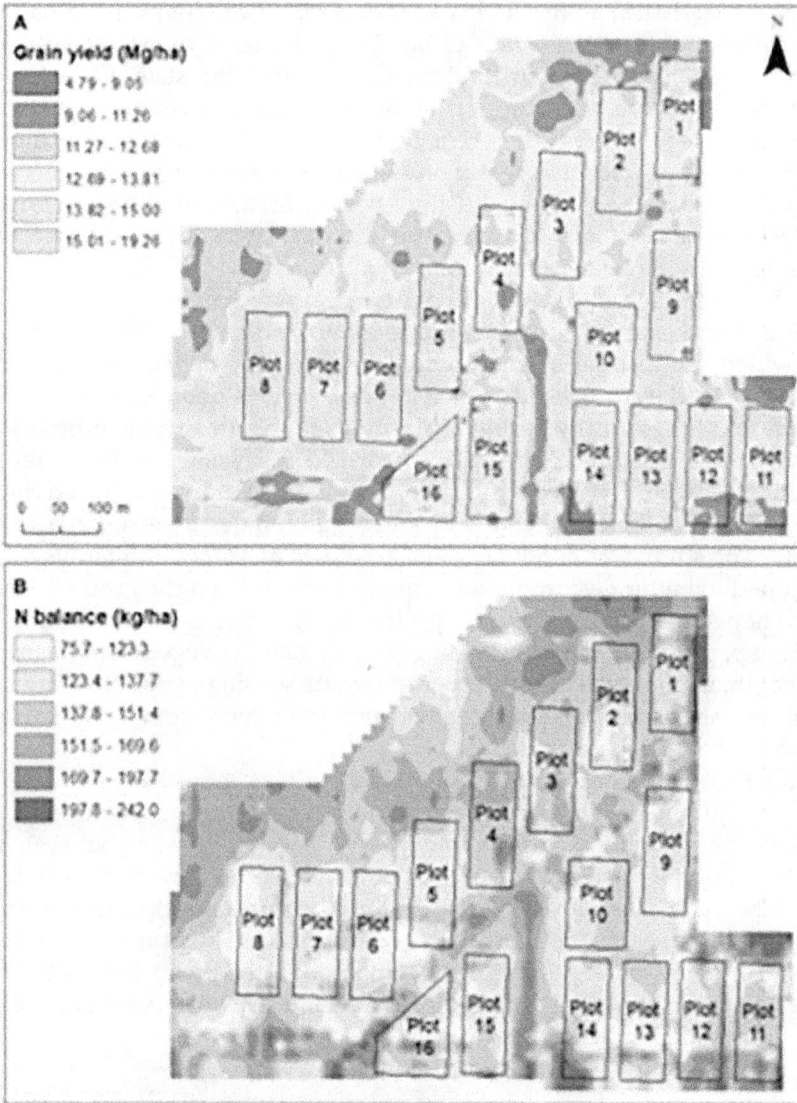

FIGURE 5.7
The change of vegetation index (NDVI) depending on vegetation growing.

habitats should look like as well as their knowledge and understanding of how different habitats should look in these photos. The purpose of this study was to develop a method for mapping fluctuations in field conditions using high-resolution remote sensing photos and image classification. Using OBIA, it was possible to identify no vegetation and monitor its condition in an agricultural field.

The data collection phase includes multi-scale picture segmentation, ruleset development, data pre-processing, the definition of features used to describe land use, classification based on the ruleset, and reliability evaluation. E-cognition software can retrieve information based not only on transmitted signals but also on size, density, and other factors. Figure 5.9 shows the workflow of mapping spatial variability by OBIA.

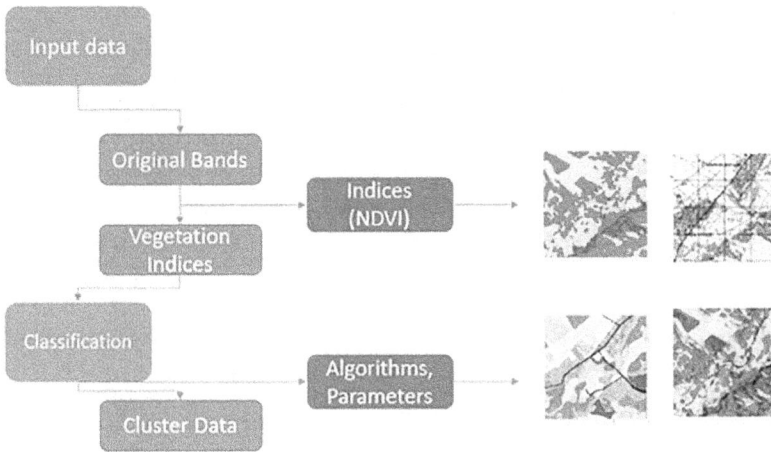

FIGURE 5.8
Workflow of mapping spatial variability by unsupervised classification methods.

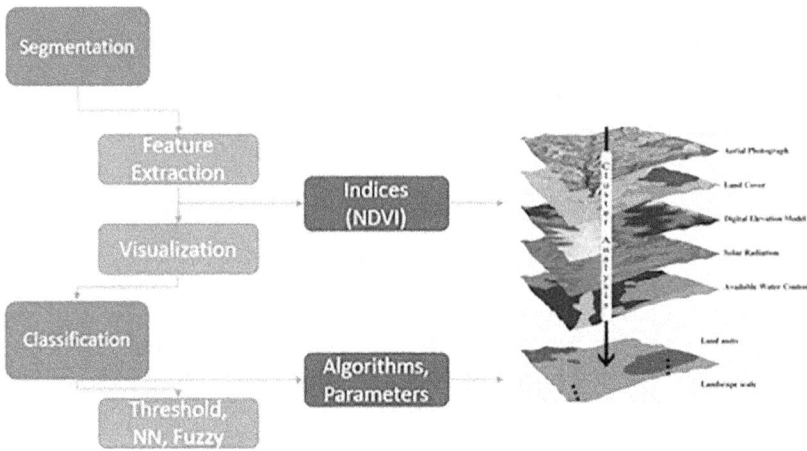

FIGURE 5.9
Workflow of mapping spatial variability by OBIA.

5.5 Conclusion

A remote monitoring system is useful for tracking and calculating the fluctuations in agricultural production. During the vegetative stage, images can be used to monitor crop growth and identify any issues that need to be addressed. Multi-temporal imaging can also be used to create production maps. These maps can be used to track yield fluctuation over time. Vegetation indices are useful in crop monitoring and crop estimation: Problems are identified; stressed plants are detected; irrigation requirements within a field change and fertilizer and pesticide requirements are identified, and prospective control zones are noted. This analysis should take into account that the reliability of such surveys depends on many factors, such as the type of index, the time of data gathering, and the stage of the plant. With the increasing availability of satellite and airborne imagery, more research is required to create the most appropriate algorithms for spatial prediction and other smart farming activities. Geospatial data is highlighted as an important component of the IoT. It uncovers critical tactics for dealing with environmental catastrophes such as floods, droughts, cyclones, pollution, climate change, and blights, among others, using geospatial data, as well as raising awareness about the use of PA. The proposal's goal is to raise awareness of PF and available control application software tools on the market and to encourage people to use them. A farmer can now immediately contact the area management office to update crop information and receive awareness and regulation data like crop diseases, soil nutrition, and soil irrigation. The agricultural sector would surely undergo major changes if the proposed model were developed as a prototype and executed in reality.

References

Adeyemi, O., Grove, I., Peets, S., & Norton, T. (2017). Advanced monitoring and management systems for improving sustainability in precision irrigation. Sustainability, 9(3), 353.

Akkaş, M. A., & Sokullu, R. (2017). An IoT-based greenhouse monitoring system with Micaz motes. Procedia Computer Science, 113, 603–608.

Alahi, M. E. E., Nag, A., Mukhopadhyay, S. C., & Burkitt, L. (2018). A temperature-compensated graphene sensor for nitrate monitoring in real-time application. Sensors and Actuators A: Physical, 269, 79–90.

Ayaz, M., Ammad-Uddin, M., Sharif, Z., Mansour, A., & Aggoune, E. H. M. (2019). Internet-of-Things (IoT)-based smart agriculture: Toward making the fields talk. IEEE Access, 7, 129551–129583.

Barik, R. K., Dubey, H., Misra, C., Borthakur, D., Constant, N., Sasane, S. A., Lenka, R. K., Mishra, B. S. P., Das, H., & Mankodiya, K. (2018). Fog assisted cloud computing in era of big data and internet-of-things: Systems, architectures, and applications. In Cloud Computing for Optimization: Foundations, Applications, and Challenges (pp. 367–394). Springer, Cham.

Bhanumathi, V., & Kalaivanan, K. (2019a). Application specific sensor-cloud: Architectural model. In Computational Intelligence in Sensor Networks (pp. 277–305). Springer, Berlin, Heidelberg.

Bhanumathi, V., & Kalaivanan, K. (2019b). The role of geospatial technology with IoT for precision agriculture. In Cloud Computing for Geospatial Big Data Analytics (pp. 225–250). Springer, Cham.

Bhardwaj, A., Sam, L., Bhardwaj, A., & Martín-Torres, F. J. (2016). LiDAR remote sensing of the cryosphere: Present applications and future prospects. Remote Sensing of Environment, 177, 125–143.

Campos, I., González-Gómez, L., Villodre, J., Calera, M., Campoy, J., Jiménez, N., Plaza, C., Sánchez-Prieto, S., & Calera, A. (2019). Mapping within-field variability in wheat yield and biomass using remote sensing vegetation indices. Precision Agriculture, 20(2), 214–236.

Fang, B., Lakshmi, V., Bindlish, R., & Jackson, T. J. (2018). AMSR2 soil moisture downscaling using temperature and vegetation data. Remote Sensing, 10(10), 1575.

Foughali, K., Fathallah, K., & Frihida, A. (2018). Using Cloud IOT for disease prevention in precision agriculture. Procedia Computer Science, 130, 575–582.

Hammoudi, S., Aliouat, Z., & Harous, S. (2018). Challenges and research directions for Internet of Things. Telecommunication Systems, 67(2), 367–385.

Im, J., Park, S., Rhee, J., Baik, J., & Choi, M. (2016). Downscaling of AMSR-E soil moisture with MODIS products using machine learning approaches. Environmental Earth Sciences, 75(15), 1–19.

López-Martínez, J., Blanco-Claraco, J. L., Pérez-Alonso, J., & Callejón-Ferre, Á. J. (2018). Distributed network for measuring climatic parameters in heterogeneous environments: Application in a greenhouse. Computers and Electronics in Agriculture, 145, 105–121.

Mulla, D. J. (2013). Twenty five years of remote sensing in precision agriculture: Key advances and remaining knowledge gaps. Biosystems Engineering, 114(4), 358–371.

Nagarajan, G., & Minu, R. I. (2018). Wireless soil monitoring sensor for sprinkler irrigation automation system. Wireless Personal Communications, 98(2), 1835–1851.

Pajares, G. (2015). Overview and current status of remote sensing applications based on unmanned aerial vehicles (UAVs). Photogrammetric Engineering & Remote Sensing, 81(4), 281–330.

Pradhan, C., Das, H., Naik, B., & Dey, N. (2018). Handbook of Research on Information Security in Biomedical Signal Processing. IGI Global, Hershey PA.

Sahani, R., Rout, C., Badajena, J. C., Jena, A. K., Das, H., & others. (2018). Classification of intrusion detection using data mining techniques. In Progress in Computing, Analytics and Networking (pp. 753–764). Springer, Singapore.

Sarkar, J. L., Panigrahi, C. R., Pati, B., & Das, H. (2016). A novel approach for real-time data management in wireless sensor networks. Proceedings of 3rd International Conference on Advanced Computing, Networking and Informatics, 599–607.

Srinivasan, R., Kavitha. M, Shashank Reddy, D., & Naga Harshitha, C. (2019), Precision agriculture using fog-edge computing, International Journal of Innovative Technology and Exploring Engineering, 8(7), 2539–2543.

Thorp, K. R., Hunsaker, D. J., French, A. N., Bautista, E., & Bronson, K. F. (2015). Integrating geospatial data and cropping system simulation within a geographic information system to analyze spatial seed cotton yield, water use, and irrigation requirements. Precision Agriculture, 16(5), 532–557.

Tzounis, A., Katsoulas, N., Bartzanas, T., & Kittas, C. (2017). Internet of Things in agriculture, recent advances and future challenges. Biosystems Engineering, 164, 31–48.

6

Machine Learning Approaches for Agro-IoT Systems

S. A. Jadhav and A. Lal

Vellore Institute of Technology
Vellore, India

CONTENTS

6.1 Introduction 89
6.2 IoT for Agriculture 90
    6.2.1 Role of IoT Data in ML-Based Agro-IoT System 90
        6.2.1.1 Data Collected by Smart Agriculture Sensors 91
        6.2.1.2 Data Collection by Agricultural Drones 92
        6.2.1.3 Predictions by Processing Raw Data 92
    6.2.2 Need of ML in IoT 92
    6.2.3 Assimilate ML with IoT 92
6.3 ML in Agriculture 95
    6.3.1 ML Learning Types and Models 95
        6.3.1.1 Supervised Learning 95
        6.3.1.2 Unsupervised Learning 98
        6.3.1.3 Semi-Supervised Learning 99
        6.3.1.4 Reinforcement Learning 100
6.4 Applications on ML Agro-IoT System 100
    6.4.1 Pest Control Management 100
    6.4.2 Resource Management 101
    6.4.3 Safeguarding Crops from Animals, Birds, and Human Attack 101
    6.4.4 Livestock Management 102
    6.4.5 Yield Management 102
    6.4.6 Quantifying the Emission of Greenhouse Gases 102
6.5 Conclusion 103
References 103

6.1 Introduction

Technological advancement sees no limits. One of the leading technologies Internet of Things (IoT) has made lots of advancements in making the environment around us smart, like smart homes, smart office, smart factory, and smart farming. Agriculture IoT and precision farming are some of the booming topics that inculcate the responsiveness of smart farming. The IoT has overpowered challenges like irrigation, scarcity, soil quality, weather issues, and recurrent infection to plants due to pests and diseases. Machine learning (ML)

DOI: 10.1201/9781003185413-6

is a technique of statistics scrutiny that automates investigative replica structure. It is a branch of artificial intelligence (AI) based on the initiative that systems can learn from statistics, classify patterns and formulate decisions with a nominal human intrusion. Whereas AI is the wide knowledge of mimicking human beings' abilities, ML is a precise detachment of AI that trains an apparatus how to be taught. Arthus Samuel masters in the field of AI invented ML in 1959 and quoted that "it gives computers the ability to learn without being explicitly programmed" (Lee et al., 2017).

6.2 IoT for Agriculture

The IoT platform is a collection of software tools that facilitates the systematic storage and process of raw data received from sensors and actuators. Cloud computing is one the most popular technologies used from the last decade for speedy and skillful delivery of processed data to end-users and to manage centralized storage of massive data. Data analytics is performed on the cloud to predict output for decision-making. IoT sensors connected or mounted on agricultural land will provide a wide variety of data based on which analysis is done. Thus integration of cloud computing and IoT is the most used pattern for problem-solving and application development. But recent trend shows lots of research has been done using the combination of IoT with ML. Figure 6.1 shows the timeline of various revolutions that took place and clearly indicates the difference between then and now.

6.2.1 Role of IoT Data in ML-Based Agro-IoT System

Data, whether it is raw or processed, labeled or unlabeled, is the most vital element for solving any ML problem. It can be in images, videos, audio, or even text files. Not only for ML or AI but data is needed for most of the technologies such as big data, analytics in IoT, blockchain, and edge computing. IoT sensors are a rich source of generating data that can be later used by various technologies for analysis. The result of any application or

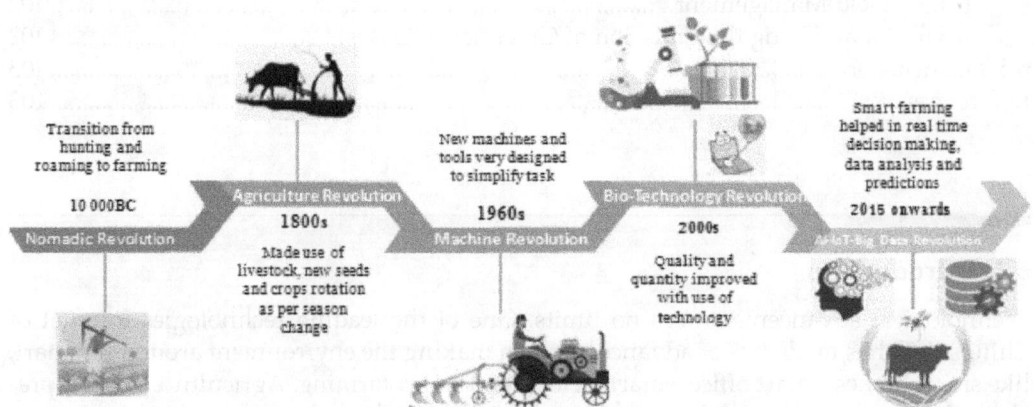

FIGURE 6.1
Timeline of agriculture and technology advancements.

problem-solving technique relies on the correctness of data because wrong data will give the wrong output. Cloud offers scalable on-demand services to the IoT devices for effective communication and knowledge sharing (Saravanan and Srinivasan, 2018).

IoT enables communication between the real and virtual worlds. Objects with virtual IDs known as sensors or actuators have the ability to collect the data from the real-time environment and the smart interfaces help in connectivity and communication. The collected data is then stored, analyzed, and viewed on a computer or mobile device. When it comes to the agriculture domain most used sensors are for quantifying environmental parameters like temperature, humidity, soil salinity, water turbidity, water level, etc.

ML also relies on lots of data for data analysis. Many research authors (Wang et al., 2021; Hasan et al., 2021; Yashodha et al., 2021) have proved that ML accuracy and performance are less for most problems if dataset size is small. Similarly, the author (Laure et al., 2018) tried to explain why massive data is needed for ML and techniques to manage the same. Based on the application requirement data is split into "training data, validation data, and test data." Whenever we prepare an ML model for some application it is trained first using training data and the learning process is initiated. Validating the data set is optional but recommended as it helps in the evaluation of the model during the training phase. Test data is used for real-time evaluation purposes, where input value from the user is fed to the model to predict the results, and later that result is compared with authentic output in testing data. Figure 6.2 shows how data is processed in ML for the prediction of output value.

6.2.1.1 Data Collected by Smart Agriculture Sensors

Data is gathered using mobile phones or stationary equipment like sensors, robotics, bots, autonomous vehicles, automated hardware, camera, and wearable devices. The wireless sensor network is used for connecting the initially low level; here raw data is collected from various sensors and passed to the next connected intermediate level sensor, and so on. The end layer is generally the decision layer and is deployed using cloud computing. The collected data is stored centrally over the cloud and used for analysis and decision-making. Prediction of certain decisions is also made possible by technology like ML and deep learning.

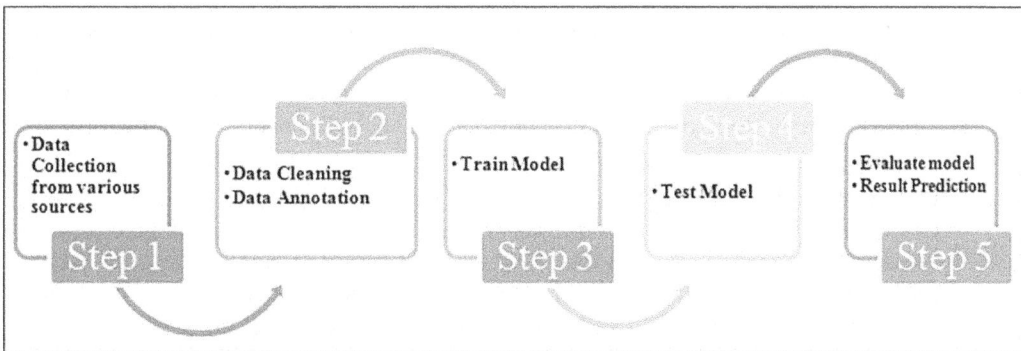

FIGURE 6.2
Machine learning process (Wood et al., 2020).

6.2.1.2 Data Collection by Agricultural Drones

Surface-station and aerial station drones have been utilized in the cultivation domain. Crop health monitoring, spraying of fertilizers and pesticides, water management, and soil analysis are a few of the common tasks performed by drones. Drones efficiently collect data in the form of images and transfer it to the server connected.

6.2.1.3 Predictions by Processing Raw Data

To forecast the prediction results artificial networks utilize data or information obtained by sensors from the cultivation land. This includes parameters such as soil, water, warmth, precipitation, moisture, etc. Using all these parameters and efficient algorithms early detection of anomalies and prediction of crop yield is possible.

6.2.2 Need of ML in IoT

Initially, ML was assumed to be a powerful tool for designing prediction models only, but with advancements in research, and it has been proved that ML is more beneficial in combination with wireless sensor networks (WSNs) and IoT. ML models have several advantages in solving IoT challenges (Akpakwu et al., 2017) like quality of service (White et al., 2017), network congestion and overload (Haroon et al., 2016), interoperability and heterogeneity, security and privacy (Sicari et al., 2015), and network mobility and coverage (Kishore Ramakrishnan et al., 2014). Table 6.1 contains the information of few notable research papers where researchers have works covering various agriculture issues and providing a solution using IoT and ML (see also Figure 6.3).

6.2.3 Assimilate ML with IoT

It is known that IoT is a layered framework, where data flows from one layer to another layer to pass the information, which helps in solving prediction problems. Research is being performed for placing ML algorithms at different layers of the IoT frameworks. Here the decision of placement totally depends on application/business requirement. Three choices can be given to integrating ML in IoT as shown in Figure 6.4.

- Choice 1: ML at IoT edge/endpoint, i.e., physical layer
  The foremost paradigm about IoT devices is that sensors are dumb and all the intelligence resides inside the cloud, but contradictory to this author (Gopinath et al., 2019) proved the accuracy of running ML algorithms locally on small, resource-constrained devices. The task of classification, regression, etc., for the prediction can be done on the microcontroller itself, so no need for connection to the cloud.

- Choice 2: ML at IoT gateway, i.e., network layer
  A gateway is a unit where IoT endpoints are connected and the data collected from these devices are transferred to the cloud. ML can be easily deployed on the local server or gateway where ML algorithms for classification, regression, or clustering can be computed. The basic idea behind this is to reduce the computation cost of transferring data to the cloud, bandwidth, processing power and reduce execution time too. If deployed at this layer it can control or give the command to all the end devices connected to the gateway.

TABLE 6.1

Research Paper on IoT and ML Algorithm in Agriculture Domain

Area	Model/Algorithm	Reference
Yield prediction	Sugarcane yield prediction using multilayer perceptron (MLP) in IoT agriculture. Accuracy – 99%, Precision – 95%, Recall – 96% Minimum mean absolute error (MAE) – 0.04% Root mean square error (RMSE) – 0.006%	Wang et al. (2021)
Pest management	Identifying tea leaf disease using neural network	Yashodha et al. (2021)
Crop, soil, water, and pest management	Survey on how IoT technology, UAV systems, and machine learning algorithms will benefit the agriculture domain	Hasan et al. (2021)
Crop management	Using deep neural network model in hydroponics system to monitor plant growth	Vanipriya et al. (2021)
Soil management	Soil moisture forecasts for potato crop using SVM, random forest, and neural network	Dubois et al. (2021)
Soil and water management	K-means and support vector regression (SVR) for forecasting droughts	Kaur et al. (2021)
Water management	Designed smart irrigation system, digital soil assessment (DSA), sustainable intensification (SI), and smart earth technologies using IoT and machine learning	Goel et al. (2021)
Crop, soil, water, and pest management	Survey on 5G technology in the agricultural using IoT and machine learning	Tang et al. (2021)
Water management	Predict irrigation patterns using support vector regression and random forest regression	Vij et al. (2020)
Water management	Using logistic regression, SVM, averaged perceptron, and fast forest for smart water management system	Singh et al., 2020
Crop, soil, water, and pest management	Reveal responsibility of machine learning in the WSN and IoT technology	Messaoud et al. (2020)
Livestock management	Early-stage lameness detection using random forest with accuracy of 91% and K-nearest neighbors (KNN) with 87% accuracy	Byabazaire et al. (2019)
Faulty sensor detection	Using ML models like KNN, isolation forest for detecting abnormal sensors	Rossi et al. (2020)
Water management	Deep reinforcement learning for smart irrigation system	Bu et al. (2019)
Livestock management	IoT framework for disease detection in cows using neural network	Vyas et al. (2019)

(Continued)

TABLE 6.1 *(Continued)*

Research Paper on IoT and ML Algorithm in Agriculture Domain

Area	Model/Algorithm	Reference
Crop management	Crop quality improvement using IoT drones and SVM	Saha et al. (2018)
Water management	Smart irrigation using SVR and K-means Accuracy – 96%	Goap et al. (2018)
Water management Livestock management	IoT-based hydroponics system using deep neural networks Cloud IoT-based Livestock Management System for animal health monitoring	Mehra et al. (2018) Saravanan and Saraniya (2018)
Crop management	Monitoring and data prediction in rose greenhouse farm using linear regression, neuronal networks, and SVM	Rodríguez et al. (2017)
Robotic grander	Autonomous gardening robotic vehicle which identifies and classifies the plant variety using neural network	Kumar et al. (2016)
Pest management	Early Detection of grape diseases using machine learning and IoT	Patil et al. (2016)

- Choice 3: ML on IoT cloud, i.e., application layer
  The most common and popular place where machine algorithms can be executed is on a cloud-based platform. Generally, data storage and analysis take place cloud-only. If an ML model is deployed at this layer it can control and pass commands to all the endpoints and gateways.

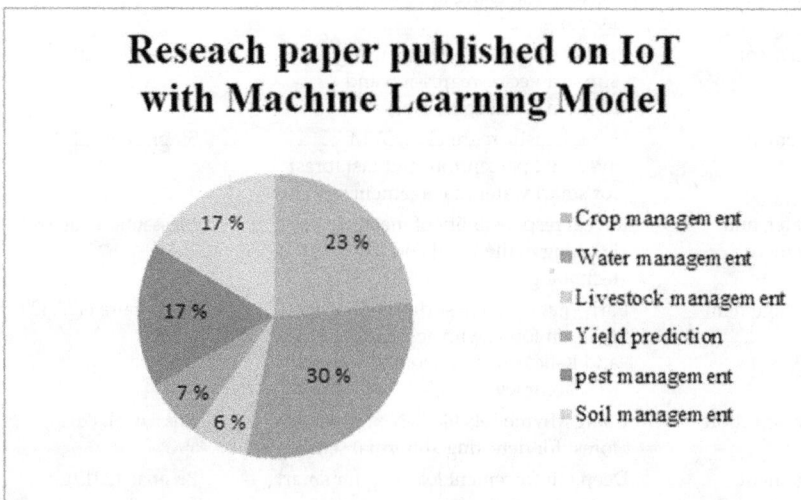

FIGURE 6.3
Recent research paper published on ML with IoT.

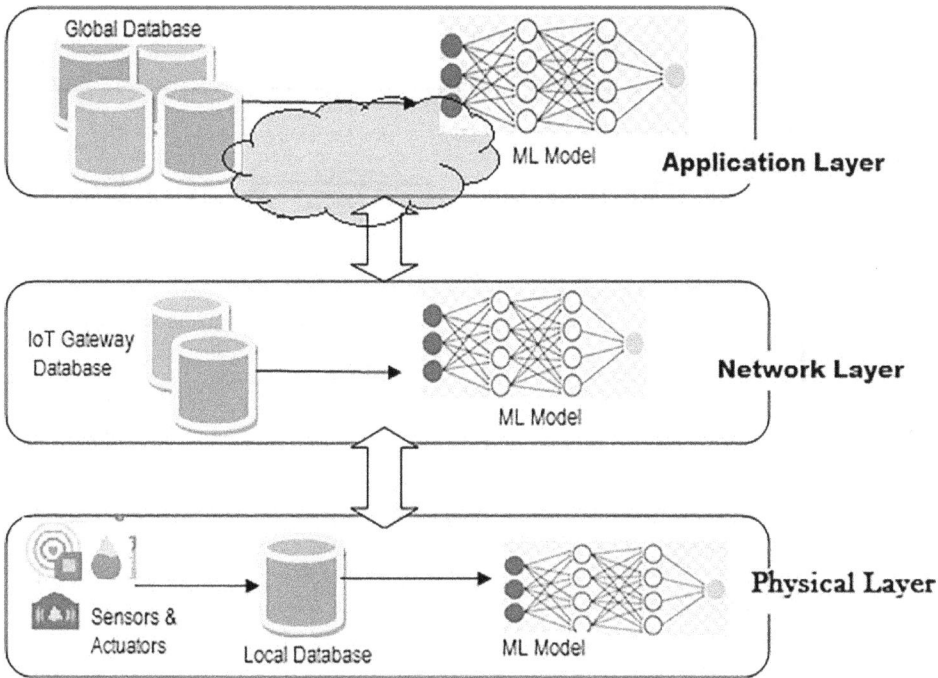

FIGURE 6.4
Integrating ML in IoT layers.

6.3 ML in Agriculture

ML models can be used for thing identification, review, prediction, categorization, and grouping of objects. Numerous models have been invented and a lot of research is present for various reasons in the agriculture domain nowadays. Figure 6.5 shows the list of models popularly used.

6.3.1 ML Learning Types and Models

The learning task can be ML categorized into "supervised," "unsupervised," and "semi-supervised." Depending upon the problem statement different models are used. The problem can be broadly divided in terms of identification, classification, quantification, and prediction.

6.3.1.1 Supervised Learning

If a model is trained using some labels for classification or prediction purposes it is termed supervised learning. It travels around classification and regression algorithms and learns about techniques for feature selection, feature transformation, and hyper-parameter tuning.

Regression is an arithmetical method to decide the relationship between one dependent variable and a chain of other variables or independent variables. In ML, we allow

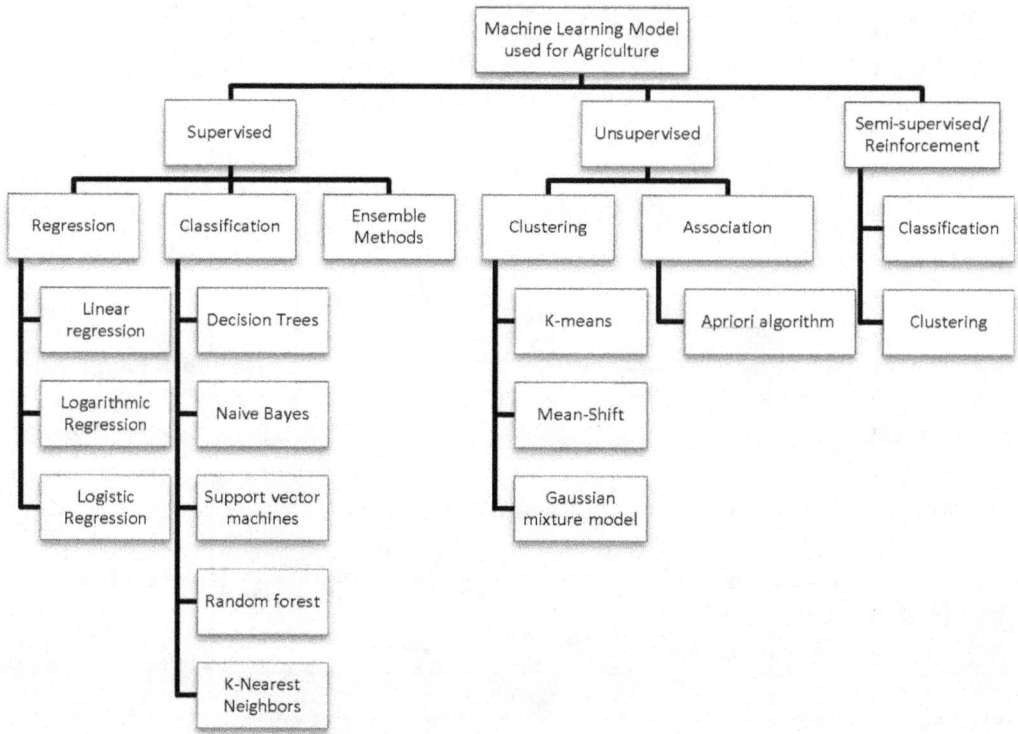

FIGURE 6.5
Machine learning models used in agriculture application.

machines to learn the relationships from the raw data provided and predict the output value. Regression techniques are not very popular in the agriculture domain but are used in solving resource management problems. Soil moisture quantity prediction system based on hybrid regression model was designed by author (Chatterjee et al., 2019), who did better than other models. Soil moisture and soil temperature will provide real-time values based on which the ML model can predict and decide if more watering is needed, thus improving the soil quality and having proper water management. Regression models are suitable for solving prediction problems.

Classification is a method to decide to which class a given object will belong. "Support vector machines" (SVMs) were pioneered in the work of numerical learn assumption. SVM is mainly a dual classifier for linear unraveling hyper-plane to categorize data instances and has been utilizing for categorization, "regression," and "clustering." SVM is one of the most popular models among all the others because of its capability of solving various issues including identification, classification, quantification, and prediction. This model is best suitable for any IoT-designed agriculture smart system, for example, in resource management, capitulate forecast, tidy discovery, ailment recognition, harvest superiority analysis, and farm animals management.

Bayesian models (BM) can be utilized for "classification" or "regression problems." Naive Bayes and Gaussian naive Bayes is the most well-known algorithms in the literature. For IoT smart agriculture systems this model can be used in the identification and classification of crop disease and its type. Also, it is suited for yield prediction where the author (Amatya et al., 2016) designed computerize quivering and grabbing cherries system during harvest, where the sensor provides information regarding color and quantity of fruits on trees and are based on ML decision system action is taken to shake the tree or not. Here too mechanical sensors are used for shaking. Livestock management can also take advantage of the Bayesian model for animal tracking and behavior explanation.

Most of the time a single learning algorithm cannot give expected accurate results, so multiple learning algorithms can be combined for the best prediction results. Thus ensemble learning (EL) focuses on such a combination of learning algorithms to train and test the ML models. Supervised machine learning algorithms used in agricultural applications are shown in the Table 6.2.

TABLE 6.2

Supervised ML Algorithm Used in Agricultural Applications

Application	Type	Results/Finding	Article
Water management	Multivariate adaptive regression spline (MARS)	Assessment of the amount of water evaporation and transpiration on a monthly basis	Mehdizadeh et al. (2017)
Plant-stress prediction	Dirichlet aggregation regression	Abiotic stress, namely drought, was predicted on barley crops well in advance before it became visible to human eyes	Kersting et al. (2012)
Plant-stress prediction	Multiple regression	Rice blast disease was predicted on rice crop	Kaundal et al. (2006)
Counting fruits	Linear regression	Based on RGB images apples and oranges were tallied	Chen et al. (2014)
Mapping crop and soil	Logistic regression	Soil typology and land distribution	Piccini et al. (2019)
Soil salinity prediction	Random forest regression (RFR) and support vector regression (SVR)	Accuracy: RFR 93.4–94.2% SVR 85.2–89.4% "Normalized root mean square error" (NRMSE): RFR 6.10–7.69% SVR 10.29–10.52%	Wu et al. (2018)
Plants water stress detection and drought conditions analysis	Linear and exponential regression	Skeleton for understanding and prediction of plant mechanisms under drought conditions	Sun et al. (2020)
Weed detection	Logistic regression	Accurately differentiated between crop and weed	Potena et.al. (2016)
Greenhouse monitoring	Linear and exponential regression	Framework designed for early fault detection and diagnosis	Lakhiar et al. (2018)
Stress	SVM variant	Nitrogen, phosphorus, and potassium (NPK) stress was identified for rice crop	Chen et al. (2014)
Plant disease	SVM	*Cercospora* leaf spot, sugar beet rust, and powdery mildew disease identified and classified for beetroot plant	Rumpf et al. (2010)
Water stress detection	Decision tree	Multiple trained decision trees are combined to provide an improved prediction performance	Castillo-Guevara et al. (2020)
Packaged food products	Decision tree	Identification of contaminants, diseases or defects or bruise detection	Lu et al. (2011)

(Continued)

TABLE 6.2 *(Continued)*

Supervised ML Algorithm Used in Agricultural Applications

Application	Type	Results/Finding	Article
Plant disease	Bayesian classifier	Identification and classification of powdery mildew disease in tomato plants	Hernández-Rabadán et al. (2014)
Plant disease	Naïve Bayes	Recognition and categorization of *Alternaria alternata, A. brassicae, A. brassicicola,* and *A. dauci* diseases in rapeseed-mustard oilseed plant	Baranowski et al. (2015)
Plant disease	Bayesian classifier	Classification of beetroot plant disease like Uromycesbetae, Cercosporabeticola	Bauer et al. (2011)
Weed detection	Naïve Bayes	Accurate detection, location, and classification of weeds	Hasan et al. (2021)
Plant disease classification	Bayesian classifier	With no overfitting problem achieved an accuracy rate of 98.9%	Sachdeva et al. (2021)
Livestock management	Bayesian	Modeling cattle movements	Lindström et al. (2013)
Plant disease	KNN	Identification of Huanglongbing disease in citrus plant	Sankaran et al. (2011)
Yield prediction	SVM	Successfully detected coffee beans and their count from a particular branch. Also predicted their ripeness	Ramos et al. (2017)
Fruit counting	SVM	Identified the ripeness of citrus fruit Accuracy 80%	Amatya et al. (2016)
Livestock management	SVM	Early detection and warning of problems related to the production curve for hens with 98% accuracy	Morales et al. (2016)
Yield prediction	Ensemble model	Identified number of tomatoes with Recall: 0.6066 Precision: 0.9191 F-Measure: 0.7308	Senthilnath et al. (2016)
Quality management	Ensemble model	Prediction and classification of geographical origin of a rice sample with 93.83% accuracy	Maione et al. (2016)
Livestock management	Ensemble model	Classification of cattle behavior with an accuracy of 96%	Dutta et al. (2015)

6.3.1.2 Unsupervised Learning

In this type of learning, the model learns patterns without any supervision. It is generally used when the input data is not labeled. This type of algorithm learns patterns from untagged data. Here no supervision is needed by the uses or any expert person and permits the algorithm to follow its own way to predict patterns. These models are the most popular for clustering problems. This type of model is mostly needed in situations where it is difficult to translate domain knowledge into feature crafting; this may be useful for high-dimensional data.

K means clustering is a form of unsupervised learning which forms "k" clusters of data. It is a form of top-down clustering algorithm and follows centroid-based clustering. The data in a cluster are more similar to each other than the other clusters. The measure of similarity can be calculated using the Silhouette coefficient. Distance between data points can be calculated using distances such as Manhattan and Euclidean.

TABLE 6.3

Unsupervised ML Algorithm Used in Agriculture Application

Application	Type	Results/Finding	Article
Plant disease	K-means	3D to distinguish between *Cercospora beticola* and *Uromyces betae* disease in beetroot crop	Bauer et al. (2011)
Plant disease	K-means	Classified into "healthy" and "injured" classes of clover plants	Atas et al. (2012)
Weed detection	K-means	Accuracy: 92.89% Designed weed identification model for soybean crop	Tang et al. (2017)
Image background preprocessing	Principal component analysis (PCA)	Image cropping, contrast enhancement, and removal of background was done efficiently using PCA	Atas et al. (2012)
Plant fungus	PCA	Identified the presence of aflatoxins which is a harmful substance produced by fungus in chilli pepper plant	Atas et al. (2012)
Yield prediction	Clustering	Identified number of tomatoes with Recall: 0.6066 Precision: 0.9191 F-Measure: 0.7308	Senthilnath et al. (2016)
Disease detection	Self-organizing map (SOM)	Accuracy 99.27% in the detection of yellow rust in wheat crop	Moshou et al. (2005)
Weed detection	SOM	Recognition and discrimination of *Zea mays* and weed species with an accuracy of 100%	Pantazi et al. (2016)

Association rules are the next category under unsupervised learning, which helps to find relationships between variables. A priori algorithm is used to find rules which are satisfied in a particular dataset. Based on how frequently the number of items occurs and their correlation, this algorithm uses a bottom-up approach to get the desired rules. This algorithm makes use of "support" and "confidence" to figure out which items to consider or eliminate. The Unsupervised machine learning algorithms used in agriculture applications are shown in Table 6.3.

6.3.1.3 Semi-Supervised Learning

It can be said as semi-supervised is a combination of both supervised and unsupervised techniques. It takes advantage of both the learning type for presenting accurate results. It makes use of the tiny quantity of labeled data along with the huge quantity of unlabeled data. This is basically done large size of the labeled dataset is not available for problem solving. It is popularly used in classification and clustering problems. Semi-supervised machine learning algorithms used in agriculture applications are shown in Table 6.4.

TABLE 6.4

Semi-Supervised and Reinforcement ML Algorithm Used in Agriculture application

Application	Type	Results/Finding	Article
Weed detection	Semi-supervised learning	Accuracy: 82.13%	Shorewala et al. (2021)
Soil spectroscopy	Laplacian support vector regression	A robust model for soil spectroscopy was developed and soil samples from five different countries were studied successfully	Tsimpouris et al. (2021)
Water management, plant growth	Reinforcement learning	Framework for smart agriculture is designed	Bu et al. (2019)
Yield prediction	Reinforcement learning	Efficiently predicts the crop yield with an accuracy of 93.7%.	Elavarasan et al. (2020)
Water management	Reinforcement learning	Framework for irrigation control technique is designed which is suitable for various geographical locations	Sun et al. (2017)

6.3.1.4 Reinforcement Learning

This type of ML represents the way humans learn, i.e., by action and reaction. Here a mediator gains knowledge of how to perform in the neighboring surroundings. Deploy this algorithm; the apparatus is skilled in formulating precise decisions. This mechanism, this technique the apparatus is uncovered to a situation where it trains itself incessantly by means of trial and error. This apparatus acquires from past occurrences and it tries to incarcerate the best probable comprehension to compose precise commerce decisions. The instance of Reinforcement Learning is Markov Decision Process. Reinforcement machine learning algorithm used in agriculture applications are shown in Table 6.4.

6.4 Applications on ML Agro-IoT System

6.4.1 Pest Control Management

With an increasing demand for organic food items, farmers are more focused on cultivating crops that are pesticide-free or with minimum use of fertilizers. With proper use of some gas sensors and image sensors farmers can remotely monitor pest population and, if the levels reach the desired threshold value, can remotely release chemicals to reduce or stop the multiplication of pests. Also if an anonymous pest is detected it can be given to the agriculture department for studies giving scope for research activities. Climate change can also give rise to pest occurrence; with data made available by the meteorological department as well as temperature sensors and humidity sensors installed in agricultural land, farmers can get alter notifications in the form of SMS on mobile devices. It will help to avoid future losses and taking necessary precautionary measures. Pest has the worst effects on production as well as on human health, so controlling it becomes a prime motive. IoT technology aids overcome food safety challenges. Image sensors can be used to detect pests in a real-time environment. An infra-red sensor can also be used to perceive body heat based on which pest can be spotted. A four-layer IoT framework is used to build a pest monitoring system, whereas ML will be used for identification, classification, or prediction of type/class of pests. Ramalingam et al. (2020) designed a distant entrap supervision framework using IoT and the "Faster RCNN (Region-based Convolutional Neural Networks)" and "Residual neural Networks 50 (ResNet50)" for object recognition with an average of 94% accuracy. Identification and detection of plant diseases are some of the most important factors for agriculture. Crops can be monitored throughout the lifecycle with the help of IoT technology. Image sensors can be used to closely observe coloring patterns. Based on color values diseases can be detected easily. It can also facilitate understanding fruiting and flowering patterns. Image sensors can also be used to keep a watch on any pest or bug infecting the crop. Early detection can help in early prevention. Soil moisture sensors and water level sensors can help to calculate salinity and water need. The author (Yahata et al., 2017) suggested using ML models for plant breeding using biological features as well as environmental stress (Garg et al. 2020).

6.4.2 Resource Management

The scarcity of various natural resources, mainly water and soil, makes it important to wisely utilize these resources. The smart irrigation system is the best example for this where sensors can sense the water level and moisture level and accordingly supply water to crop, thus reducing water wastage and also preserving the soil nutrition level. Over-watering and under-watering both will have ad worse effect on plants and crops so have to use it properly. Water pumps and pipes can be remotely controlled by farmers giving them the freedom from manual efforts needed for watering the huge land. Sensors like pH (potential of hydrogen or "power of hydrogen"), ORP (oxidation–reduction potential), EC (electrical conductivity), and turbidity have demonstrated precise water quality status. The quality of soil will decide the types of crop cultivated and also the quantity of yield. Every time farmers make an additional effort during the land preparation phase to ensure and increase the quality of the soil. Land preparation includes different stages like clearing weeds, plowing, harrowing, flooding, and leveling. The traditional method for testing soil quality was to send samples of soil to the geological department and wait for results. Most accurate results were provided by this technique but the disadvantage were that it was time-consuming, required more testing equipment, was not economical for poor farmers, and fewer laboratories were set up, making it difficult for every farmer to connect to it. IoT provides an easy solution to this problem too. Soil sensors like NPK sensors, pH sensors, EC sensors, moisture sensors, etc., will provide information of soil nutrition level and quality to farmers on mobile phones through SMS or by some mobile designed application. The image sensor can also be used to detecting weeds and uneven leveling of land. Similarly, water level senor will be useful to maintain intensity during the flooding stage, especially for paddy land.

6.4.3 Safeguarding Crops from Animals, Birds, and Human Attack

Once the crop starts to give a good yield the next danger the farmers have to face is various types of attacks. It is a very common practice to stay awake day and night to look after crops till they are harvested. Famers have to protect their crops from various animals and birds; furthermore, humans are also harmful as they can destroy the crops by burring out the land out of any personal rivalry or by stealing the crops. Traditionally farmers used electric fencing for preventing animal attach but it was made illegal as it had a high risk of animal or human death. Image sensor along with proximity sensor is the best to answer for this, were the proximity sensor can detect the presence of any object within marked area and image sensor will help to identify that object, thus making farmer attentive and ready to take quick action. A flame detector, a type of fire sensor, is moreover accurate and responds faster, making it more reliable to identify a fire in the field.

6.4.4 Livestock Management

Raising livestock is not an economic chore. It mainly involves two tasks, firstly monitoring the health of their livestock and tracking their location. Health supervision is performed by placing appropriate wearable on livestock and evaluating body temperature, heat, pressure, heart rate, respiratory rate, digestive level, etc. It will avoid illness and lend a hand for early diagnose of diseases. It is an ordinary behavior of livestock to roam anywhere and get separated from the group. IoT wearable can track the livestock if lost in minimum time and also provides information about livestock movement patterns. Depending on the type of livestock different IoT-based applications can be designed, for example, monitoring cattle behavior as food consumed and steps walked for analyzing health as well as quantity of milk it may produce. An electronic shepherd guides and allows sheep to feed unwanted weeds only, thus safeguarding the vegetation and help to eliminate the weed. Many such applications based on needs are designed and deployed with IoT as a support system. The amalgamation skeleton of IoT and ML for livestock behavior and disease prediction was designed by the author (Lee 2018).

6.4.5 Yield Management

Predicting the crop capitulates to be healthy ahead of the yield time would help the strategists and farmers to captivate appropriate procedures for selling and storeroom. Precise forecast of crop expansion stages plays a pivotal position in a crop manufacturing organization. Such predictions will also sustain the allied industries for the device the strategy of logistics of their business. The crop yield forecast is a vital agricultural crisis. Each and every cultivator has forever tried to recognize how much he will get from his anticipation. In the precedent, capitulate forecast was intended by analyzing farmer's preceding occurrence on a meticulous crop. The Agricultural capitulate is mainly relies on climate circumstances, vermin, and preparation of harvest maneuver. The precise figure about the history of crop capitulation is a significant thing for conception decisions associated with agricultural hazard organization. A greenhouse is constructed with walls and roofs made mostly of see-through material like glass, in which plants that want regulated climatic conditions are developed. Skilled laborers are needed to work in such an environment, but with IoT technology, manual work is eliminated as most tasks are automated. Smart greenhouses deployed with the support of IoT will intelligently monitor and control the inside climate.

6.4.6 Quantifying the Emission of Greenhouse Gases

Climate change has affected the agriculture domain badly, and rises in the atmospheric concentration of greenhouse gas (GHG) is the main factor for this which future influences Global warming. Last three decades the warming effect has increased by 37%, which is highly alarming. "Water vapor, carbon dioxide (CO_2), methane (CH_4), ozone and nitrous oxide (N_2O)" are the five most important GHG. Care must be taken to control the increase in the amount of emission of these gases. Industrial and metropolitan pollution is the foremost reason for the increase in the level of GHG but also agriculture is one of the reasons. Paddy lands reveal the fact that they are also one of the sources for the release of greenhouse gases such as methane (CH_4) and nitrous oxide (N_2O), which are caused due to excessive use of fertilizer and poor climate conditions. GHG affects not only the crop's growth but also affect the health of farmers. Most farmers work in paddy land where the concentration of GHG s high is suffering from lung diseases. Author (Jadhav et al., 2019) IoT can give the solution by finding harmful gases by deploying various gas sensors like CO_2 sensor, MQ-2, MQ-4 or MQ-5 for methane gas, MQ-7 or MQ-9 for Carbon Monoxide (CO), MQ-135 for ammonia, etc., at different locations to monitor, detect and prevent GHG emission.

6.5 Conclusion

Affluent insight for decision making and improving action in order to protect the crops or to increase yield can be facilitated by integrating ML to sensor data. Most research shows that IoT challenges like Quality of Service, Network congestion and Overload, Interoperability and heterogeneity, Security and privacy, and Network Mobility and Coverage can be easily addressed by ML techniques. ML models facilitate inaccurate classification, regression, or clustering for predicting output values. Thus knowledge-based agriculture systems can be designed by incorporating ML models into the IoT framework. Figure 6.3 shows a count of research papers published based on the integration of these two massive technologies in the agriculture domain. The numbers are not huge and thus encourage us to explore more in all the directions for finding solutions to numerous problems.

References

Akpakwu, G. A., Silva, B. J., Hancke, G. P., & Abu-Mahfouz, A. M. (2017). A survey on 5G networks for the Internet of Things: Communication technologies and challenges. *IEEE access, 6*, 3619–3647.

Amatya, S., Karkee, M., Gongal, A., Zhang, Q., & Whiting, M. D. (2016). Detection of cherry tree branches with full foliage in planar architecture for automated sweet-cherry harvesting. *Biosystems engineering, 146*, 3–15, doi: 10.1016/j.biosystemseng.2015.10.003

Atas, M., Yardimci, Y., & Temizel, A. (2012). A new approach to aflatoxin detection in chili pepper by machine vision. *Computers and electronics in agriculture, 87*, 129–141.

Baranowski, P., Jedryczka, M., Mazurek, W., Babula-Skowronska, D., Siedliska, A., & Kaczmarek, J. (2015). Hyperspectral and thermal imaging of oilseed rape (*Brassica napus*) response to fungal species of the genus Alternaria. *PloS one, 10*(3), e0122913.

Bauer, S. D., Korč, F., & Förstner, W. (2011). The potential of automatic methods of classification to identify leaf diseases from multispectral images. *Precision agriculture, 12*(3), 361–377.

Bu, F., & Wang, X. (2019). A smart agriculture IoT system based on deep reinforcement learning. *Future generation computer systems, 99*, 500–507, doi: 10.1016/j.future.2019.04.041

Byabazaire, J., Olariu, C., Taneja, M., Davy, A. (2019). Lameness Detection as a Service: Application of Machine Learning to an Internet of Cattle, doi: 10.1109/CCNC.2019.8651681

Castillo-Guevara, M. A., Palomino-Quisne, F., Alvarez, A. B., & Coaquira-Castillo, R. J. (2020, October). Water stress analysis using aerial multispectral images of an avocado crop. In *2020 IEEE Engineering International Research Conference (EIRCON)* (pp. 1–4). IEEE, doi: 10.1109/ EIRCON51178.2020.9254011

Chatterjee, S., Kumar, S., Saha, J., & Sen, S. (2019, March). Hybrid regression model for soil moisture quantity prediction. In *2019 International Conference on Opto-Electronics and Applied Optics (Optronix)* (pp. 1–5). IEEE, doi: 10.1109/OPTRONIX.2019.8862329

Chen, D., Neumann, K., Friedel, S., Kilian, B., Chen, M., Altmann, T., & Klukas, C. (2014). Dissecting the phenotypic components of crop plant growth and drought responses based on high-throughput image analysis. *The plant cell, 26*(12), 4636–4655.

Dubois, A., Teytaud, F., & Verel, S. (2021). Short term soil moisture forecasts for potato crop farming: A machine learning approach. *Computers and electronics in agriculture, 180*, 105902, ISSN 0168-1699, doi: 10.1016/j.compag.2020.105902

Dutta, R., Smith, D., Rawnsley, R., Bishop-Hurley, G., Hills, J., Timms, G., & Henry, D. (2015). Dynamic cattle behavioural classification using supervised ensemble classifiers. *Computers and electronics in agriculture, 111*, 18–28.

Elavarasan, D., & Vincent, P. D. (2020). Crop yield prediction using deep reinforcement learning model for sustainable agrarian applications. *IEEE Access, 8*, 86886–86901, doi: 10.1109/ ACCESS.2020.2992480

Garg, D., Khan, S., & Alam, M. (2020). Integrative use of IoT and deep learning for agricultural applications. In *Proceedings of ICETIT 2019* (pp. 521–531). Springer, Cham.

Goap, A., Sharma, D., Shukla, A. K., & Krishna, C. R. (2018). An IoT based smart irrigation management system using Machine learning and open source technologies. *Computers and electronics in agriculture, 155*, 41–49, doi: 10.1016/j.compag.2018.09.040

Goel, R. K., Yadav, C. S., Vishnoi, S., & Rastogi, R. (2021). Smart agriculture – Urgent need of the day in developing countries, *Sustainable Computing: Informatics and Systems, 30*, 100512, ISSN 2210-5379, doi: 10.1016/j.suscom.2021.100512

Gopinath, S., Ghanathe, N., Seshadri, V., & Sharma, R. (2019, June). Compiling kb-sized machine learning models to tiny IoT devices. In *Proceedings of the 40th ACM SIGPLAN Conference on Programming Language Design and Implementation* (pp. 79–95), doi: 10.1145/3314221.3314597

Haroon, A., Shah, M. A., Asim, Y., Naeem, W., Kamran, M., & Javaid, Q. (2016). Constraints in the IoT: The world in 2020 and beyond. *Constraints, 7*(11), 252–271, http://dx.doi.org/10.14569/ IJACSA.2016.071133

Hasan, A. M., Sohel, F., Diepeveen, D., Laga, H., & Jones, M. G. (2021). A survey of deep learning techniques for weed detection from images. *Computers and electronics in agriculture, 184*, 106067, doi: 10.1016/j.compag.2021.106067

Hernández Rabadán, D., Ramos, F., & Guerrero Juk, J. (2014). Integrating SOMs and a Bayesian classifier for segmenting diseased plants in uncontrolled environments. *The scientific world journal, 2014*, 214674, doi: 10.1155/2014/214674

Jadhav, S. A., & Lal, A. M. (2019). Analysis of methane (CH4) and nitrous oxide (N2O) emission from paddy rice using IoT and fuzzy logic. *International journal of cloud computing, 8*(3), 258–265, doi: 10.1504/IJCC.2019.103933

Kaundal, R., Kapoor, A. S., & Raghava, G. P. (2006). Machine learning techniques in disease forecasting: A case study on rice blast prediction. *BMC bioinformatics, 7*(1), 1–16, doi: 10.1186/1471-2105-7-485

Kaur, A., & Sood, S. K. (2021). Energy efficient cloud-assisted IoT-enabled architectural paradigm for drought prediction. *Sustainable computing: Informatics and systems, 30*, 100496, =doi: 10.1016/j. suscom.2020.100496

Kersting, K., Xu, Z., Wahabzada, M., Bauckhage, C., Thurau, C., Roemer, C., … & Pluemer, L. (2012, July). Pre-symptomatic prediction of plant drought stress using Dirichlet-aggregation regression on hyperspectral images. In *Proceedings of the AAAI Conference on Artificial Intelligence* (Vol. 26, No. 1).

Kishore Ramakrishnan, A., Preuveneers, D., & Berbers, Y. (2014). Enabling self-learning in dynamic and open IoT environments. *Procedia computer science, 32*, 207–214.

Kumar, V. S., Gogul, I., Raj, M. D., Pragadesh, S. K., & Sebastin, J. S. (2016). Smart autonomous gardening rover with plant recognition using neural networks. *Procedia computer science, 93*, 975–981, doi: 10.1016/j.procs.2016.07.289

Lakhiar, I. A., Jianmin, G., Syed, T. N., Chandio, F. A., Buttar, N. A., & Qureshi, W. A. (2018). Monitoring and control systems in agriculture using intelligent sensor techniques: A review of the aeroponic system. *Journal of sensors, 2018*, 1–19.

Laure, B. E., Angela, B., & Tova, M. (2018, April). Machine learning to data management: A round trip. In *2018 IEEE 34th International Conference on Data Engineering (ICDE)* (pp. 1735–1738). IEEE, doi: 10.1109/ICDE.2018.00226

Lee, M. (2018). IoT livestock estrus monitoring system based on machine learning. *Asia-pacific journal of convergent research interchange, SoCoRI, 1*, 119–128.

Lee, A., Taylor, P., Kalpathy-Cramer, J., & Tufail, A. (2017). Machine learning has arrived!. *Ophthalmology, 124*(12), 1726–1728.

Lindström, T., Grear, D., Buhnerkempe, M., Webb, C., Miller, R., Portacci, K., & Wennergren, U. (2013). A Bayesian approach for modeling cattle movements in the United States: Scaling up a partially observed network. *PloS one, 8*, e53432, doi: 10.1371/journal.pone.0053432

Lu, J., Liu, Y., & Li, X. (2011, September). The decision tree application in agricultural development. In *International Conference on Artificial Intelligence and Computational Intelligence* (pp. 372–379). Springer, Berlin, Heidelberg, doi: 10.1007/978-3-642-23881-9_49

Maione, C., Batista, B. L., Campiglia, A. D., Barbosa Jr, F., & Barbosa, R. M. (2016). Classification of geographic origin of rice by data mining and inductively coupled plasma mass spectrometry. *Computers and electronics in agriculture, 121*, 101–107.

Mehdizadeh, S., Behmanesh, J., & Khalili, K. (2017). Using MARS, SVM, GEP and empirical equations for estimation of monthly mean reference evapotranspiration. *Computers and electronics in agriculture, 139*, 103–114.

Mehra, M., Saxena, S., Sankaranarayanan, S., Tom, R. J., & Veeramanikandan, M. (2018). IoT based hydroponics system using deep neural networks. *Computers and electronics in agriculture, 155*, 473–486, doi: 10.1016/j.compag.2018.10.015

Messaoud, S., Bradai, A., Bukhari, S. H. R., Qung, P. T. A., Ahmed, O. B., & Atri, M. (2020). A Survey on machine learning in Internet of Things: Algorithms, strategies, and applications. *Internet of Things*, 100314, doi: 10.1016/j.iot.2020.100314

Morales, I. R., Cebrián, D. R., Blanco, E. F., & Sierra, A. P. (2016). Early warning in egg production curves from commercial hens: A SVM approach. *Computers and electronics in agriculture, 121*, 169–179.

Moshou, D., Bravo, C., Oberti, R., West, J., Bodria, L., McCartney, A., & Ramon, H. (2005). Plant disease detection based on data fusion of hyper-spectral and multi-spectral fluorescence imaging using Kohonen maps. *Real-time imaging, 11*(2), 75–83.

Pantazi, X.-E., Moshou, D., & Bravo, C. (2016). Active learning system for weed species recognition based on hyperspectral sensing. *Biosystems engineering, 146*, doi: 10.1016/j.biosystemseng.2016.01.014

Patil, S. S., & Thorat, S. A. (2016, August). Early detection of grapes diseases using machine learning and IoT. In *2016 Second international conference on cognitive computing and information processing (CCIP)* (pp. 1–5). IEEE, doi: 10.1109/CCIP.2016.7802887

Piccini, C., Marchetti, A., Rivieccio, R., & Napoli, R. (2019). Multinomial logistic regression with soil diagnostic features and land surface parameters for soil mapping of Latium (Central Italy). *Geoderma, 352*, 385–394, doi: 10.1016/j.geoderma.2018.09.037

Potena, C., Nardi, D., & Pretto, A. (2016, July). Fast and accurate crop and weed identification with summarized train sets for precision agriculture. In *International conference on intelligent autonomous systems* (pp. 105–121). Springer, Cham.

Ramalingam, B., Mohan, R. E., Pookkuttath, S., Gómez, B. F., Sairam Borusu, C. S. C., Wee Teng, T., & Tamilselvam, Y. K. (2020). Remote insects trap monitoring system using deep learning framework and IoT. *Sensors, 20*(18), 5280, doi: 10.3390/s20185280

Ramos, P. J., Prieto, F. A., Montoya, E. C., & Oliveros, C. E. (2017). Automatic fruit count on coffee branches using computer vision. *Computers and electronics in agriculture, 137*, 9–22.

Rodríguez, S., Gualotuña, T., & Grilo, C. (2017). A system for the monitoring and predicting of data in precision agriculture in a rose greenhouse based on wireless sensor networks. *Procedia computer science, 121,* 306–313, doi: 10.1016/j.procs.2017.11.042

Rossi, F., Souza, P., Ferreto, T., Lorenzon, A., Caggiani Luizelli, M., Rubin, F., & Hohemberger, R. (2020). Detecting abnormal sensors via machine learning: An IoT farming WSN-based architecture case study. *Measurement, 162,* doi: 10.1016/j.measurement.2020.108042

Rumpf, T., Mahlein, A. K., Steiner, U., Oerke, E. C., Dehne, H. W., & Plümer, L. (2010). Early detection and classification of plant diseases with support vector machines based on hyperspectral reflectance. *Computers and electronics in agriculture, 74(1),* 91–99.

Saha, A. K., Saha, J., Ray, R., Sircar, S., Dutta, S., Chattopadhyay, S. P., & Saha, H. N. (2018, January). IOT-based drone for improvement of crop quality in agricultural field. In *2018 IEEE 8th Annual Computing and Communication Workshop and Conference (CCWC)* (pp. 612–615). IEEE. doi: 10.1109/CCWC.2018.8301662

Sachdeva, G., Singh, P., & Kaur, P. (2021). Plant leaf disease classification using deep convolutional neural network with Bayesian learning. *Materials today: Proceedings,* doi: 10.1016/j.matpr.2021.02.312

Sankaran, S., Mishra, A., Maja, J. M., & Ehsani, R. (2011). Visible-near infrared spectroscopy for detection of Huanglongbing in citrus orchards. *Computers and electronics in agriculture, 77(2),* 127–134.

Saravanan, K., & Saraniya, S. (2018). Cloud IOT based novel livestock monitoring and identification system using UID. *Sensor review,* doi: 10.1108/SR-08-2017-0152

Saravanan, K., & Srinivasan, P. (2018). Examining IoT's applications using cloud services. In *Examining cloud computing technologies through the Internet of Things* (pp. 147–163). IGI Global, doi: 10.4018/978-1-5225-3445-7.ch008

Senthilnath, J., Dokania, A., Kandukuri, M., Ramesh, K. N., Anand, G., & Omkar, S. N. (2016). Detection of tomatoes using spectral-spatial methods in remotely sensed RGB images captured by UAV. *Biosystems engineering, 146,* 16–32.

Shorewala, S., Ashfaque, A., Sidharth, R., & Verma, U. (2021). Weed density and distribution estimation for precision agriculture using semi-supervised learning. *IEEE access, 9,* 27971–27986, doi: 10.1109/ACCESS.2021.3057912

Sicari, S., Rizzardi, A., & Grieco, L. A., Coen-Porisini, A. (2015). Security, privacy and trust in internet of things:, The road ahead. *Computer networks, 146,* doi: 10.1016/j.comnet.2014.11.008

Singh, M., & Ahmed, S. (2020). IoT based smart water management systems: A systematic review. *Materials today: Proceedings,* doi: 10.1016/j.matpr.2020.08.588

Sun, Y., Wang, C., Chen, H. Y., & Ruan, H. (2020). Response of plants to water stress: A meta-analysis. *Frontiers in plant science, 11,* 978, doi: 10.3389/fpls.2020.00978

Sun, L., Yang, Y., Hu, J., Porter, D., Marek, T., & Hillyer, C. (2017, December). Reinforcement learning control for water-efficient agricultural irrigation. In *2017 IEEE International Symposium on Parallel and Distributed Processing with Applications and 2017 IEEE International Conference on Ubiquitous Computing and Communications (ISPA/IUCC)* (pp. 1334–1341). IEEE, doi: 10.1109/ISPA/IUCC.2017.00203

Tang, Y., Dananjayan, S., Hou, C., Guo, Q., Luo, S., & He, Y. (2021). A survey on the 5G network and its impact on agriculture: Challenges and opportunities. *Computers and electronics in agriculture, 180,* 105895, doi: 10.1016/j.compag.2020.105895

Tang, J., Wang, D., Zhang, Z., He, L., Xin, J., & Xu, Y. (2017). Weed identification based on K-means feature learning combined with convolutional neural network. *Computers and electronics in agriculture, 135,* 63–70, doi: 10.1016/j.compag.2017.01.001

Tsimpouris, E., Tsakiridis, N. L., & Theocharis, J. B. (2021). Using autoencoders to compress soil VNIR–SWIR spectra for more robust prediction of soil properties. *Geoderma, 393,* 114967, ISSN 0016-7061, doi: 10.1016/j.geoderma.2021.114967

Vanipriya, C. H., Malladi, S., & Gupta, G. (2021). Artificial intelligence enabled plant emotion xpresser in the development hydroponics system. *Materials today: Proceedings,* doi: 10.1016/j.matpr.2021.01.512

Vij, A., Vijendra, S., Jain, A., Bajaj, S., Bassi, A., & Sharma, A. (2020). IoT and machine learning approaches for automation of farm irrigation system. *Procedia computer science, 167,* 1250–1257, doi:10.1016/j.procs.2020.03.440

Vyas, S., Shukla, V., & Doshi, N. (2019). FMD and mastitis disease detection in cows using Internet of Things (IOT). *Procedia computer science, 160,* 728–733, doi: 10.1016/j.procs.2019.11.019

Wang, P., Hafshejani, B. A., & Wang, D. (2021). An improved multilayer perceptron approach for detecting sugarcane yield production in IoT based smart agriculture. *Microprocessors and microsystems, 82,* 103822, doi: 10.1016/j.micpro.2021.103822

White, G., Nallur, V., & Clarke, S. (2017). Quality of service approaches in IoT: A systematic mapping. *Journal of systems and software, 132,* 186–203.

Wu, W., Zucca, C., Muhaimeed, A. S., Al-Shafie, W. M., Fadhil Al-Quraishi, A. M., Nangia, V., ... & Liu, G. (2018). Soil salinity prediction and mapping by machine learning regression in Central Mesopotamia, Iraq. *Land degradation & development, 29*(11), 4005–4014, doi: 10.1002/ldr.3148

Wood, N., & Lam, C. (2020). Preparing and Architecting for Machine Learning White Paper Preparing and architecting for machine learning 2, doi: 10.13140/RG.2.2.23893.58080

Yahata, S., Onishi, T., Yamaguchi, K., Ozawa, S., Kitazono, J., Ohkawa, T., ... & Tsuji, H. (2017, May). A hybrid machine learning approach to automatic plant phenotyping for smart agriculture. In *2017 International Joint Conference on Neural Networks (IJCNN)* (pp. 1787–1793). IEEE, 1787–1793, doi: 10.1109/IJCNN.2017.7966067

Yashodha, G., & Shalini, D. (2021). An integrated approach for predicting and broadcasting tea leaf disease at early stage using IoT with machine learning–A review. *Materials today: Proceedings, 37,* 484–488, doi: 10.1016/j.matpr.2020.05.458

7

A Survey on Internet of Things (IoT)-Based Precision Agriculture

Aspects and Technologies

R. Srinivasan, M. Kavitha, and R. Kavitha

VelTech Rangarajan, Dr Sagunthala R&D Institute of Science and Technology
Chennai, India

K. Saravanan

Anna University Regional Campus
Tirunelveli, India

CONTENTS

7.1 Internet of Things 109
7.2 Introduction 110
    7.2.1 Types of Sensors in PA 113
    7.2.2 Layers Design for IoT in PA 113
        7.2.2.1 The Perception Layer 113
        7.2.2.2 The Transmission Layer 113
        7.2.2.3 The Application Layer 114
    7.2.3 Specifications for Energy and Power in PA 114
7.3 Precision Agriculture Requirement 114
7.4 Multispectral Remote Sensing in Precision Agriculture 115
    7.4.1 Hyperspectral Remote Sensing 116
    7.4.2 Hyperspectral Data in Agriculture 116
7.5 Global Positioning System 116
7.6 Technologies for IoT-Based PA 116
7.7 Challenges of IoT in PA in India 117
7.8 Conclusions 118
References 118

7.1 Internet of Things

Figure 7.1 depicts the IoT concept's sensing, monitoring, planning, analysis, and control. The Internet of Things (IoT) is an electronics-oriented integrated system in which sensors, controllers, various forms of software, and network interconnectivity are used to

DOI: 10.1201/9781003185413-7

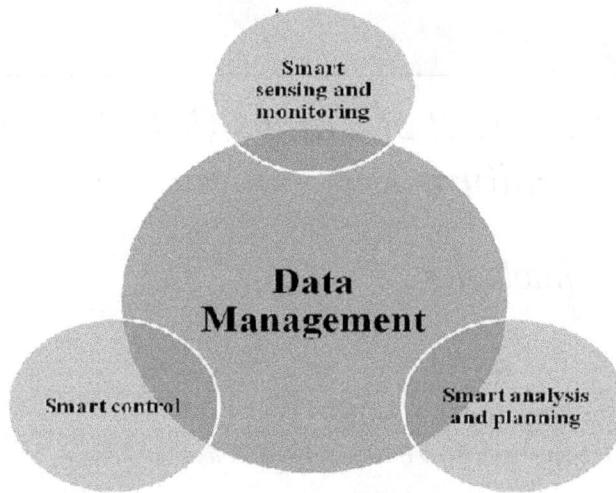

FIGURE 7.1
Concept of IoT.

acquire and exchange data between actual electronics circuits and devices. IoT's accuracy and efficiency are the major benefits for real-time applications to handle a variety of problems. Agriculture is a major manufacturing application of IoT to use the accelerated growth of IoT controllers for the overall rate, and subsistence is solely based on agricultural goods.

IoT concepts are enhanced by the huge number of electronic devices connected through the Internet. With it, we can look, at any time, anyone or any connection paradigm [1], with applications for different fields of work, such as travel, intellectual property, education, marketing, logistics, transformation, industrial and environmental production, and smart agriculture [2].

Figure 7.2 and 7.3 depicts the various states and applications of the Internet of Things in the field of agriculture.

7.2 Introduction

Table 7.1 depicts the many types of agricultural expansion. To maintain economic development and stability, agriculture plays a vital role in countries' production [3, 4]. Overcoming the gap between population growth and grain yield is a big challenge for agriculture.

Figure 7.4 shows how sensors work in the field. The main objective of precision agriculture (PA) is to increase the production of the crops, decrease labor time, proper irrigation processes, and effective consumption of fertilizer and provide higher productivity and use of resources when compared with traditional methods. PA is used for efficient use of various inputs like the efficient use of seeds; pesticide; fertilizer; and water, fuel, land, and soil.

IoT offers suitable solutions for several applications such as agriculture, security, traffic congestion, smart cities, and industrial control. Wireless sensor networks (WSNs), along

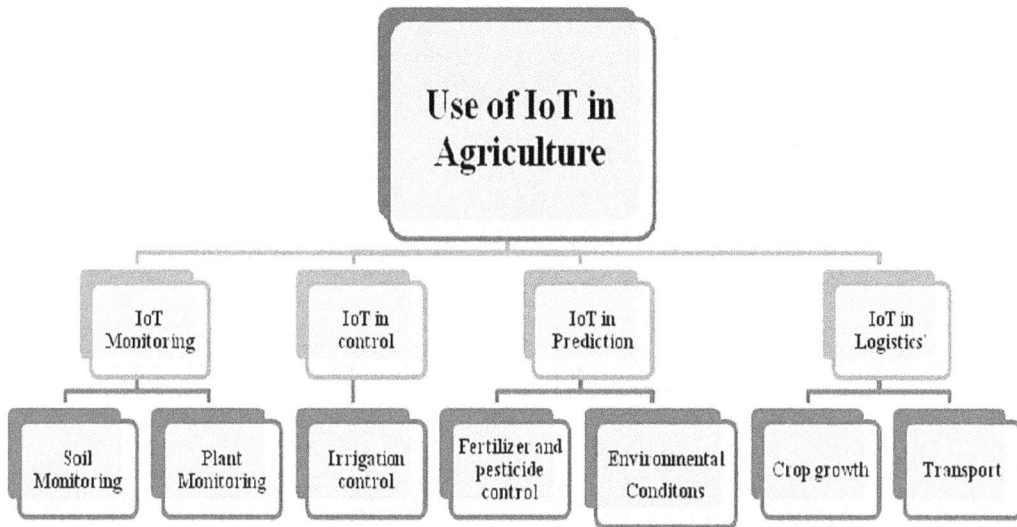

FIGURE 7.2
Hierarchical structure of usage of IoT in precision agriculture.

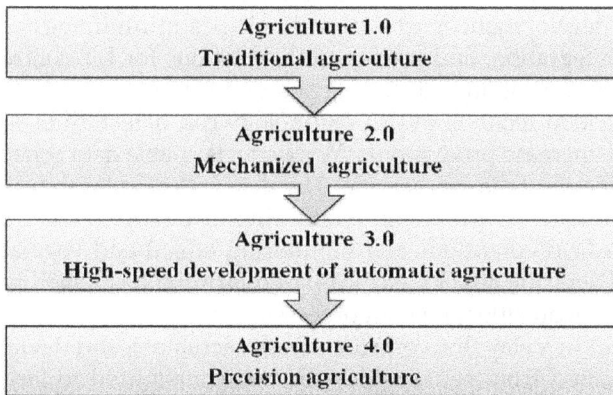

FIGURE 7.3
Characteristics of agriculture growth and how to deal with it (Agriculture 1.0 to Agriculture 4.0).

TABLE 7.1

Different Types Agricultural Growth with Issues and Periods of Time

Agriculture Growth			
Agriculture 1.0	Traditional agriculture	From 1784 to 1870	Operation efficiency is low
Agriculture 2.0	Mechanized agriculture	In the 20th century	Inefficient use of resources
Agriculture 3.0	High-speed development of automatic agriculture	From 1992 to 2017	Intelligence accuracy is low
Agriculture 4.0	Precision agriculture	From 2017	Security issues

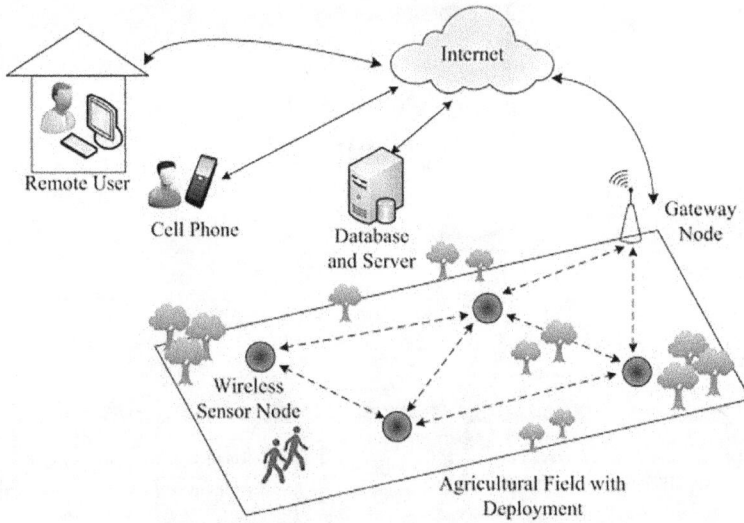

FIGURE 7.4
Wireless sensor node with agriculture field.

with IoT-based automation of PA measures, can modify the agriculture sector. Present IoT trends are security, development of network technologies, minimizing energy use, efficiency of devices, device integration, and user-friendly solutions for IoT controls. IoT is used for multiple-device communication and sharing and understanding their internal and external contexts with embedded technology. IoT technology can detect all of these problems and provide solutions to increase productivity. WSN efforts enable data collecting from sensory devices and transmission to larger servers. The data collected by the sensors will provide details about the unique environment: monitoring of environmental conditions or crop production such as field conditions and monitoring of soil and vegetation, movement of unwanted material, wildlife attacks, etc. IoT-based farming has major features: (1) physical architecture, (2) data acquisition, (3) data processing, and (4) data statistics (Figure 7.5). The whole system is built in a way that controls sensors, actuators, and devices.

Data can be collected from sensory devices and transmitted to larger servers thanks to WSN initiatives and obtain exactly what they require to optimize production and

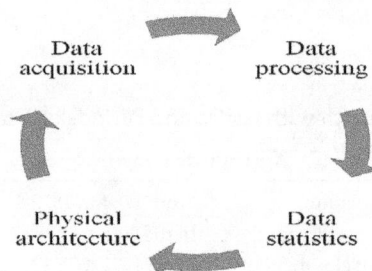

FIGURE 7.5
Major features of IoT-based PA.

sustainability. Actual information regarding the environment like weather changes, crop, and soil parameters can be retrieved from the sensor devices, which will be deployed in the yield. Many aspects and technologies of WSNs with IoT are being used presently in the PA for effective irrigation, fertilization, and pest control. PA manages the production of each crop, herbicide, seed, insecticide, effort fertilizer, etc. PA involves five stages: (1) collecting the data, (2) diagnosis, (3) analyzing the data, (4) precision field operation, and (5) evaluation. PA is mainly used for supervision plans to utilize information knowledge to improve quality and manufacture. PA systems aim to (1) reduce the cost, (2) decrease the time and effort, (3) save water and energy, and (4) provide a user-friendly interface for farmers. Some of the challenges in the technology implementation are a high investment, inexpert labor, fear of new technology, coverage and connectivity, split market, etc.

7.2.1 Types of Sensors in PA

To measure the different types of crops, PA plays an important role in segregating it [5], as shown in Table 7.2.

7.2.2 Layers Design for IoT in PA

IoT's core architecture is segregated into the perception layer, transmission layer, and application layer.

7.2.2.1 The Perception Layer

The perception layer is a top layer used to build the environment for various sensors, which collect data from various environments. This helps brief the current state of the environments [6].

7.2.2.2 The Transmission Layer

The transmission layer includes all types of network communication protocols. It will collect data from the perception layer and transmits it to the application layer based on network protocols [7].

TABLE 7.2

Different Types of Precision Agriculture Sensors

Agriculture Sensors	Functional Description
Location sensors	Location sensors are used to sense the latitudinal and longitudinal position of the area. To improve the accuracy of the sensing position, sensors use GPS satellite technology.
Optical sensors	Optical sensors are used to measure the structure of the soil. These sensors are mounted on robots, drones, and satellites to detect the organic matter and soil moisture content.
Electro-chemical sensors	Electro-chemical sensors help collect chemical data from the soil by detecting certain ions in the soil. They provide information on the pH status and nutrient levels of the soil.
Mechanical sensors	To analyze the soil compaction or mechanical resistance.
Dielectric soil moisture sensors	To measure the humidity levels with the electrolytic instability of the soil.
Airflow sensors	To analyze the air inflow from mobile mode or fixed mode.

TABLE 7.3

Specifications for Energy and Power for Wireless Communication in PA

Type of Data Communication	Application Possibilities	Size of Data	Depletion of Energy
The size of the data is minimal, and it consumes lesser energy	• Air temperature/direction/ humidity speed • Humidity and temperature of soil • Color and thickness of leaf • Thickness of trunk • Size of fruit	100 bytes	Less than an mA
The size of data was medium, and the energy consumption also medium	• Still camera • Multiple cameras • Sensors that detect sound	13 mega bytes	13 mA
The size of data was large and high energy consumption	Video camera streaming	Ten seconds of Mb and a minute	50 A

7.2.2.3 The Application Layer

The application layer plays a major role in IoT architecture. It may be a cloud-based or local system–based function.

- Data storage – e.g., cloud-based platform and Hadoop Distributed File System for quick and secure access to data [8];
- Data management – e.g., Supervisory Control And Data Acquisition (SCADA) for real-time data monitoring [7];
- Data statistics – e.g., decision-making process, production modes, and crop controls for automatic control in agricultural production [9]; and
- Data marketing – e.g., data detection, tracking capabilities of agricultural products for new business models, ownership, and privacy [10].

7.2.3 Specifications for Energy and Power in PA

The energy and power in PA play a major role in processing data and analysis. Table 7.3 shows the various specification of energy and power level for communication.

7.3 Precision Agriculture Requirement

In PA a set of rules must be followed to attain better yield. Related activities to agriculture are:

1. Before going to yield the crop in the field, a field survey is important. Soil sample as well as study of soil conductivity, soil moisture, and pH according to the soil for the chosen agricultural plant are done using sensors.

2. Remove unnecessary plants growing with crops and avoid unnecessary competition between them.

3. Monitor the plant growth and health, periodically examining nutrient status for phosphorous, potassium, nitrogen, etc.

4. An important factor to be considered after crops are planted is detecting diseases in the crops.

5. In the time of growth of crops, check the water level and soil moisture of the yield.

6. The finding of lodging is also a vital part of PA.

7.4 Multispectral Remote Sensing in Precision Agriculture

Remote hearing is vital to the PA component. It uses multispectral satellites to collect high-resolution images for agriculture practices. The multispectral imaging camera sensors mounted on agricultural drones allow farmers to monitor crops, soil, parasites, fertilizers, and water, the data that they need more accurately. Consequently, such drones have proven to be helpful in terms of increased yield and other benefits. Multispectral sensors use four bands, namely red, green, red-edge, and near-infrared (NIR) bands, to capture images of crops and vegetation in the visible and invisible regions (Figure 7.6) [1].

Modern PA is designed to increase yields and resources such as reducing environmental impacts such as over-fertilization and the use of pesticides. Many benefits of using more images or data include higher accuracy, simplicity, and lower costs. This helps to balance yield, crop growth, and soil quality [11, 12]. A wide variety of multispectral camera sensors are used in agricultural practices such as [13, 14].

Different types of sensors used in multispectral remote sensing are

- Sentera Quad
- Parrot Sequoia

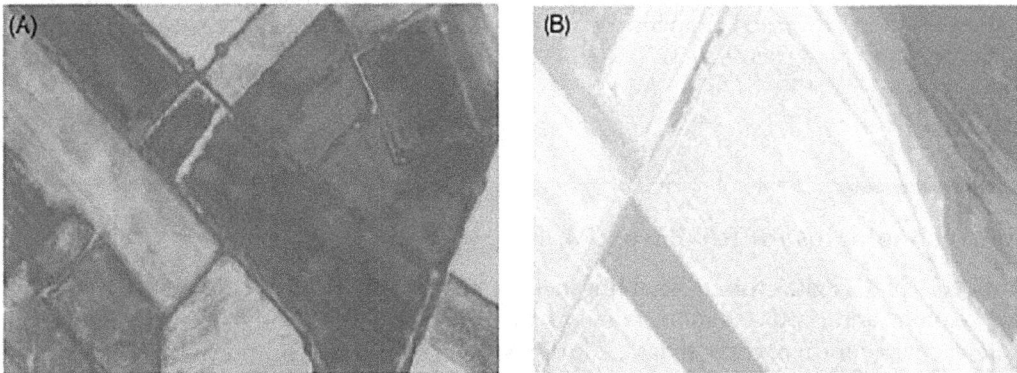

FIGURE 7.6
(A) Image of the field in multispectral images; (B) image of the field from unmanned aerial vehicles.

- ACD light sensor – Tetracam
- MicaSense Rededge Sensor
- Airinov multiSPEC 4C Agronomic Sensor

7.4.1 Hyperspectral Remote Sensing

Hyperspectral imaging sensors have more advantages than multispectral sensors for classification and discrimination. They provide detailed information on any item due to the availability of band information. Hyperspectral sensors have good spectral correction. High-resolution spectral processing of hyperspectral data has the advantage of PA capture and monitoring, but it also includes unwanted data, which affects the level of accuracy [15].

7.4.2 Hyperspectral Data in Agriculture

Hyperspectral data involves the acquirement of images in hundreds of narrow adjoining spectral bands to get high-resolution information for each pixel of an exact scene. Extracted spectral signatures from hyperspectral images are used to recognize and categorize the characteristics of objects. Several applications of hyperspectral and multispectral imaging are being confirmed in different types of agriculture techniques, together with excellence in organizing, classifying, and categorizing farming products and in the classifying insect and contaminants as well as in food protection [16].

7.5 Global Positioning System

GPS technology is used to monitor the growing situation. It takes in parameters such as air, water, soil, pesticides, and fertilizers. The GPS structure is used to locate the exact place in a farming field and check a range of farming parameters by using wireless communication networks. It interfaces with Acorn RISC Machine (ARM) like an intelligent monitoring system to attain functions like SMS or MMS to make an alarm to the farm manager when unwanted changes occur. It is also used for the maintenance and monitoring of the crop for agriculture [17].

7.6 Technologies for IoT-Based PA

Protocols play a vital role in enabling network connectivity in IoT devices. Combining applications and protocols allows devices to exchange data over the network, define the data exchange format, encode data, address schemes for devices, and route packets from sources to destination, and protocols include functions like flow control, sequence control, and retransmission of lost packets. Agri-IoT data gaining component consists of several protocols such as Hypertext Transfer Protocol (HTTP), Message Queuing Telemetry

TABLE 7.4

Wireless Communication Technologies with Their Standards

Parameters	Zigbee	LoRa
Standard	IEEE 802.15.4	IEEE 802.15.4g
Channel bandwidth	2 MHz	<100 Hz
Data rate	20, 40, and 250 kbps	100 Mbps
Network size	65,000	1,000,000
Application	WPANs agriculture	Agriculture Environment

Transport (MQTT), Data Distribution Service (DDS), and Advanced Message Queuing Protocol (AMQP) and also communication wireless protocols such as IEEE 802.11 Wi-Fi, LoRaWAN, WiMax, Bluetooth, Zigbee, and 2G/3G/4G Mobile Communications Standards.

Table 7.4 depicts the many types of wireless communications technologies and standards. Table 7.5 depicts the many forms of precision agricultural routing protocols.

7.7 Challenges of IoT in PA in India

- Lack of knowledge among farmers about the advantages of PA
- Extra manual work
- Frequent changing of weather
- Expensive for machinery work
- No interest in PA among young and educated professionals
- More expensive
- Difficulty in understanding the technology among farmers
- IoT devices and smart farming require breakage-free Internet connectivity. It creates challenges in developing countries.

TABLE 7.5

Routing Protocol Schemes in PA

Parameters	Sink Mobility	Multipath	Cluster Head	Routing Metric
Wireless protocols/ devices	Simulation	Zigbee	Zigbee/simulation	Zigbee/IEEE 802.15.4
Power savings/ battery lifetime	High power	1825 min	20 times established without cluster heads	28.4 days
Application	Forest area	Irrigation system	Crop farming	Precision agriculture
Limitations	Packet losses lead to more energy consumption	Consumes a lot of power at low communication distance	Unreliable communication beyond 80 m	Short battery life

7.8 Conclusions

PA and agricultural systems based on IoT have proven to be incredibly beneficial to farmers, as less irrigation is beneficial to agriculture. Sensor coefficients like temperature, data collecting through sensors, humidity, and wetness could be set dependent on the state of the agriculture field. The proposed approach will create optimal resource usage and solve the problem of irrigation scarcity. An important ability of wireless networks was better represented graphically than prior technologies that could be recovered and statistically analyzed. Using IoT technology, real-time field monitoring is conceivable. The presented method closely monitors the waste of agricultural resources. PA is the science of art to improve crop yield and to support management via high technology sensors and analysis tools [18]. PA is the application of technology to manage spatial and temporal unpredictability of inputs to improve productivity and environmental quality. PA is a practical approach that reduces the risk and variables in agriculture. The growth of technologies in the 21st century led to the development of the PA concept [19]. PA is used for the efficient use of various inputs like effective use of fertilizer, seed, pesticide, fuel, land, data, and water. Agriculture and the agricultural industry in a remote area can benefit from the WSNs and cloud server–based vast networks with IoT.

References

1. S. Liaghat, and S.K. Balasundram, "A review: the role of remote sensing in precision agriculture," Am. J. Agric. Biol. Sci., vol. 5, no. 1, pp. 50–55, 2010.
2. R. Srinivasan, and E. Kannan, "A review: Precision agriculture (PA) using energy-efficient wireless sensor networks," J. Comput. Theor. Nanosci., vol. 15, pp. 1–4, 2018.
3. J. Wolfert, C. Srensen, and D. Goense, "A future internet collaboration platform for safe and healthy food from farm to fork," in: 2014 Annual SRII. IEEE, 2014, pp. 266–273.
4. X. Yang, et al. "A survey on smart agriculture: Development modes, technologies, and security and privacy challenges," IEEE/CAA J. Automatica Sinica, vol. 8. no. 2, pp. 273–302, 2020.
5. Kumar,"https://www.rfwireless-world.com/Terminology/Advantages-and-uses-of-Agriculture-Sensors.html"
6. S. Wang, Y. Lin, Y. Qin, and C. Chen, "Security enhancement of internet of things using service level agreements and light weight security," in: Advances in Information and Communication Networks, Springer, 2018, pp. 221–235.
7. A. Tzounis, N. Katsoulas, T. Bartzanas, and C. Kittas, "Internet of things in agriculture, recent advances and future challenges," Biosyst. Eng., vol. 164, pp. 31–48, 2017.
8. Z. Zong, R. Fares, B. Romoser, and J. Wood, "Faster to improving the performance of a large scale hybrid storage system via caching and prefetching," Cluster Comput., vol. 17, no. 2, pp. 593–604, 2014.
9. D. Ko, Y. Kwak, and S. Song, "Real time traceability and monitoring system for agricultural products based on wireless sensor network," Int. J. Distrib. Sens. Netw., vol. 10, no. 6, pp. 832510, 2014.
10. S. Kang, X. Hao, T. Du, L. Tong, X. Su, H. Lu, X. Li, Z. Huo, S. Li, and R. Ding, "Improving agricultural water productivity to ensure food security in China under changing environment: From research to practice," Agric. Water Manag., vol. 179, pp. 5–17, 2017.
11. N. Bagheri, H. Ahmadi, S.K. Alavipanah, and M. Omid, "Multispectral remote sensing for site-specific nitrogen fertilizer management," Pesqui. Agropecuária Brasileira, vol. 48, no. 10, pp. 1394–1401, 2013.

12. M. Wójtowicz, A. Wójtowicz, and J. Piekarczyk, "Application of remote sensing methods in agriculture," Commun. Biometry Crop Sci., vol. 11, no. 1, pp. 31–50, 2016.

13. Corrigan, F., 2018. Multispectral Imaging Camera Drones in Farming Yield Big Benefits. DroneZon., https://www.dronezon.com/learn-about-drones-quadcopters/multispectral-sensor-drones-in-farming-yield-big-benefits/. (accessed 01.06.19).

14. L. Deng, Z. Mao, X. Li, Z. Hu, F. Duan, and Y. Yan, "UAV-based multispectral remote sensing for precision agriculture: A comparison between different cameras," ISPRS J. Photogramm. Remote Sens., vol. 146, pp. 124–136, 2018.

15. P.C. Pandey, K. Manevski, P. Srivastava, and G. Petropoulos, "The use of hyper spectral earth observation data for land use/cover classification: Present status, challenges and future outlook," in: P. Thenkabail (ed.), Hyperspectral Remote Sensing of Vegetation, (1st ed.), 2018a, pp. 147–173.

16. P.S. Thenkabail, and J.G. Lyon, Hyperspectral Remote Sensing of Vegetation, CRC Press, 2016.

17. V. Satyanarayana and S. D. Mazaruddin, "Wireless sensor based remote monitoring system for agriculture using Zig Bee and GPS," Proc. Conf. Adv. Commun. Control Syst., Apr. 2013, pp. 1–5.

18. D. Schimmelpfennig, and J. Lowenberg-DeBoer. "Precision agriculture adoption, farm size and soil variability." Precision Agriculture, 21, Wageningen Academic Publishers, 2021, pp. 769–776.

19. L. García, et al. "IoT-based smart irrigation systems: An overview on the recent trends on sensors and IoT systems for irrigation in precision agriculture," Sensors, vol. 20, no. 4, pp. 1042, 2020.

8

Novel Semantic Agro-Intelligent IoT System Using Machine Learning

C. S. Saravana Kumar

Robert Bosch Engineering and Business Solutions
Coimbatore, India

S. Vinoth Kumar

Vel Tech Rangarajan Dr. Sangunthala R&D Institute of Science and Technology
Chennai, India

CONTENTS

8.1 Introduction 121
8.2 Related Works 123
8.3 Challenges with the Existing Approaches 123
    8.3.1 Issues and Challenges 124
        8.3.1.1 Hardware 124
        8.3.1.2 Infrastructure 124
        8.3.1.3 Networking 124
        8.3.1.4 Reliability 124
        8.3.1.5 Data Security 124
8.4 Problem Statement 124
8.5 Proposed Approach 125
    8.5.1 Improving the Training Data 125
    8.5.2 Agro-Intelli Algorithm 125
        8.5.2.1 Prerequisites 126
        8.5.2.2 Dimensionality Vector 126
        8.5.2.3 Intelli-Group Algorithm 128
8.6 Results and Discussion 129
8.7 Conclusion and Future Work 131
References 131

8.1 Introduction

Agriculture is the backbone of any country. We are in the digital world. Technology has a bigger role in the field of agriculture and its importance is increasing day by day. Technology has not limited itself only to crop diagnosis but also changing the very old practices to improve the yield and promote better monitoring ways. Various technologies are now being used in agriculture. The ecosystem in various countries is getting affected due to the dip in the agriculture yield. It is high time that we put agriculture back on track and technology is going to play a bigger role here. A large-scale agricultural system requires a lot of maintenance, knowledge, and supervision.

The volume of devices getting connected to the Internet is growing at an enormous rate. The primary concept of IoT is to connect all the physical devices in the world virtually and sharing the information between them. It is a revolutionary technology that represents how a connected data can be computed and analyzed in a better way: for example, traffic systems, whereby performing the dynamic analysis information can be sent to the nearby police officials on any trouble in the surrounding. Here different systems are connected through Internet and exchange the data for analysis. Recently IoT has been booming up and it has been applied in various fields. IoT aid to build some of the most efficient and cost-effective solutions [1] to humans in the field of agriculture and some of its highlights are as follows,

- IoT involves the connection of any devices over the Internet [2, 3]. For example, Android, IoT sensors, etc.
- Soil nature and soil transformation can be monitored remotely [4, 5].
- Fail Fast technique – to predict issues at an early stage. For example, prediction of infection in the crop by sensing the growth rate and proposing the prevention steps at an early stage.
- With the help of remote monitoring we save a lot of time and human power.
- Connecting different physical devices as well as involving the cloud makes communication and data sharing efficient.
- IoT aids in performing a wide range of analytics.

Smart farming [6, 7] is achieved by involving new technologies like IoT, the cloud, and machine learning (ML). Today's world population is increasing at a tremendous pace and the greatest challenge that is put forth before mankind is feeding all people. This can be achieved in the best possible way by combining the above said latest techniques. There are numerous challenges faced in agriculture today like degradation of soil fertility, insect attacks, and any recent development in various plant diseases. Farmers do not possess the knowledge of the existing trends for specific crop cultivation nor do they have a proper chance to get an alert if there is a change in the nature of the soil. This chapter explains how the farmland can be continuously monitored for any changes in the soil involving a model using ML techniques where the ML system continuously monitors and provides feedback on soil change and what adaptation or proactive action should be taken to attain the expected yield.

ML techniques can be involved for model creation, where the system can be created with the vector space containing the attributes for soil monitoring. With the help of IoT the sensors connected to the soil can send values of different parameters like soil nature, weather changes, and crop growth, and when the system senses any abnormal situation or any deviations from what has been expected, an alert will be sent to the farmer. Also the system can be connected with the cloud [8, 9] system so that recent trends of crop growth can be inferred by referring to the data collected globally. Also a statistical opinion can be provided to the farmer to improve the crop yield. This chapter discusses the same, where Section 8.2 explains the existing techniques of ML in the field of agriculture combined with IoT. Section 8.3 explains our proposed approach, explaining how the ML system connects the data received from IoT devices semantically and its response to the user. Section 8.4 details out the results and discussion. Section 8.5 gives the conclusion.

8.2 Related Works

There are many techniques followed in different countries with respect to the subject "Smart Farming." The important ones are as follows:

Malaysia has developed an IoT system for fruit traceability. It is called as Mi-Trace [10]. This system provides traceability to sellers and buyers for any fruit.

Taiwan has introduced an IoT system to monitor soil conditions. This is utilized especially in turmeric cultivation, where with the help of the IoT system the quality of the turmeric is increased from 40 to 60% as well as 70% of the water is saved.

Thailand has developed an IoT system for water control, where based on different vegetations the system helps in guiding the water irrigation to the farmers.

China has built an IoT system to send the instruction to remote places, thereby saving the labor cost by 60% and reducing fertilizer and pesticide use rate to 60% [10].

8.3 Challenges with the Existing Approaches

To meet the rapidly growing population's needs it is very important to increase the yield. To increase the yield the soil has to be monitored continuously. Upon any change in the nature of the soil, the details should be collected by the sensors and the farmer has to be updated or notified immediately. Even the plant growth rate is the one to watch where an insect attack or any recent developing plant disease might hinder the plant growth rate. Water beds are drying up day by day and the weather is unpredictable. To cope with all these conditions we need a system that can monitor and send feedback continuously to the user to improve the yield. Hence the only way to achieve these is to adapt to smart farming, where the latest technologies like ML with IoT would aid to achieve the target. Recent research works have been focused on Agro-IoT, where an IoT-based framework monitors the data continuously; it performs analysis and reasoning. SmartFarmNet is one such IoT-based platform for collecting environmental, soil, fertilizer, and irrigation data. The availability of open-source smart agriculture software is not high. Some of them are explained below:

Farm At Hand is a multi-user agricultural software that has features such as farm equipment management, inventory, and sales.

FarmRexx is an online agricultural software that can be used any place with Internet access. It features maintaining livestock records, monitoring livestock movements, farm equipment management, weather, and managing farm chemicals.

FarmOS [11] is another web-based farm management application that uses a web application framework called Drupal. The same can be used for planning, recording, and agricultural management activities [12].

Trimble is yet another web-based farm management application where its basic plan is free of cost to farmers. Farmers can use this application from anywhere having Internet access. It features the maintenance of yield records and weather prediction.

Tania is an open-source farm management application that is built on Go, VueJS, and SQLite. It allows us to connect with various sensors and manages them while we are on the go.

8.3.1 Issues and Challenges

8.3.1.1 Hardware

IoT devices are exposed to harsh environmental conditions like rain, high temperature, and humidity, which may destroy the same. These devices are bound to high industrial standards and thus are costlier [13, 14].

8.3.1.2 Infrastructure

Assuring smart agriculture [15] environment is challenging because smart agriculture demands continuous monitoring and feedback. Open-source tools and frameworks should be made available to agriculture application developers to build sustainable and affordable applications.

8.3.1.3 Networking

Wireless technology [16] is preferred more in smart agriculture due to the wiring cost and maintenance. Due to the physical obstacles the signal becomes weak when they reach the transceivers, which leads to the demand for more robust networking technologies.

8.3.1.4 Reliability

In smart agriculture more IoT devices are connected to send the data to the cloud. There are many external impacts where those systems can fail to send the data. A large number of gateways are needed to make sure the data has been sent to the server or cloud systems without any hindrance.

8.3.1.5 Data Security

In smart agriculture a lot of data needs to be sent every day. We need to make sure the data that has been sent over is safe. The data that has been sent has to be encrypted. New security and privacy problems arise such as authentication, integrity, and access control.

8.4 Problem Statement

One factor that is more concerned in agriculture is "high yield." Recently crop yield has faced a great deal of downtime due to incorrect predictions. Inability to understand the nature and demand of the soil leads to incorrect decisions. Incorrect decisions could have been prevented by predicting the change in the soil nature, crop growth, climate, recent plant diseases, appropriate fertilizer, etc. To aid these we need a system that can feedback continuously and act as a fail-fast system to inform the farmer if there is a deviation in the expected behavior in the farmland. The system not only should provide feedback of changes in the nature of soil or plant growth or the plant structure to the farmers but also refer to the global data available on the cloud. The objective is to develop an IoT system involving ML techniques where the system not only exchanges data between the devices but also analyzes the same semantically and reasons on it based on the reference with the cloud data. Hence it not only provides feedback to farmers about the yield prediction but also provides suggestions to the farmers on tackling any new issues.

8.5 Proposed Approach

8.5.1 Improving the Training Data

The data fed into training the model cannot depend only on the data from the cloud; instead the training data received can be multiplied into "N" no. of valid data with the help of multipliers, where each word attribute extracted is cross-multiplied and analyzed to determine if it forms a meaningful sentence that can be fed as a training sentence to the model [17, 18]. If the sentence is a valid sentence then the same will be added as a training sentence to the data set.

8.5.2 Agro-Intelli Algorithm

From Figure 8.1, the complete architecture of Agro-Intelli algorithm is depicted. Individual segment involved in the architecture are discussed as follows.

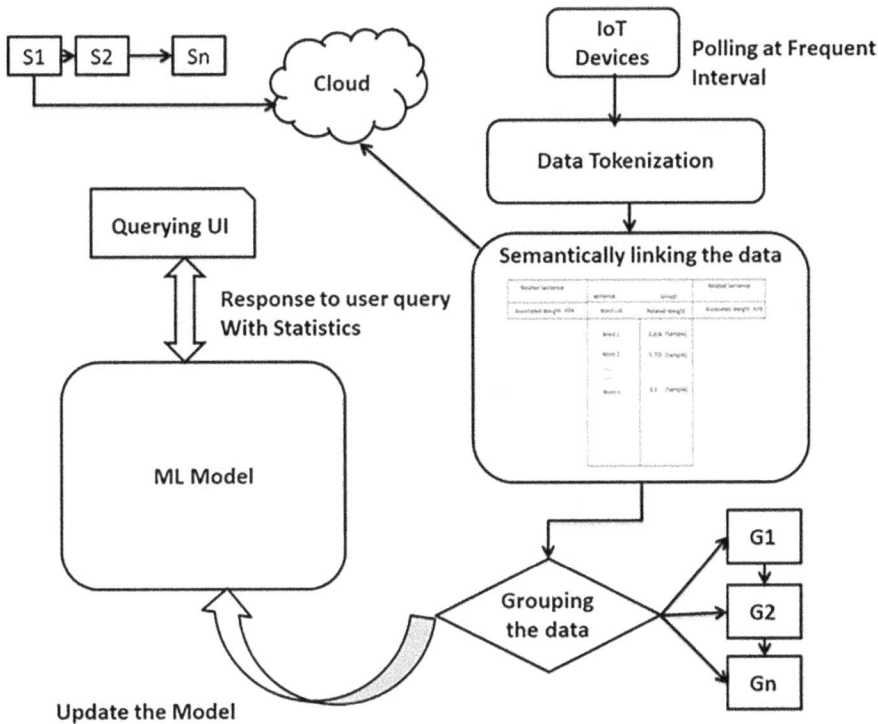

FIGURE 8.1
Architecture of Agro-Intelli Algorithm.

8.5.2.1 *Prerequisites*

- IoT devices like wireless sensors and actuators placed in the agriculture field to scan and send the field information.
- Sensors are connected to the soil to determine the change in the nature of the soil, in particular the soil behavior when there is an overdose of fertilizer or a change in the soil behavior when the chemical combination in the fertilizer overreacts with the soil nature.
- Sensors to scan the plant thickness and structure. This is to study the plant growth rate and structure to know about insect attacks and other factors.
- In the template user has to fill out the plan details and the target month.

8.5.2.2 *Dimensionality Vector*

Before we get into the algorithm, we will walk you through the attributes involved in dimensionality vector of the Agro-Intelli Algorithm.

- Humidity Rate – This attribute points to soil humidity. Soil humidity changes based on the weather and other factors.
- PW (plant width), PH (plant height) – These attributes point to the plant width and height, respectively. This attribute is taken to calculate whether the growth of the plant at a particular stage is on track or not.
- Weather – Through this attribute the weather is converted into a numeric value, which is rated against the forecast weather. The current weather conditions will aid the plant growth of not is calculated.
- Market factor – This attribute is calculated based on the influence of other attributes as well as based on the current projected situation regarding how the market factor for a particular crop is going to be when it is out for market after harvest.

From Figure 8.2, the steps involved in the Agro-Intelli Algorithm are as follows:

STEP 1 – Collect the data from the IoT devices connected in the field at periodic intervals.

STEP 2 – Tokenize the data collected such that the feature data is extracted and tagged.

Note: The feature data is passed from the sensors where the information in numeric is appended with the plain texts.

The data from the cloud is also referred to periodic intervals, where the data globally received in the form of text has been tokenized to differentiate and group conditions based on each specific crop cultivation.

STEP 3 – From the tagged data the dimensionality vector is formed, where soil humidity rate is placed as one of the attributes. The other attributes considered in the dimensionality vector are the plant width and height, current weather, plant growth rate, and the market factor.

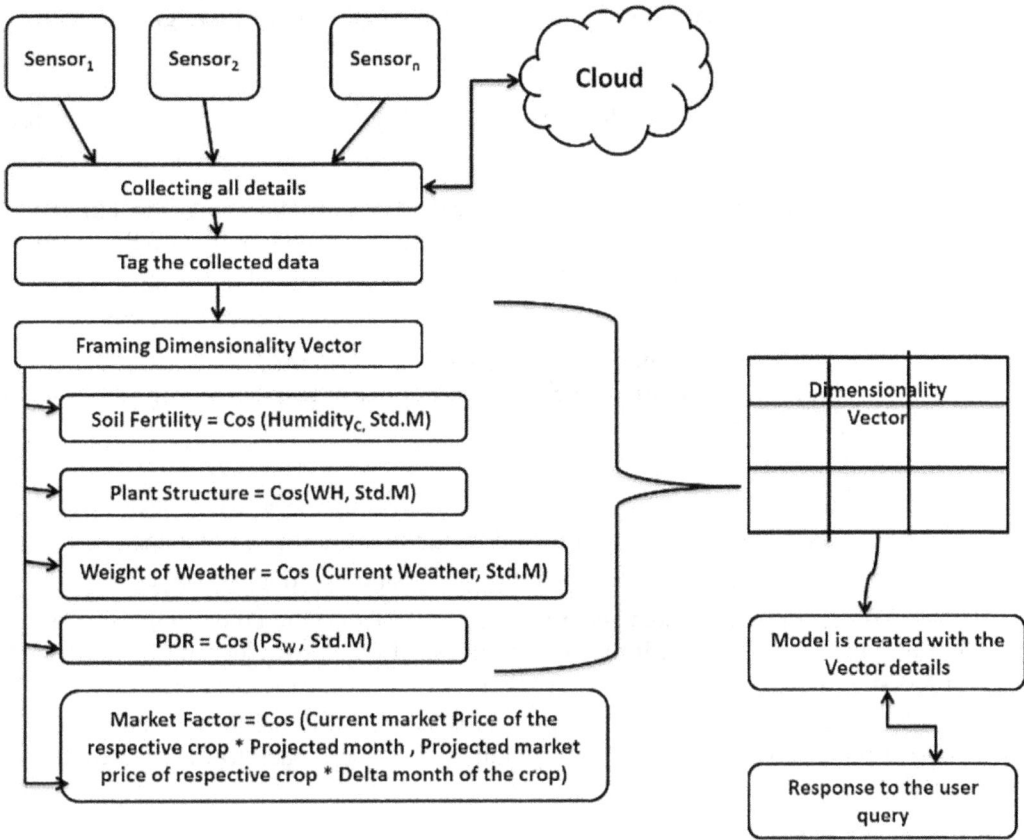

FIGURE 8.2
Agro-Intelli Algorithm.

STEP 4 – The soil fertility rate compared against the weather condition is calculated and the formula is given as

$$S_{FW} = Cos\left(Humidity_C, Std.M\right) \tag{8.1}$$

where S_{FW} (Soil Fertility Weight) = Cosine similarity between current Humidity and Standard Measure.

STEP 5 – Plant structure is calculated by using its width and height and the weight is calculated by comparing it with Standard Measure as follows.

$$PS_W \left(Plant\,Structure\,Weight\right) = Cos\left(WH, Std.M\right) \tag{8.2}$$

where WH = Width + Height of Current Crop and Std.M = Standard Measure.

STEP 6 – Weather weight is calculated by comparing the current weather with the standards

$$W_W \left(Weather\,Weight\right) = Cos\left(Current\,Weather, Std.M\right) \tag{8.3}$$

STEP 7 – PDR (Plant Damage Rate) is calculated by comparing the plant growth with the Standard Measure.

$$PDR = Cos(PS_W, Std.M) \tag{8.4}$$

STEP 8 – Market Factor Weight (MF_W) is calculated by knowing the current market price of the respective crop multiplied with Projected month with Projected market price of respective crop multiplied with Delta month of the crop.

$$MF_W = Cos\text{(Current market Price of the respective crop*}$$
$$\text{Projected month, Projected market price of respective crop*} \tag{8.5}$$
$$\text{Delta month of the crop)}$$

STEP 9 – With the dimensionality vector updated with the attribute values it is mapped to the output class determined during the grouping of the data.

STEP 10 – Now the model is trained with the dimensionality data available.

STEP 11 – As and when the data gets refreshed in the cloud then the dimensionality vector will be refreshed with the newer values, which in turn leads to the update of the ML Model.

STEP 12 – As soon as the query from the user is input, a response with suggestions, current plant statistics, and forecast details will be sent back to the user by the system.

Note: The system model will get refreshed at periodic intervals with the input from IoT devices and it will have the updated information.

8.5.2.3 Intelli-Group Algorithm

When we have any data from the cloud that data should be grouped based on the crop. It is done using the Intelli-Group Algorithm as shown in Figure 8.3. The steps of the same is as follows:

STEP 1 – From the input data, the data is tokenized based on the attributes. Here the attributes are the filtering conditions of each feature.

STEP 2 – Using the bag-of-words model the tagged data is grouped by referring to the group table. The group table is formed where conditions pointing to a specific cause are collectively represented as one group. When the tokenized data points to a similar grouping condition then it will be tagged as the corresponding group.

STEP 3 – Semantic association is established between the groups. The semantic relationship is established by involving cosine similarity. Cosine similarity is calculated to know the semantic relationship between the feature of a group with the neighbor group.

STEP 4 – The group is linked in order based on the semantic weight between them.

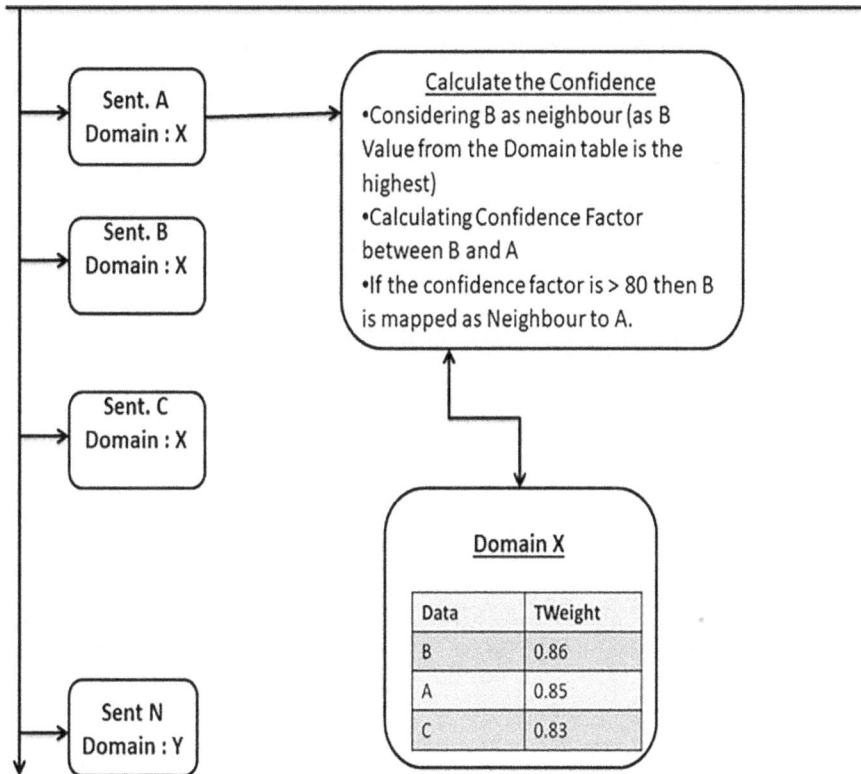

FIGURE 8.3
Intelli-Group Algorithm.

STEP 5 – The chain from a group to another group will not exist if the threshold value weight of 80% is not met.

STEP 6 – The link between the groups is refreshed at frequent intervals so that when there is more related association is formed for a group then the linking chain will be re-established.

8.6 Results and Discussion

The datasets are extracted from https://data.gov.in, and for modeling, Weka jar and Pandas (for data processing) are involved. The integrated development environments (IDEs) used to develop the model are Eclipse and Jupyter Notebook as shown in Figure 8.4. For the Agro-Intelli Algorithm using Jupyter Notebook along with ML techniques Python libraries are used to test our theory and the results are as follows.

Figure 8.5 depicts the accuracy difference in group extraction between the SmartAgri framework and the Agro-Intelli algorithm when worked upon data from the cloud.

Figure 8.6 depicts the response of statistics when the user queries about the plant growth and predictions to the model trained with standard data between AgriSys and Agro-Intelli.

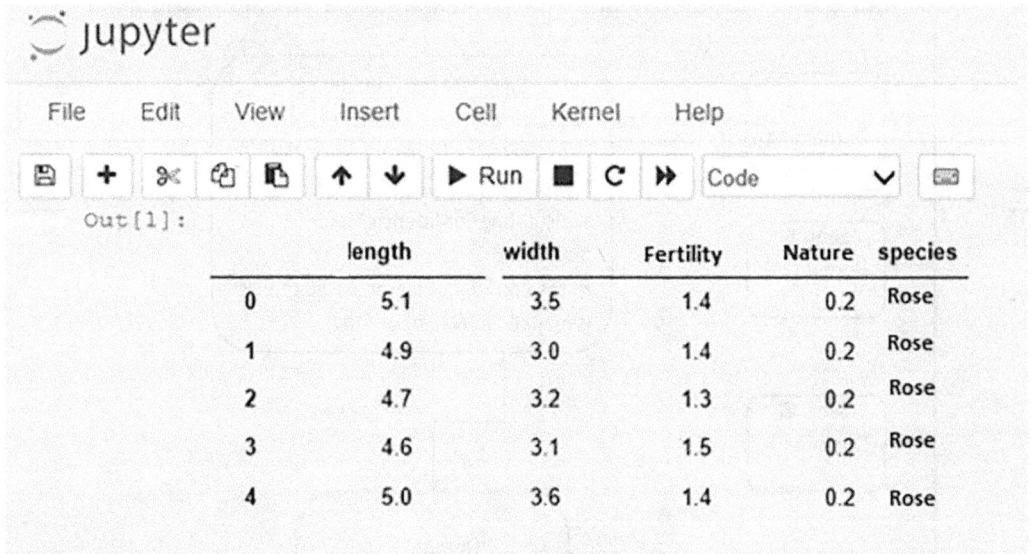

FIGURE 8.4
Jupyter Notebook involving plant structure details in the feature vector.

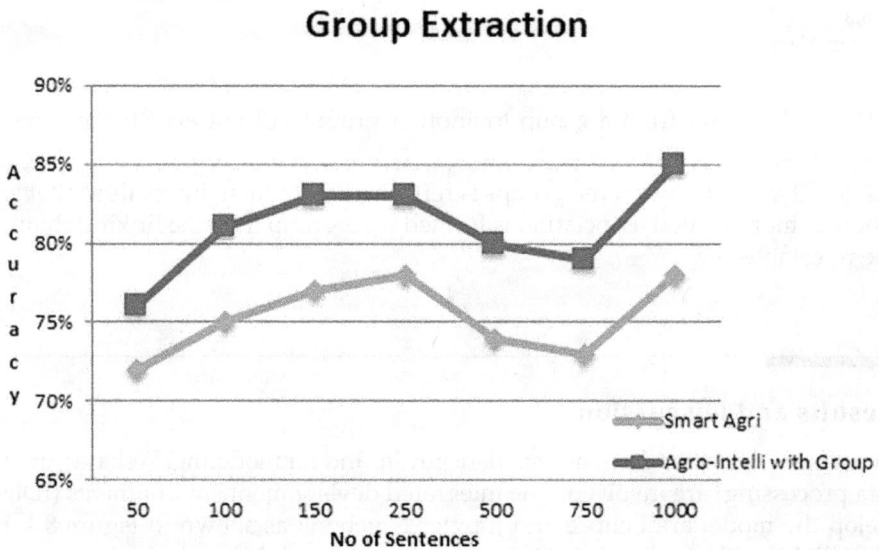

FIGURE 8.5
Accuracy comparison of group extraction between the SmartAgri framework and Agro-Intelli with grouping logic.

Statistics Accuracy

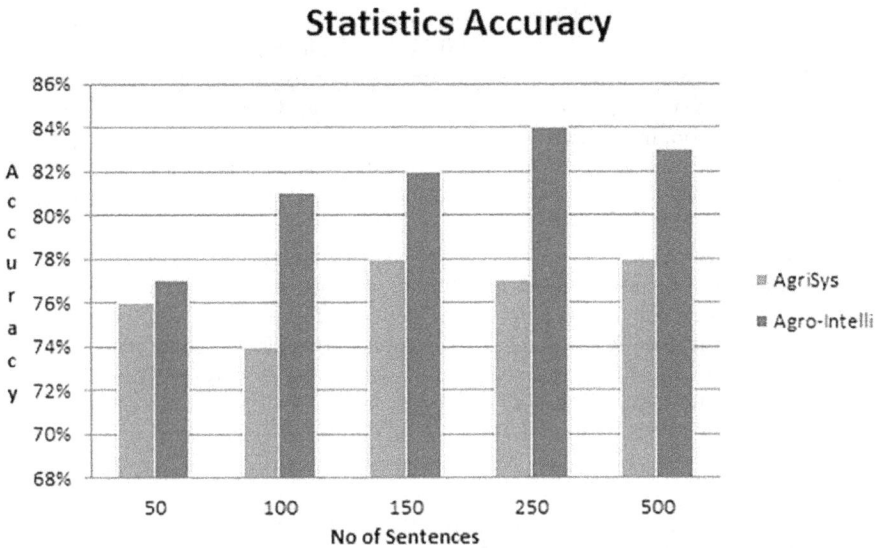

FIGURE 8.6
Comparison of statistics of accuracy between the AgriSys application and the Agro-Intelli algorithm.

8.7 Conclusion and Future Work

As we said in the beginning agriculture is the backbone of any country. To satisfy the population's need the yield has to be improved with good profit, and it is high time that we shift to modern agriculture involving the latest techniques. By making a system intelligent enough to predict the failures early comes as a real savior for the farmer. The system predicts the growth rate of the plant not only with the help of data from IoT devices but also with the knowledge it has received globally with the help of the cloud. The system can provide solutions not only to prevent the recent plant diseases developing for a specific crop but also able to control them, thus making the system a good and efficient feedback system. The work should also be extended to consider additional factors like rainfall and animal movement to make it more efficient and effective.

References

1. Sonam Tenzin, Satetha Siyang, Theerapat Pobkrut, Teerakiat Kerdcharoen. "Low cost weather station for climate-smart agriculture," 9th International Conference on Knowledge and Smart Technology, Chonburi, Thailand, 2017.
2. Sriveni Namani, Bilal Gonen. "Smart agriculture based on IoT and cloud computing," 3rd International Conference on Information and Computer Technologies (ICICT), San Jose, CA, 2020.
3. Chandoul Marwa, Soufiene Ben Othman, Hedi Sakli. "IoT based low-cost weather station and monitoring system for smart agriculture," 20th International Conference on Sciences and Techniques of Automatic Control and Computer Engineering (STA), Monastir, Tunisia, 2020.

4. George Suciu, Cristiana-Ioana Istrate, Maria-Cristina Diţu. "Secure smart agriculture monitoring technique through isolation," Global IoT Summit (GIoTS), Aarhus, Denmark, 2019.
5. K. A. Patil, N. R. Kale. "A model for smart agriculture using IoT," International Conference on Global Trends in Signal Processing, Information Computing and Communication (ICGTSPICC), Jalgaon, India, 2016.
6. Rahul Dagar, Subhranil Som, Sunil Kumar Khatri. "Smart farming – IoT in agriculture," International Conference on Inventive Research in Computing Applications (ICIRCA), Coimbatore, India, 2018.
7. R Maheswari, H Azath, P Sharmila, S Sheeba Rani Gnanamalar. "Smart village: Solar based smart agriculture with IoT enabled for climatic change and fertilization of soil," IEEE 5th International Conference on Mechatronics System and Robots (ICMSR), Singapore, 2019.
8. Saravanan K, Srinivasan, P. Examining IoT's Applications Using Cloud Services. In P. Tomar, & G. Kaur (Eds.), Examining Cloud Computing Technologies Through the Internet of Things (pp. 147–163). Hershey, PA: IGI Global, 2017. doi:10.4018/978-1-5225-3445-7.ch008.
9. Saravanan K, Saraniya S. "Cloud IOT based novel livestock monitoring and identification system using UID," Sensor Review, Vol. 38, Issue: 1, pp. 21–33, 2018. https://doi.org/10.1108/SR-08-2017-0152
10. Muhammad Shoaib Farooq; Shamyla Riaz; Adnan Abid; Kamran Abid; Muhammad Azhar Naeem, "A Survey on the Role of IoT in Agriculture for the Implementation of Smart Farming," New Technologies for Smart Farming 4.0: Research Challenges and Opportunities, 2019.
11. Mohamed Rawidean Mohd Kassim. "IoT applications in smart agriculture: Issues and challenges," IEEE Conference on Open Systems, Kota Kinabalu, Malaysia, 2020.
12. Aalaa Abdullah, Shahad Al Enazi, Issam Damaj. "AgriSys: A smart and ubiquitous controlled-environment agriculture system," 3rd MEC International Conference on Big Data and Smart City (ICBDSC), Muscat, Oman, 2016.
13. Tana Krongthong, Benchalak Muangmeesri. "Modeling and simulink of smart agriculture using IoT framework," 1st International Conference on Cybernetics and Intelligent System (ICORIS), Denpasar, Indonesia, 2019.
14. Xiaomin Li, Rihong Zhang. "Integrated multi-dimensional technology of data sensing method in smart agriculture," IEEE 9th Joint International Information Technology and Artificial Intelligence Conference (ITAIC), Chongqing, China, 2020.
15. Putro S. Budi Cahyo Suryo, I. Wayan Mustika, Oyas Wahyunggoro, Hutomo Suryo Wasisto. "Improved time series prediction using LSTM neural network for smart agriculture application," 5th International Conference on Science and Technology (ICST), Yogyakarta, Indonesia, 2019.
16. G. Sushanth, S. Sujatha. "IOT based smart agriculture system," International Conference on Wireless Communications, Signal Processing and Networking (WiSPNET), Chennai, India, 2018.
17. Saravana Kumar, C S, Santhosh, Dr. R. "Effective information retrieval and feature minimization technique for Semantic Web Data," Computers and Electrical Engineering, Vol. 81, 2020. Article No.106518.
18. Kumar, Saravana C. S., Amudhavalli, P., Santhosh, R., Kalaiarasan, C. "T structured semantic weight relationship algorithm combined with decision trees for data extraction," Journal of Computational and Theoretical Nanoscience, Vol. 16, Issue: 2, pp. 735–739, 2019.

9

Yield Prediction Based on Soil Content Analysis through Intelligent IoT System for Precision Agriculture

K. Selvakumar and P. J. A. Alphonse

National Institute of Technology
Trichy, India

L. SaiRamesh

Anna University
Chennai, India

CONTENTS

9.1 Introduction...133
9.2 Related Work 135
9.3 Proposed Precision Agriculture Algorithm...137
9.4 Proposed Soil Moisture Evaluation for Location System with Crop
    Yield Prediction...137
    9.4.1 SVM-Based Ensemble Classification with Decision Tree 141
9.5 Experimental Setup and Result Analysis...141
    9.5.1 Performance of Metrics 143
        9.5.1.1 Accuracy..143
        9.5.1.2 Precision 144
        9.5.1.3 Recall (Sensitivity)...144
        9.5.1.4 F1 Score..144
9.6 Conclusion 144
References..145

9.1 Introduction

Agriculture is the critical juncture in the history of humans. About 11,000 years ago, people changed from hunting life to food producers by agriculture. It changed their lifestyle and had more time to learn and gain knowledge by studying the life that surrounds them. They settled down in life by farming. For decades, people started to cultivate a new type of crop and used animals by experimenting in their own land, which is enough to feed their families. About 1960, high-yield production of wheat and rice were developed by researchers emerged as the green revolution. Food production plays an important role in population growth and change in politics. Shortage of food arises due to uneven distribution of resources and overpopulation. The challenges in feeding the hungry depend on the type of land and water, which is overcome by agricultural science.

Agriculture is the primary source of food for humans to survive and solve hunger all over the world. Before the industrial revolution, agriculture is the main source for governments in earning profits. Agriculture plays an important role in business including biotechnology, insurance companies, farm equipment, and chemical industries. The agricultural biodiversity aspects are historical resources, ecological community, biotic factors, and social-economical culture.

In agriculture, yield is important, which is a measure of the amount of crop grown in a particular area. Due to technology development in farming methods and tools have improved crop yield. The yield of a crop is usually measured by kilograms per hectare. Today, productivity in agriculture is measured in money produced per land and yields are measured by weights of crop per land. Seed ratio is another paradigm in crop yield. To ensure food security there is a need to forecast the crop yield [1].

Healthy soil is defined by its fertility which depends on physical, chemical, and biological properties. Nowadays, people are not aware of the cultivation of crops at the right time and at the right place. Because of these cultivating techniques the climatic condition also changed against fundamental assets like soil, water, and air leads to food insecurity. By analyzing all these issues and problems like weather, temperature, and several factors, there is no proper solution and technology to overcome [2]. In India, there are several ways to increase the economic growth in the field of agriculture and improve crop yield and quality of crops.

Data mining also helps in predicting the production of crop yield. Generally, data mining is extracting hidden information from database and transform it into an understandable structure for future use. Data mining is the software that allows the user to classify analyze data from more dimensions and provides a summary of the relationship among the class labels [3]. The ultimate goal of data mining is a prediction that has many business applications.

Information from data is obtained from the patterns, associations, or relationships in data mining. It can be converted into knowledge about historical patterns and future trends. For example, review and survey about the production of crops help the farmers to identify the risk factors for poor production and preventive measures to overcome. Crop yield production is a serious agricultural problem. Farmers expect crop yield based on their own experience on a particular crop as they cultivate more crops based on climatic changes. Accurate information about crop yield history helps for making decisions related to agriculture which reduces the risk and cost management [4].

In that point, crop yield prediction based on the soil content analysis is one of the research carried over by many researchers to improve the growth of the crop. In earlier days, crops were chosen by the farmers based on the soil and the monsoon for the particular period [5]. That technique worked well most of the time previously. But, once global warming became a factor and modifies the monsoon periods, which affects the crop yield by changing the mineral quantities and qualities of soil content. In that manner, the earlier prediction model is not suited and the farmers need assistance like an analysis model for predicting the crop yield based on the soil content.

Data mining techniques are used to classify soil and analyze soil for crop production [6, 7]. The fertility of the soil is predicted by the decision tree algorithm. Clustering algorithms help in understanding the current usage of land for agriculture and future prediction of land for crop production. K-means algorithm is used to classify soil and plants before marketing. K-nearest neighbor algorithm is used to predict the weather conditions. Neural networks help in forecasting water resources and identify the ripeness of fruit. Support vector machine (SVM) is used for crop classification and climatic changes. This paper is to predict the category of analyzed soil dataset to indicate the crop yield in particular soil type and suggest suitable crop to be cultivated.

9.2 Related Work

A hybrid decision model is needed to achieve high accuracy and generality in terms of yield prediction capabilities. Analyzing the agricultural data is a challenging task to summarize the report about soil fertility, weather condition, types of crops cultivated, and high yield production. To increase the crop yield with many dimensions the data mining techniques, and multiple linear regression are used [8]. It analyzes the existing soil types and climatic parameters and optimizes the solution to increase production in agriculture.

Crop planning plays an important role in increasing crop production. There are many methods that predict and suggest suitable crops to be cultivated during a particular time period based on soil type and climate. Artificial neural networks (ANNs) and multiple linear regression [5] are used in the prediction of wheat yield based on parameters including rainfall, biomass, evaporation of soil, amount of water soil extracts, and fertilizer usage [9]. Sensors are used in prediction and control irrigation using instruments like electronic sensors, satellite imaging, hyperspectral remote sensing [10]. Many researchers proposed algorithms that increase the production of crop yield by comparing various algorithmic approaches. They analyzed the dataset with algorithms and incremental methods and provided suggestions for further use [6]. The corn and soybean production are based on weekly rainfall [7], which is predicted by ANN and optimized a solution to improve the production. Crop yield can be increased and predicted based on parameters like rainfall, atmosphere, and pH level [11].

Artificial intelligence and satellite imagery [12] are used for crop monitoring with low resolution achieves accuracy compared to CNN algorithm and satellite imagery. ANN analyzes various parameters and achieves the same accuracy as CNN. Fertilizer usage randomly leads to a decrease in the yield, which damages the top layer of soil and causes the soil to be more acidic. To overcome this, a decision tree and SVM algorithm–based system predicts the rainfall and recommends the crop to cultivate [3]. Multilinear regression and decision tree regression form the tree structure to predict yield for the next upcoming years [13]. Prediction is based on k-means clustering and poly regression. Based on the analyzes of crop yield it predicts the soil condition, availability of rainfall for the next 3 years. Lasso and ridge regression [4] were applied to estimate the crop yield. Combining decision tree, linear regression classifies the soil and cross-validates with other methods to outperform.

Prediction is based on analyzing the historical data or mining information from raw data. Harvest is predicted by SVM, RNN, SVR, KNN-R [14] based on parameters like soil, rainfall, humidity, temperature, and water extracts by soil to help management to earn profit. Statistical textural features are considered to predict whether the plant is a crop or weed [15] by using SVM and k-fold cross-validation. SVM and ellipsoid estimation techniques are used to detect the fruit, track and 3D reconstruction [16] to map the crop yield production and estimate the profits.

To detect drought condition, an unsupervised learning method SVM is used [17]. It provides the factor that causes drought conditions and preventive measures to overcome the condition, suitable crop to cultivate in such condition. It also provides a way to increase the yield production of crops based on the threshold. The volume of product to be cultivated per unit area in land is a challenging task [18]. Using the generalized linear model (GLM) and SVM it is able to estimate the amount of crop harvested per unit of land and volume of crop cultivated in a particular area. Stem breakage and root lodging are important in classifying crops to improve production. Using binary patterns, Gabor filters as features of SVM are able to classify the crop lodging with co-occurrence matrix.

The condition of soil is important for agriculture. Soil consists of heavy metals such as nitrogen, potassium, organic and heavy metals that exploits the soil. To detect the heavy metals present in the soil SVM, Random Forest (RF), Extreme Learning Machine (ELM) is used. It detects the metals and suggests the crop that suits that particular soil. Soil properties are predicted by SVM and multilayer perceptron neural networks. Sometimes water logging [19] in land exploits the crop and decreases the crop yield. To overcome this, SVM and ANN help to predict and track the water level by sensors. The moisture of soil during the winter season to cultivate wheat is achieved by machine learning methods of SVM and RF. It predicts the moisture content and suggests whether to cultivate a crop or not. Overall activity in the agricultural field can be recognized by using a smart shirt and classify the various activity using Naive Bayes and KNN.

Location-aware similarity uses spatial social union (SSU) between two users for measuring similarity. It interconnects the users, items, and locations to predict the user preferred location. Rating prediction [20] and item recommendation algorithms are used to experiment on a dataset in real time. The geographic probabilistic factor model (Geo-PFM) framework is an effective user mobility model to recommend location which captures geographical areas [21]. It combines the effects of user mobility and latent factors and influences geographical recommendation systems.

An efficient location recommendation system [22] has challenges such as predicting the user's new location and existing location. To overcome these problems, the probabilistic approach is developed for each user to predict the location based on geographical influence. It computes accurately and suggests the location. Encounter probability [23] is used to measure the similarity in behaviors of two users to identify the similar experience of choosing the location. Based on the rating, the location is recommended and similarity is a measure between random users [24]. It is calculated by applying factorization between users who are dissimilar in locating items. Distance and transitive node similarity [25] proposed algorithms to predict the similarity between the users and the relationship between the different items rated by strangers.

From the above works of literature, the importance of crop yield prediction is identified and the need for machine learning algorithms for prediction was detected. Most of the prediction algorithms use datasets that cannot provide the appropriate decision for the current scenario of crop cultivation. Moreover, the existing systems evaluated the soil nutrients content statically for the specific and the suitable crop for the available soil nutrients are not predicted. Some of the systems provide crops suitable for the soil based on the geographical features and the monsoon period. This is also considered as the static prediction, which suggests the crop cultivation based on the previous years' data collection. The proposed crop yield prediction overcomes all these concerns by providing the continuous soil nutrients level with the current soil moisture level, which is mainly helpful to maintain the required soil content for the efficient predicted crop yield cultivation. The intelligent rules applied here with ensemble classifiers assist in choosing the suitable crop for the soil based on the initial soil nutrients level.

9.3 Proposed Precision Agriculture Algorithm

The proposed IoT system is composed of pH sensors, humidity and temperature sensors, Soil moisture sensors, soil nutrient sensors (NPK) probes, microcontroller/microprocessor equipped with Wi-Fi and Cloud storage. When the sensors are implemented, they measure the corresponding characteristics and transmit time-stamped live data to the cloud server. These sensors work together and provide wholesome data to the analyst. For the Recommending system, we proposed an ensemble classifier model to get the crop suitable for the given soil data and helps to enhance the growth using an optimized farming process.

- The proposed system uses the parameters like soil type, groundwater level, local population, daily, seasonally needs of the local people, labors available of the farmer, the range of the same plantation, and the range of farming land in the locality.
- Classification of soil into low, medium, and high categories is done by adopting data mining techniques to predict the crop yield using the available dataset.

Figure 9.1 shows the overview of the prediction system for crop cultivation at a specific location. The data for prediction was collected from the agricultural field, which includes soil moisture level, nutrient level, and temperature level at a different location. The location plays a major role in crop cultivation because the nutrients level differs based on the location, and also moisture-level changes with respect to the temperature rating at the location. The collected information is transferred to the cloud storage, where the data analysis takes place after the necessary pre-processing stages. The crop yield needs continuous monitoring and, for that, the analysis is viewed through smart devices to improve the moisture level, which implicitly helps to maintain the consistency of soil nutrients.

9.4 Proposed Soil Moisture Evaluation for Location System with Crop Yield Prediction

The soil moisture at a particular depth alone is not a deciding factor for watering the plants, since the water holding capacity of each layer in the field is highly dependent on the adjacent layers. If the bottom soil layer is too dry, then the water supplied at the upper layer will be drained soon. Hence in the real-time environment, the soil moisture content to be maintained in a particular depth is highly dependent on the soil moisture content of the upper and bottom layers. The proposed scientific processing unit accesses the soil

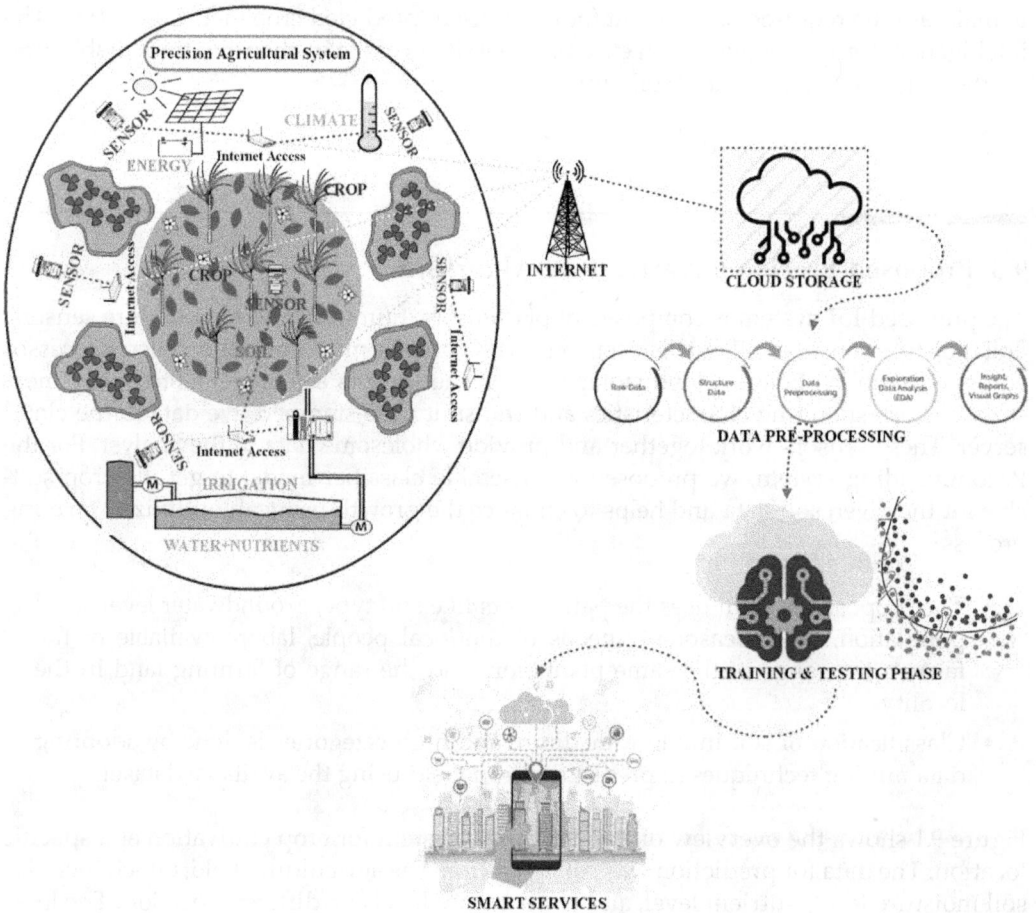

FIGURE 9.1
Prediction of crop cultivation for a specific location.

moisture data from the classified database and analyzes it for calculating the actual quantity of water to be supplied in the plant's root in different periods for cultivation (DC). Table 9.1 shows the procedure to calculate the Soil Moisture Evaluation for the location (SMEL) in different crop periods and the list of symbols used in this algorithm.

TABLE 9.1

List of Symbols Used in SMEL Algorithm

Symbol	Definition
DC	Duration for cultivation
SML_{p1}, SML_{p2}, SML_{p3}, SML_{p4}	Soil moisture values in depth d_1, d_2, d_2, respectively
RSM	Required soil moisture level
SMEL	Soil Moisture Evaluation for Location
l_s	Location at which the soil moisture measures
Temp	Temperature at a specific location

ALGORITHM FOR SMEL CALCULATION

Input: DC, SML_{p1}, SML_{p2}, SML_{p3}, SML_{p4}
Output: SMEL at depth (l_s)

Step 1:	Begin
Step 2:	if (DC <= 30 days & Temp < 25°C) then
Step 3:	if (SML_{p1} <= SML_{p2} && SML_{p2} <= SML_{p3}) then
Step 4:	SMEL = RSM
Step 5:	else if (SML_{p1} > SML_{p2} && SML_{p2} > SML_{p3}) then
Step 6:	SMEL = RSM + SML_{p1}
Step 7:	end if
Step 8:	end if
Step 9:	if (DC> 30 days and DC <= 45 days & Temp > 30°C) then
Step 10:	if (SML_{p2} >= SML_{p1}) then
Step 11:	SMEL1 = RSM + SML_{p2};
Step 12:	else if (SML_{p2} < SML_{p1}) then
Step 13:	SMEL1 = RSM;
Step 14:	end if
Step 15:	if (SML_{p2} > SML_{p3}) then
Step 16:	SMEL = SMEL1 + SML_{p3};
Step 17:	else (SML_{p2} < SML_{p3}) then
Step 18:	SMEL = SMEL1;
Step 19:	end if
Step 20:	end if
Step 21:	if (DC > 45 days and DC <= 60 days & Temp < 25°C) then
Step 22:	if (SML_{p3} > SML_{p2} && SML_{p2} > SML_{p1}) then
Step 23:	SMEL = RSM + SML_{p2} + SML_{p1};
Step 24:	else (SML_{p3} < SML_{p2} && SML_{p2} < SML_{p1}) then
Step 25:	SMEL = RSM;
Step 26:	end if
Step 27:	end if
Step 28:	if (DC > 45 days and DC <= 60 days & Temp > 30°C) then
Step 29:	if (SML_{p4} > SML_{p3} && SML_{p3} > SML_{p2}) then
Step 30:	SMEL = RSM + SML_{p3} + SML_{p2};
Step 31:	else (SML_{p4} < SML_{p3} && SML_{p3} < SML_{p2}) then
Step 32:	SMEL = RSM;
Step 33:	end if
Step 34:	end if
Step 35:	End

The scientific processing unit calculates the SMEL at a particular location in a particular crop period based on the available soil moisture content at a different location. The conventional static measurement method calculates the quantity of water required based on the soil moisture content at a fixed location, which leads to inaccuracy. The dynamic nature of the proposed algorithm accurately estimates the quantity of water and duration of water supply for the crops. This information is stored as information in the agricultural database and it will be used in intelligent rules while making a decision on crop cultivation. The farmers can log in to the farmer's web portal with a registered username and password to receive the recommendations given by the agriculture scientists. These recommendations help farmers to avoid both overwatering and shortage irrigation of plants and also increase the crop yield. The system provides recommendations to the farmers via short message service (SMS) if any updating information is available.

As a whole, the proposed system helps both the analyst and farmers to frequently examine the dynamic nature of soil moistures in the cultivation fields without frequently visiting the lands with manual interventions. Based on the dynamic nature of the lands, the analyst can guide the farmers about the duration and quantity of water supply to the cultivation land. The proposed approach significantly increases the accuracy of soil moisture measurement with the help of multi-depth sensors and hence conserves the water in dry areas. This water conservation extends the water availability, where water is a scarce resource.

Finally, a set of rules are used in the decision-making process which are listed below based on Actual soil moisture required (ASMR) and Estimated Soil Moisture (ESM),

Rule 1:

if (DC <=30 days) then
$\qquad$ if (SM_{d1} <= SM_{d2} && SM_{d2} <= SM_{d3}) then
$\qquad\qquad$ ASMR = ESM
$\qquad$ else if (SM_{d1} > SM_{d2} && SM_{d2} > SM_{d3}) then
$\qquad\qquad$ ASMR = ESM + SM_{d1}
$\qquad$ end if

Rule 2:

if (DC >30 days and DC <=45 days) then
$\qquad$ if (SM_{d2} >= SM_{d1}) then
$\qquad\qquad$ ASMR1= ESM + SM_{d2};
$\qquad\qquad$ else if (SM_{d2} < SM_{d1}) then
$\qquad\qquad\qquad$ ASMR1 = ESM;
$\qquad\qquad$ end if
if (SM_{d2} > SM_{d3}) then
$\qquad\qquad$ ASMR= ASMR1 + SM_{d3};
$\qquad\qquad$ else (SM_{d2} < SM_{d3}) then
$\qquad\qquad\qquad$ ASMR = ASMR1;
$\qquad$ end if

Rule 3:

if (DC>45 days and DC <=60 days) then
$\qquad$ if (SM_{d3} > SM_{d2} && SM_{d2} > SM_{d1}) then
$\qquad\qquad$ ASMR= ESMR + SM_{d2} + SM_{d2};
$\qquad$ else (SM_{d3} < SM_{d2} && SM_{d2} < SM_{d1}) then
$\qquad\qquad$ ASMR= ESM;
$\qquad$ end if
$\qquad$ end if

9.4.1 SVM-Based Ensemble Classification with Decision Tree

SVM-based ensemble classification with decision tree explains the complete scenario of SVM and the ensemble with decision tree for predicting the crop based on the SMEL algorithm. The SVM classifies the soil type based on the nutrient level and that classified data is considered as the input for the decision tree with SMEL output to take the decision on crop cultivation for the soil. Then the crop that can be grown in that soil is extracted from the database. Then the level of water chosen by the farmer is compared with that of the crops.

Those crops that suit the soil are found. Now the crops are separated as per the user's needs. Now depending upon the need of the users the acre measure of the farmer is found and the land is divided into parts. Based on the value of the crop recommended.

INPUT: Soil nutrient level, ASMR, crop types
OUTPUT: Suitable crop for the soil

1: Begin

2: Get the soil attributes from dataset

3: Get the input vectors N_i, Support vectors N_s

4: Get the soil types based on features (N_i) with the available N_s

5: Input the soil types to decision with ASMR

6: Identify the information gain IG for the features of ASMR and features of SMEL

7: Segregate the attributes according to IG for each soil

8: Calculate the IG for child nodes

9: Labeled the child node with more IG as next level of parent

10: If IG > bestIG

11: Then considered as the best soil for crop Else

12: Check for next child IG

13: End

In ensemble classification, the decision tree starts by selecting the type of soil. Then the crop that can be grown in that soil is extracted from the database. Then the level of water and the calculated temperature chosen by the farmer are compared with those of the crops. Those crops that suit the soil are found. Now the crops are separated as per the user's needs. Now depending upon the need of the users the acre measure of the farmer is found and the land is divided into parts. The proportional land is given to the farmers for their cultivation. And find the precision and calculate the measures with a decision tree to recommend the crop with accuracy.

9.5 Experimental Setup and Result Analysis

A methodological approach has been followed for measurement of various soil parameters. The cloud system is fed with precise data from the sensors and this can be rendered on a digital screen or as serial input. The system also allows remote monitoring of soil conditions and provides crop surveillance. The experiment was carried out in the field

FIGURE 9.2
Field setup

area 600 × 100 sq. ft. The available field area is categorized into four parts with the area 300 × 500 sq. ft and experiments are conducted with different soil moisture levels and varying temperatures based on exposure to sunlight. The nutrients are added in regular intervals based on the crop planted (lady's finger, tomato) in the field. The area with proper nutrients and soil moisture level maintenance provides high yield than the other field. The yield is not discussed in the result section because the proposed setup concentrates on the prediction of the crop yield based on this experiment conducted on the field. The experiments carried out in the field are considered as the training set and classification algorithms predict the crop yield in the test field as given in Figure 9.2.

The experiment involved the pH sensor observations using a pH meter, which classifies the pH level soil. Different crops require different pH conditions and a well-informed choice can be made. A pH of 7 is neutral, lesser than 7 is acidic and more than 7 is basic/alkaline. A variety of fertilizer mixes can then be used to alter the pH slightly to the demands of the farmer. A moisture sensor is used to measure the soil measure level in addition to temperature sensors to evaluate the humidity level of the region/location.

Another important is NPK sensors. The key/major nutrients in the soil are nitrogen, potassium, and phosphorus (NPK); these nutrient content fluctuations dictate what crop cycles are followed. These readings are especially useful at the start of the season when crop choices are decided and also the farmer can use various fertilizers as per the requirement of the soil. The NPK contents in the soil are measure using a digital soil analyzer. In most of the existing systems for crop yield prediction, the soil nutrients level is not considered, which is very much essential in crop yield. Figure 9.3 shows the initial page for the crop recommendation system, which gets the details of the soil with the location to analyze the soil nutrients level and recommend the crop as per the nutrient level of the specified region as shown in Figure 9.4. Once the user submits the user details, the land details page will ask for information like User name, country, state, District, area hectare, soil type, season, transport cost, Duration days, Market price. Once the user fills and submits the form, the data will be analyzed by applying the SMEL algorithm.

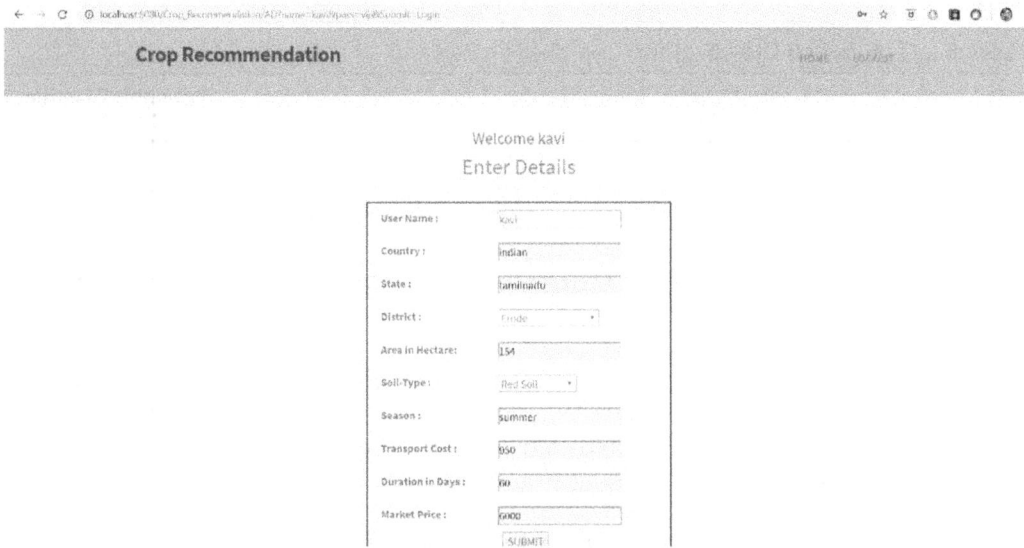

FIGURE 9.3
Crop recommendation system.

9.5.1 Performance of Metrics

9.5.1.1 Accuracy

Accuracy is the most intuitive performance measure and it is simply a ratio of correctly predicted observation to the total observations.

$$Accuracy = TP + TN \,/\, TP + FP + FN + TN$$

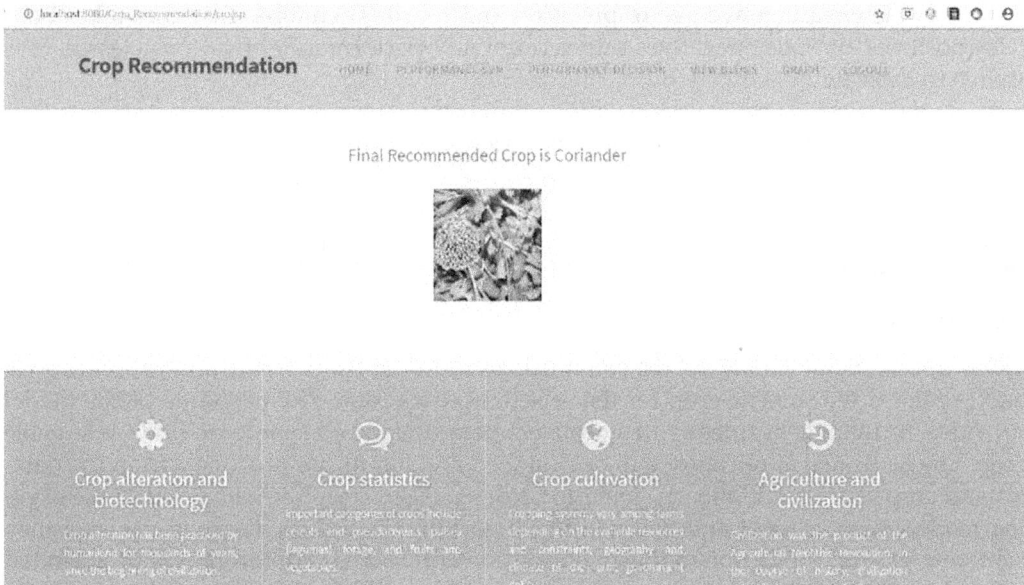

FIGURE 9.4
Crop recommendation based on SMEL-based ensemble classification algorithm.

TABLE 9.2

Performance of Various Classifiers for Crop Yield Prediction

Measures	SVM	KNN	DT	Ensemble Classifier (SVM + DT)
Accuracy	84	88	91	98
Precision	95.5	96.7	97.8	98.9
Recall	97.7	98.8	97.8	99.9
F1 Score	96.6	97.7	97.8	99.4

9.5.1.2 Precision

Precision is the ratio of correctly predicted positive observations to the total predicted positive observations.

$$\text{Precision} = TP / (TP + FP)$$

9.5.1.3 Recall (Sensitivity)

Recall is the ratio of correctly predicted positive observations to all observations in actual class.

$$\text{Recall} = TP / (TP + FN)$$

9.5.1.4 F1 Score

F1 core is the weighted average of precision and recall. Therefore, this score takes both false positives and false negatives into account and performance of various classifiers is shown in Table 9.2.

$$\text{F1 Score} = 2 * (\text{Recall} * \text{Precision}) / (\text{Recall} + \text{Precision})$$

9.6 Conclusion

The proposed model analyzed the soil nutrients with the temperature and moisture level and predicted the suitable crop for the selected soil location. The proposed SMEL model provides the intelligent rules by including temperature and soil nature to find the suitable crop. The dedicated application will be designed and inputs are processed from the dataset and crop recommendation is going to be done. Also, it will be compared to the existing models such as SVM and decision tree algorithms based on the accuracy level they achieved for crop prediction to the given soil content. Further, this work will be extended by applying fuzzy rough set or swarm intelligence-based techniques to predict the soil for future crop cultivation.

References

1. Pratheba, R., Sivasangari, A., & Saraswady, D. (2014, March). Performance Analysis of Pest Detection for Agricultural Field Using Clustering Techniques. In *2014 International Conference on Circuits, Power and Computing Technologies [ICCPCT-2014]* (pp. 1426–1431). IEEE.
2. Reshma, R., Sathiyavathi, V., Sindhu, T., Selvakumar, K., & SaiRamesh, L. (2020, October). IoT Based Classification Techniques for Soil Content Analysis and Crop Yield Prediction. In *2020 Fourth International Conference on I-SMAC (IoT in Social, Mobile, Analytics and Cloud) (I-SMAC)* (pp. 156–160). IEEE.
3. van Klompenburg, T., Kassahun, A., & Catal, C. (2020). Crop yield prediction using machine learning: A systematic literature review. *Computers and Electronics in Agriculture, 177,* 105709.
4. Kavita, M., & Mathur, P. (2020, October). Crop Yield Estimation in India Using Machine Learning. In *2020 IEEE 5th International Conference on Computing Communication and Automation (ICCCA)* (pp. 220–224). IEEE.
5. Shastry, K. A., Sanjay, H. A., & Deshmukh, A. (2016). A parameter based customized artificial neural network model for crop yield prediction. *Journal of Artificial Intelligence, 9,* 23–32.
6. van Klompenburg, T., Kassahun, A., & Catal, C. (2020). Crop yield prediction using machine learning: A systematic literature review. *Computers and Electronics in Agriculture, 177,* 105709.
7. Kaul, M., Hill, R. L., & Walthall, C. (2005). Artificial neural networks for corn and soybean yield prediction. *Agricultural Systems, 85*(1), 1–18.
8. Majumdar, J., Naraseeyappa, S., & Ankalaki, S. (2017). Analysis of agriculture data using data mining techniques: Application of big data. *Journal of Big data, 4*(1), 1–15.
9. Khairunniza-Bejo, S., Mustaffha, S., & Ismail, W. I. W. (2014). Application of artificial neural network in predicting crop yield: A review. *Journal of Food Science and Engineering, 4*(1), 1.
10. Lee, W. S., Alchanatis, V., Yang, C., Hirafuji, M., Moshou, D., & Li, C. (2010). Sensing technologies for precision specialty crop production. *Computers and Electronics in Agriculture, 74*(1), 2–33.
11. Dahikar, S. S., Rode, S. V., & Deshmukh, P. (2015). An artificial neural network approach for agricultural crop yield prediction based on various parameters. *International Journal of Advanced Research in Electronics and Communication Engineering, 4*(1), 94–98.
12. Priyanka, T., Soni, P., & Malathy, C. (2018). Agricultural crop yield prediction using artificial intelligence and satellite imagery. *Eurasian Journal of Analytical Chemistry, 13*(7), 6–12.
13. Medar, R., Rajpurohit, V. S., & Shweta, S. (2019, March). Crop Yield Prediction Using Machine Learning Techniques. In *2019 IEEE 5th International Conference for Convergence in Technology (I2CT)* (pp. 1–5). IEEE.
14. Chandraprabha, M., & Dhanaraj, R. K. (2020, November). Machine Learning Based Pedantic Analysis of Predictive Algorithms in Crop Yield Management. In *2020 4th International Conference on Electronics, Communication and Aerospace Technology (ICECA)* (pp. 1340–1345). IEEE.
15. Jiang, J., Xing, F., Zeng, X., & Zou, Q. (2019). Investigating maize yield-related genes in multiple omics interaction network data. *IEEE Transactions on Nanobioscience, 19*(1), 142–151.
16. Moonrinta, J., Chaivivatrakul, S., Dailey, M. N., & Ekpanyapong, M. (2010, December). Fruit Detection, Tracking, and 3D Reconstruction for Crop Mapping and Yield Estimation. In *2010 11th International Conference on Control Automation Robotics & Vision* (pp. 1181–1186). IEEE.
17. Gaikwad, N., Palivela, H., Chavan, G., & Prathap, P. (2015, March). Applications of Unsupervised Auto Segmentation on Dhule Area Hyperspectral Image for Drought and Yield Prediction. In *2015 International Conference on Innovations in Information, Embedded and Communication Systems (ICIIECS)* (pp. 1–5). IEEE.
18. Gamboa, A. A., Cáceres, P. A., Lamos, H., Zárate, D. A., & Puentes, D. E. (2019, April). Predictive Model for Cocoa Yield in Santander Using Supervised Machine Learning. In *2019 XXII Symposium on Image, Signal Processing and Artificial Vision (STSIVA)* (pp. 1–5). IEEE.
19. Yang, Y. W., Cao, H. X., Zhang, W., Xu, L., Wan, Q., Ke, Y., … & Huang, B. (2018). Hyperspectral identification and classification of oilseed rape waterlogging stress levels using parallel computing. *IEEE Access, 6,* 57663–57675.

20. Hao, F., Li, S., Min, G., Kim, H. C., Yau, S. S., & Yang, L. T. (2015). An efficient approach to generating location-sensitive recommendations in ad-hoc social network environments. *IEEE Transactions on Services Computing, 8*(3), 520–533.
21. Liu, B., Xiong, H., Papadimitriou, S., Fu, Y., & Yao, Z. (2014). A general geographical probabilistic factor model for point of interest recommendation. *IEEE Transactions on Knowledge and Data Engineering, 27*(5), 1167–1179.
22. Zhang, J. D., Chow, C. Y., & Li, Y. (2014). iGeoRec: A personalized and efficient geographical location recommendation framework. *IEEE Transactions on Services Computing, 8*(5), 701–714.
23. Qiao, X., Yu, W., Zhang, J., Tan, W., Su, J., Xu, W., & Chen, J. (2014). Recommending nearby strangers instantly based on similar check-in behaviors. *IEEE Transactions on Automation Science and Engineering, 12*(3), 1114–1124.
24. Pirasteh, P., Hwang, D., & Jung, J. J. (2015). Exploiting matrix factorization to asymmetric user similarities in recommendation systems. *Knowledge-Based Systems, 83,* 51–57.
25. Cha, S. H. (2007). Comprehensive survey on distance/similarity measures between probability density functions. *City, 1*(2), 1.

10

Fuzzy-Based Intelligent Crop Prediction over Climate Fluctuation Using IoT

S. Subramani

SNS College of Technology
Coimbatore, India

C. Chandru Vignesh

Veltech Rangarajan Dr Sagunthala R&D Institute of Science & Technology
Chennai, India

J. Alfred Daniel

SNS College of Technology
Coimbatore, India

C. B. Sivaparthipan and Balaanand Muthu

Adhiyamaan College of Engineering
Hosur, India

N. Suganthi

Kumaraguru College of Technology
Coimbatore, India

CONTENTS

10.1 Introduction 147
10.2 Background 148
10.3 Crop Prediction Framework 149
10.4 Prediction Using Fuzzy Rules 149
    10.4.1 Fuzzy-Based Crop Prediction 151
    10.4.2 Crop Prediction Using Fuzzy Equation 152
    10.4.3 Semantic Extraction in Big Data 154
10.5 Results and Discussion 155
10.6 Conclusion 160
References 161

10.1 Introduction

Nowadays, the Internet of Things (IoT) has predominant position in the future generation. Using the IoT, the arena round is getting automatic with the aid of using changing guide procedures when you consider that it's miles strength green and entails minimum guy power. The Indian authorities have taken initiatives to grow a stable and clever gadget primarily based totally on country's want to use IoT. Smart city is the most important issue targeted through authorities which includes clever parking, women safety, waste management, water management, and agriculture. IoT has a massive effect on smart farming for the reason that agricultural lands are destroyed because of the loss of employees in the field. In sensible farming, in real time, the knowledge required by the farmers is disseminated in numerous destinations. This paper proposes machine-controlled prediction of the crop with the rewards of getting ICT in a global setup. It forecast the route for rural agriculturalists towards exchanging various of the standard methods. This analysis paper overcomes the restrictions of ancient farming methods by exploiting aquatic supply expeditiously and additionally by reducing the labor salary. The agricultural system is additionally machine-controlled frequently to observe the watering (Mohanraj et al. 2016). IoT sensors have the capability to produce data regarding crop yields, downfall to the farmers. In the future, IoT will change the way food grows.

10.2 Background

IoT node combines three fundamental additives including intelligence, sensing, and wireless communications. Some of the real-time applications of IoT are precision agriculture, smart home, waste management, healthcare, and transportation.

A wireless sensor network (WSN) consists of various sensor nodes deployed in huge numbers at different locations to monitor environmental factors. The data sensed by the sensor are analyzed and processed to compare with the knowledgebase. The crop forecast and animal prediction drive up the competence of crop production (Sriharsha et al. 2012; Zhu et al. 2011).

The fact is to know how data are gathered from various heterogenous sources over numerous years. The heterogeneous resources consist of dependent and unstructured records from agricultural researchers, email, net crawling, and farmers' profiles, which lead to big information (Guo et al. 2015). Big data analyzes and examines both the structured and unstructured large data sets of the agriculture field. Semantic extraction from large information the use of ontology is particularly to reduce the searching complexity and improve the accuracy, which in turn would boom the performance of the machine. The crop has been labeled primarily based on weather, soil kind, and life span of the crop. Based on the ontology category of the crop and the use of knowledgebase, the crop which is yielding sufficient quality will be recognized (Nengfu and Wensheng 2007). Chaudhary et al. (2015) proposed a recommended machine using cotton ontology for cotton crops further improves internetworking and cell utility in recommended system.

A monitoring system became proposed by means of Keshtgari and Deljoo (2012), Leona and Jalaob (2013), Othman and Shazali (2012), which makes the farmers be greater worthwhile and sustainable, in the meantime it offers improved water administration. During rainfall circumstances, the land proprietors essential now not irrigate the land due to the fact humidity of the soil receives changed. Water may be stored, which in turn consumes electricity.

A comparative study was done by Mohanraj et al. amid numerous applications accessible with the present established system by considering several features such as knowledgebase, monitoring modules, efficiency, and reliability. The system proposed by them overcomes the restrictions of old-style agricultural measures by exploiting water supply proficiently and also with plummeting labor costs. Table I gives the technology and its objective in the existing system. In the existing method, several authors proposed several methods for automatic crop monitoring but none addressed the issues of big data. In the proposed method, rules framed for the sensed values are compared with the information extracted semantically from various heterogeneous networks to predict high yield crops.

K. Sriharsha et al. focuses particularly on sensing and tracking the Temperature, humidity and water stage of the paddy crop discipline and gives diverse sensing analyzes in the paddy crop field.

Song et al. (2012) classify the attributes of crop usage of agricultural ontology. Jimoh et al. (2013) and Agboola et al. (2013) predict rainfall for the use in fuzzy good judgment. J. C. Kang and J. L. Gao (2013) proposed in the ontology era in farming data recovery to improve accurateness and consistency of the farming records recovery.

Sherine M. Abd El-kader and Basma M. Mohammad El-Basioni (2013) used WSN in cultivating the potato crop in Egypt and additionally stated its software in precision farming and its significance for improving the agriculture in Egypt. N. Sakthipriya (2014) proposed an effective technique for crop monitoring the use of a Wi-Fi sensor network which controls water sprinkling by way of sensing soil moisture.

Hemlata Channe et al. (2015) proposed multiple processing systems for intelligent farming using mobile computing, cloud computing, IoT, and immense data evaluation to enhance agricultural manufacturing and control the cost of agro-merchandise.

10.3 Crop Prediction Framework

The efficient workflow of effective crop prediction using a semantic knowledgebase is depicted in Figure 10.1.

Various sensors such as humidity sensors, temperature sensors, and pH sensors are deployed at farmers' land. The sensed values from various sensors are collected using IoT and sent to the fuzzification function for prediction.

There are two phases of prediction. In Phase 1 based on temperature and humidity and pH values, the list of crops to be grown is predicted using the fuzzy model. Since the temperature keeps changing over time, the fuzzy model best suits crop prediction. In Phase 2, the list of crops predicted by Phase 1 will be tested with the crop knowledgebase and the effective list of crops to be grown will be predicted. The methodology used in this work is discussed in the following section.

10.4 Prediction Using Fuzzy Rules

The overall architecture of the fuzzy-based crop prediction is shown in Figure 10.2.

There are two stages in fuzzy prediction. Stage 1 predicts the climate based on temperature and humidity. Temperature sensors measure the temperature of the atmosphere over time and the data is stored in buffer.

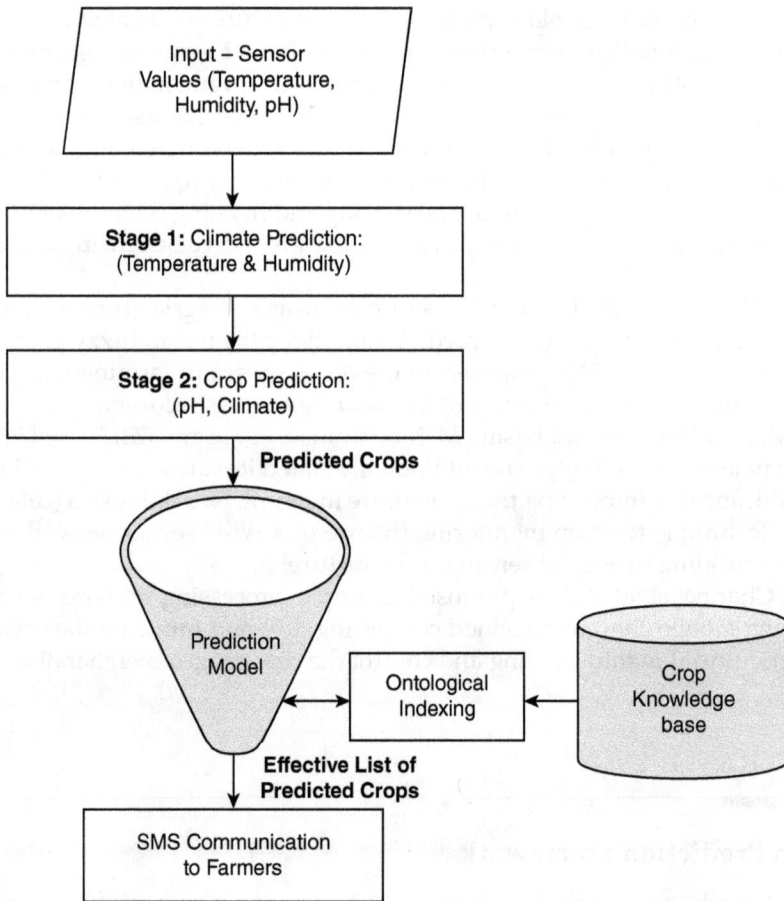

FIGURE 10.1
Effective crop prediction model.

ALGORITHM: CROP PREDICTION USING FUZZY RULES

1. Convert Temperature in Celsius to %.
   - *Split the input temperature into two boundary values as Min & Max values.*
   - *Find average of minimum and maximum values.*
   - *Find temperature in % using the following formula,*

   Temperature = Original value – minimum value/Average value

2. Normalize temperature and humidity values from 0 to 1 using the formula:

   Normalization of (x) = (x–min)/(max–min)

3. Convert it into linguistic variables.
4. Evaluate it using fuzzy rules to find climate.
5. Predict a list of crops to cultivate using weather and soil pH value.

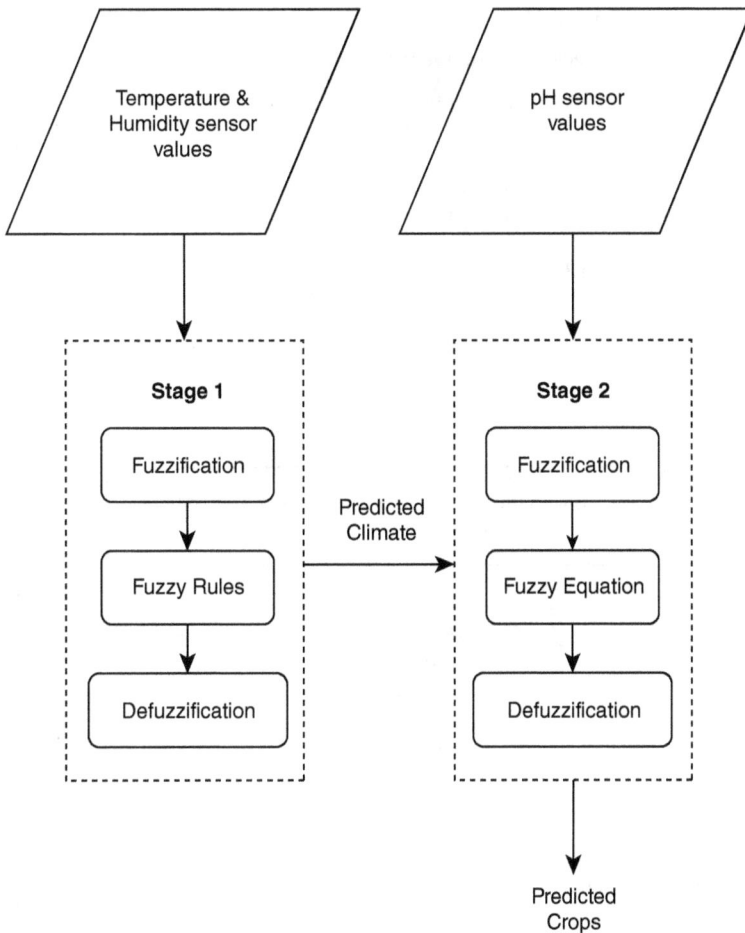

FIGURE 10.2
Crop prediction using fuzzy rules.

Similarly, the humidity sensor measures the humidity of the soil stored in buffer. Temperature and humidity values are given as input for the climate prediction. Since sensors measure different values for a time interval, the measured leads to fuzzy. Climate prediction involves various steps such as fuzzification, fuzzy rule formation using FAM, defuzzification.

Stage 2 predicts the list of crops to cultivate based on predicted weather at Stage 1 and soil pH value. The Crop prediction involves fuzzification, fuzzy equation, and defuzzification. The list of crops predicted by Stage 2 will be tested against the crop knowledgebase and the effective list of crops to be grown will be predicted.

10.4.1 Fuzzy-Based Crop Prediction

Table 10.1 illustrates the FAM (Fuzzy Associate Memory) uses matrix form map fuzzy rules.

TABLE 10.1

Forecasting the Weather through Sample Rule

Rule 1:	If (temp==very high) AND (humidity==high) THEN *(season==summer)*
Rule 2:	If (temp==moderate) AND (humidity==low) THEN *(season==spring)*
Rule 3:	If (temp==very low) AND (humidity==low) THEN *(season==winter)*

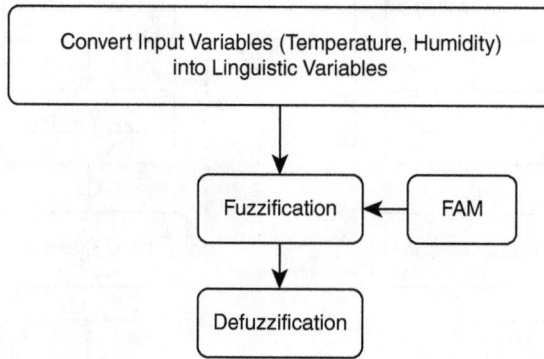

FIGURE 10.3
Flow diagram for climate prediction.

The flow diagram for the prediction of climate using fuzzy rules is shown in Figure 10.3.

Fuzzification converts the measured value from the sensors (temperature and humidity) to linguistic variables of fuzzy sets, which consists of five tuples: Very Low, Low, Medium, High, and Very High. Fuzzy rules are formed and evaluated by Fuzzy Associative Mapping (FAM), which is shown in Table 10.2. Fuzzy Associative Mapping is used to reduce the rate of false negatives (Sherine et al., 2013; Wood et al., 2005).

The output of fuzzification is then evaluated by defuzzification to generate an accurate prediction of climate. Defuzzification uses the mean of maxima defuzzification method to produce the climate (Jiang et al., 2010). The climate-related variables are shown in Table 10.3, which gives the list of crops to be grown in respective climates predicted by the defuzzification process (Sangeetha et al., 2015).

10.4.2 Crop Prediction Using Fuzzy Equation

Crop to cultivate is predicted effectively constructed on weather and soil pH value. pH sensors are used to measure the pH value of the soil.

The type of soil can be identified by analyzing its acidic nature. The list of crops to be grown in respective pH values and soil types is shown in Tables 10.4 and 10.5, respectively.

A fuzzy equation is used for evaluation. It takes climate and pH as two input variables.

TABLE 10.2

Fuzzy Associative Mapping for Climate

Temperature/Humidity	Very High	High	Moderate	Low	Very Low
Very High	Summer	Summer	Summer	Autumn	Autumn
High	Summer	Summer	Autumn	Autumn	Autumn
Moderate	Summer	Autumn	Autumn	Spring	Spring
Low	Spring	Spring	Spring	Autumn	Winter
Very Low	Spring	Spring	Winter	Winter	Winter

TABLE 10.3

Weather-Associated Variables

Characteristic	Forms	Comparative Temperature	Comparative Humidity	Crops to Be Grown
Climate	Summer	Very hot (32°C–40°C)	Very high to moderate	Millets, maize, red chilies, cotton, paddy, soya bean, sugarcane, turmeric, moong, groundnut, barley
	Autumn	Warm days (>30°C) Cool nights (21°C–29°C)	Low	Maize, oats
	Spring	Warm days (>30°C) Cool nights (25°C–29°C)	Low to moderate	Wheat, mustard, barley, peas
	Winter	Cold (10°C–15°C)	High	Oats

TABLE 10.4

pH Value and Suitable Crops

pH Value	Suitable Crops
5.5–6.5	Apple, strawberry
5.5–7.5	Tomato, corn, peas, grapes, carrot
6.0–7.5	Onion, cabbage, beans

Therefore, the list of crops to be grown is predicted based on four tuples (T, H, pH, S), where T – temperature, H – humidity, pH – pH of soil, S – type of soil.

$$S_i = \{A, B, C, D, E, F\}$$

$$X = \{A_i, B_i, C_i, D_i, E_i, F_i\},$$

where A, B, C, D, E, and F – desert soil, black soil, laterite soil, red soil, alluvial soil, and mountain soil, respectively.

X – Set of crops and $A_i, B_i, C_i, D_i, E_i, F_i$ – group of crops to cultivate in procured soil.

Climate is predicted based on fuzzy rules with the attributes such as temperature and humidity using which is represented as,

$$f_1(x) = f(t, h), \tag{10.1}$$

TABLE 10.5

Soil Associated Variables

Attribute	Soil Types	Crops to Be Grown
Soil	Alluvial	Cotton, jute, wheat, rice, sugarcane
2	Black	Oilseeds, sugarcane, millet, rice, wheat, cotton, groundnut
2	Red	Potatoes, cotton, maize, pineapples
2	Laterite	Cashew, tea, coffee, tropical crops, rubber, coconut
2	Mountain	Tea, tropical fruits, coffee, spices
2	Desert	Millet, barley

where t – temperature and h – humidity, x is a crop to cultivate and $x \in X$. Crop is predicted based on soil pH value using fuzzy equation and is represented as,

$$f_2(x) = f(f_1(x), pH), \qquad (10.2)$$

where pH – pH value of soil. $f_2(x)$ is a fuzzy equation that predicts the crops to be grown. The following fuzzy rule gives the effective list of crops to be grown. The sample rule is given as follows.

$$f_2(x) = \text{Summer AND pH} > 7.0 \text{ then } f_1(x) \cap pH_i$$

The predicted crops using Equation 10.2 will be used for predicting an accurate list of crops to be grown using crops knowledgebase. The data in the crop knowledgebase are prolonged through domain ontology, further grouped on the basis of ontology and the participation query is scrutinized for semantics. It is then compared with the predicted list of crops from Equation 10.2 to produce an effective list of crops to be grown.

10.4.3 Semantic Extraction in Big Data

The agricultural data from various heterogeneous resources are collected and stored in the database over several years. The heterogeneous resources consist of information from agricultural researchers, Email, Web Crawling, farmers Profile, which leads to Big Data. These data must be preprocessed to remove unwanted data. During preprocessing, it removes HTML and XML labels, and scripting reports from the basic documents.

In big data, the semantic analysis using ontology allows structuring the data from the unstructured data of various resources such as web pages and historical data, which serves as a basis for the implementation of the proposed system for decision support (Zhang, 2011; Mohanraj et al., 2016). The semantic analysis gets initiated with the ontology establishment for the information in the database through keyword comparison by natural language processing. Figure 10.4 shows the semantic extraction from Big Data.

The crop is classified based on ontology, which involves three features: climate, life span, and soil, which is shown in Figure 10.5.

In order to measure the quantity of water droplets in the air the proposed system uses humidity sensors. Further, to determine the temperature level of air from moisture and

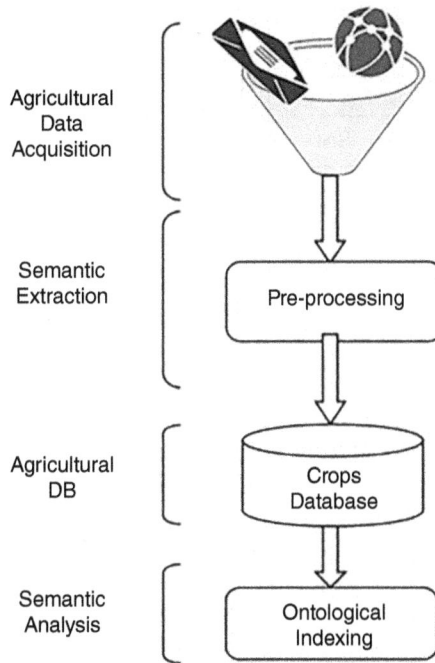

FIGURE 10.4
Semantic extraction from the knowledgebase.

radiation temperature sensors are deployed. Therefore, climate consists of temperature and humidity. The pH sensors analyze the acid level of the soil, and they measure the pH value. Henceforth, through soil pH value the type of soil can be identified. The data in the crop knowledgebase are prolonged through domain ontology, further grouped on the basis of ontology and the participation query is scrutinized for semantics. It is then compared with the semantically extracted data to produce an effective list of crops to be grown.

10.5 Results and Discussion

In order to determine climatic conditions and to predict the season change periodic measurement of humidity and temperature is recorded periodically. The periodic measurement is for a year. The acquired values are analyzed and plotted. The following Figure 10.6 depicts the periodic measurement of humidity and temperature. The values are plotted for the year 2018–2019.

Based on soil pH value the crop to be cultivated is determined. Figure 10.7 shows the pH value of the soil, which varies depending on the type of soil.

The semantic retrieval using ontology has been evaluated based on accuracy and time complexity. This proposed system using Fuzzy logic shows superior performance when compared to the existing system, which is shown in Figure 10.8. The time complexity is

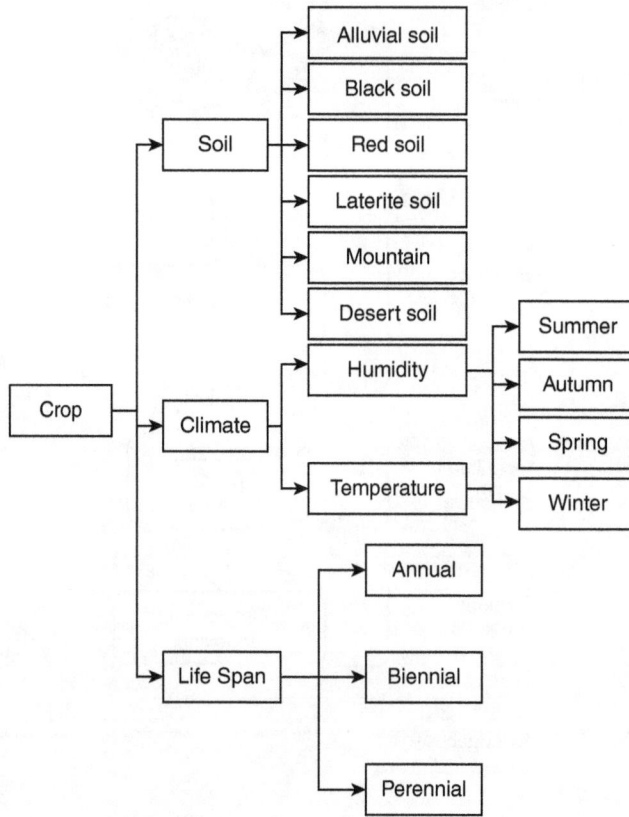

FIGURE 10.5
Ontology classification of crops.

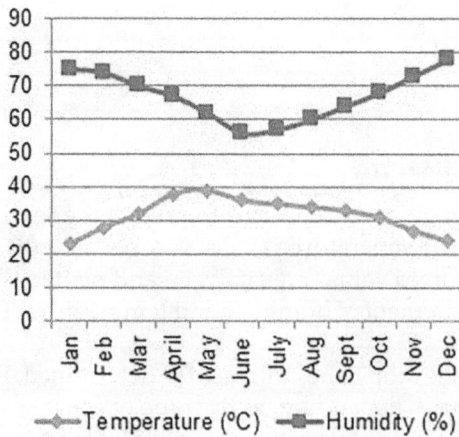

FIGURE 10.6
Humidity and temperature.

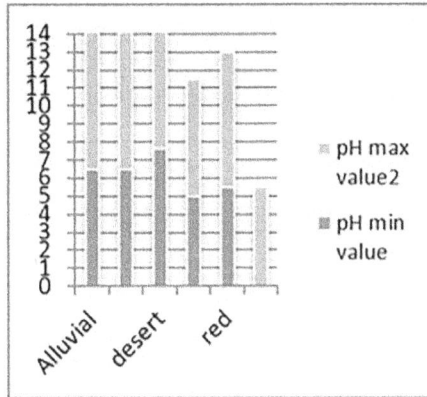

FIGURE 10.7
pH value analysis.

analyzed, which is 45 s for the proposed system and is significantly low compared to existing one (Sherine et al., 2013), which has 82 s; thus proposed system is more efficient.

The performance analysis of the proposed algorithm using the fuzzy rule is estimated based on precision, sensitivity, accuracy, and Matthew Correlation Coefficient (MCC). The accuracy of the system is premeditated by the number of accurate positive (+VE) forecasts by the entire number of positive (+VE) forecasts. Further Figure 10.9 illustrates the overall positive forecasts of the system model. The crop prediction using fuzzy rules sustains the accuracy rate of 80%, correspondingly.

In the following Figure 10.9, performance analysis of the proposed Crop prediction using fuzzy rules-based sensitivity is evaluated. Further, sensitivity is determined by the number of precise positive (+VE) forecasts by overall positive (+VE) forecasts. The proposed system sustains a sensitivity of above 84%. Further, for the Crop prediction using fuzzy rules, accuracy is considered. Figure 10.10 discusses the exactness maintained at above 80% in all measures, namely False Negative (FN), False Positive (FP), True Negative (TN), and True Positive (TP).

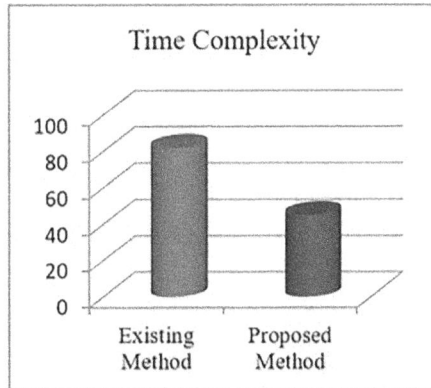

FIGURE 10.8
Time complexity.

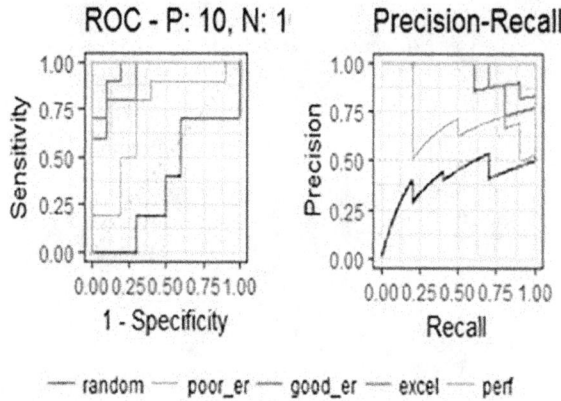

FIGURE 10.9
Sensitivity vs. specificity and precision vs. recall.

Figure 10.10 portrays the manner in which the proposed system model fluctuates for diverse input. Further Figure 10.10 depicts the comparison of numerous attributes through various outputs, namely excel, random, perf, poor_er, and good_er. The number of samples attained is assigned as ten (10), which is positive and the remaining ten are unfavorable. Finally, after analysis the system model ends up in error which is originated through serval process and resulting in consider reduction of the errors in future.

Figure 10.11 depicts the rate of recall and precision after reducing the error. In the proposed model through means of assessment the error gets reduced. In the system model the errors are carefully handled.

Figure 10.12 concentrates the regularized rank as a fixed variable to display the difference between error and precision for diverse values. The rate of accuracy is calculated using the randomized rating. As a result, the accuracy rate is greater than the error rate. This shows that the system model is effective in determining accuracy and error.

In the proposed model, after calculating the precision and error, a comparison is made to determine the specificity, sensitivity, and precision, as depicted in Figure 10.13. The system model uses a similar dataset to determine the performance.

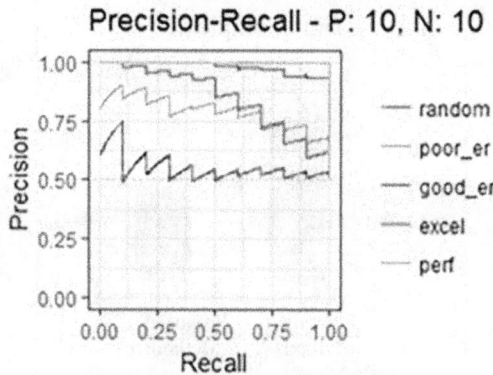

FIGURE 10.10
Recall vs. precision.

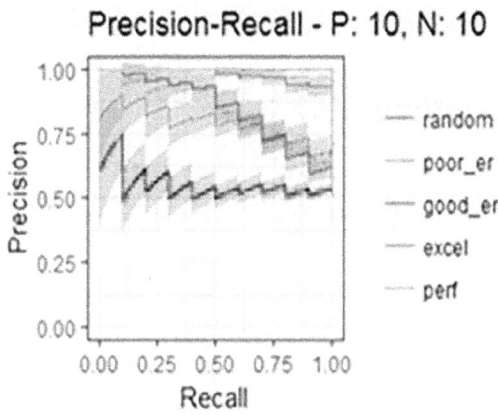

FIGURE 10.11
Subsequently error inference.

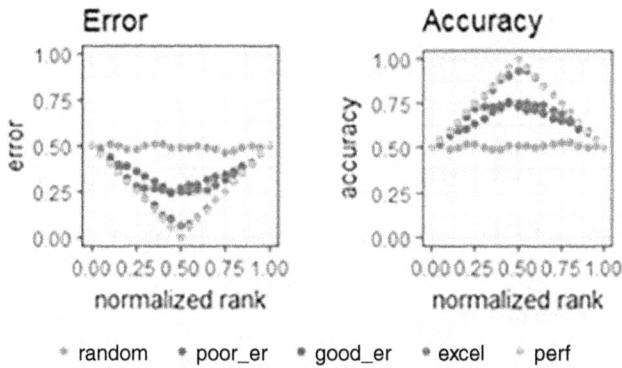

FIGURE 10.12
Error vs. accuracy.

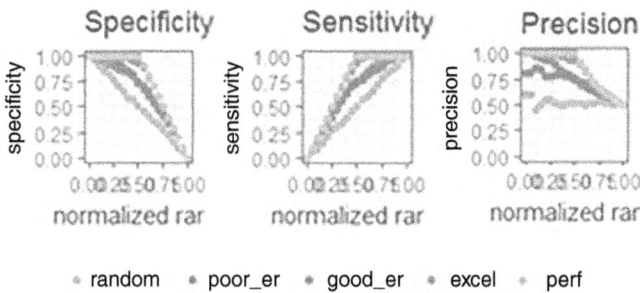

FIGURE 10.13
Comparison of specificity, sensitivity, and precision.

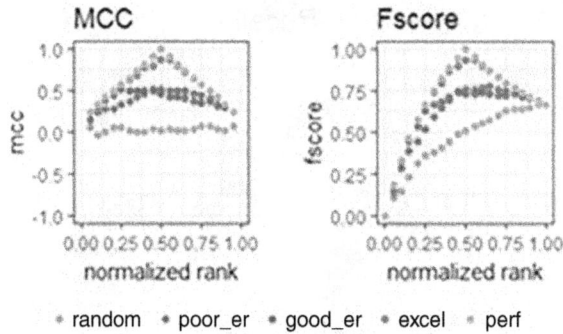

FIGURE 10.14
MCC and F-score.

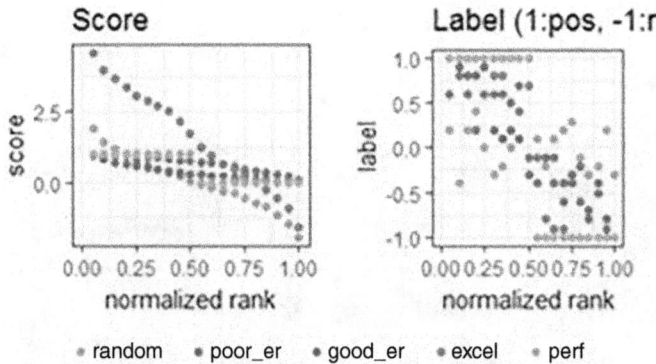

FIGURE 10.15
Normalized and random normalized rank.

The F-score and MCC are calculated, and the performance rate shows high fluctuation comparatively. F-score is the ratio between the recall and precision, as mentioned in Figure 10.14. MCC is used to determine if the error rate is high or low.

Finally, Figure 10.15 shows the comparison between the Normalized and Random normalized for the given dataset. Our proposed model illustrates progressive performance percentages in comparison with the existing system.

In addition to that, MCC is calculated in the proposed methodology. The MCC accuracy is maintained above 80% in all the criteria, namely True Positive (TP). True Negative (TN), False Positive (FP), and False Negative (FN) as mentioned in Figure 10.9.

10.6 Conclusion

This paper uses the IoT era for crop cultivation in large information, which offers low cost and green garage and consumes much less power. The ontology primarily based totally crop farming forecasts the crop to be grown with the excessive harvest. The semantic extraction in large information will increase the overall performance and in flip reduces the looking time and garage capacity. FAM is employed to predict the climate condition, which is principally to cut back the speed of false negatives. Thus, it produces an effective prediction of the crop to be used. Crop monitoring is additionally automated to avoid paucity of water. The performance analysis of the proposed algorithm using the fuzzy rule is assessed based on sensitivity, accuracy, precision, and MCC. Therefore, the proposed method shows better performance and also overcomes the restrictions of traditional agricultural procedures by utilizing water resources efficiently and also by reducing the labor cost.

References

A.H. Agboola, A.J. Gabriel, E.O. Aliyu, B.K. Alese, "Development of a Fuzzy Logic Based Rainfall Prediction Model," *International Journal of Engineering and Technology*, vol. 3, Issue 4, pp. 427–435, April 2013.

Gelian Song, Maohua Wang, Xiao Ying, Rui Yang, Binyun Zhang, "Study on Precision Agriculture Knowledge Presentation with Ontology," *AASRI Conference on Modelling, Identification and Control, AASRI Procedia*, vol. 3, pp. 732–738, 2012.

Hemlata Channe, Sukhesh Kothari, Dipali Kadam, "Multidisciplinary Model for Smart Agriculture using Internet-of-Things (IoT), Sensors, Cloud-Computing, Mobile-Computing & Big-Data Analysis," *The Int. J. Computer Technology & Applications*, vol. 6, Issue 3, pp. 374–382, 2015.

I. Mohanraj, Kirthika Ashokumar, J. Naren, "Field Monitoring and Automation Using IOT in Agriculture Domain," *The Procedia Computer Science*, vol. 93, pp. 931–939, 2016.

J. C. Kang, J. L. Gao, "Application of Ontology Technology in Agricultural Information Retrieval," *Advanced Materials Research*, vol. 756–759, pp. 1249–1253, 2013.

J. D. Wood, C. E. O'Connell-Rodwell, S. Klemperer, "Using Seismic Sensors to Detect Elephants and Other Large Mammals: A Potential Census Technique," *Journal of Applied Ecology*, vol. 42, pp. 587–594, 2005.

K. Sriharsha, T. V. Janardhana Rao, A. Pravin, K. Rajasekhar, "Monitoring the Paddy Crop Field Using Zigbee Network," *International Journal of Computer and Electronics Research*, vol. 1, Issue 4, December 2012.

Kehua Guo, Wei Pan, Mingming Lu, Xiaoke Zhou, Jianhua Ma, "An Effective and Economical Architecture for Semantic-based Heterogeneous Multimedia Big Data Retrieval," *Journal of Systems and Software*, vol. 102, pp. 207–216, 2015.

Manijeh Keshtgari, Amene Deljoo, "A Wireless Sensor Network Solution for Precision Agriculture Based on ZigBee Technology," *Scientific Research Journal on Wireless Sensor Network*, vol. 4, pp. 25–30, 2012.

Maria Rossana C. de Leona, Eugene Rex L. Jalaob, "A Prediction Model Framework for Crop Yield Prediction," *Asia Pacific Industrial Engineering and Management System*, vol. 6, Issue 4, 2013.

Mohd Fauzi Othman, Khairunnisa Shazali, "Wireless Sensor Network Applications: A Study in Environment Monitoring System," *International Symposium on Robotics and Intelligent Sensors, Procedia Engineering*, vol. 41, pp. 1204–1210, 2012.

N. Sakthipriya, "An Effective Method for Crop Monitoring Using Wireless Sensor Network," *Middle-East Journal of Scientific Research*, vol. 20, Issue 9, pp. 1127–1132, 2014.

R. G. Jimoh, M. Olagunju, I. O. Folorunso, M. A. Asiribo, "Modeling Rainfall Prediction using Fuzzy Logic," *International Journal of Innovative Research in Computer and Communication Engineering*, vol. 1, Issue 4, pp. 929–936, June 2013.

S. Sangeetha, M. K. Dharani, B. Gayathridevi, R. Dhivya, P. Sathya, "Prediction of Crop and Intrusions using WSN," *Proceedings of 3rd International Conference on Advanced Computing, Networking and Informatics in the series of Smart Innovation, Systems and Technologies*, vol. 44, pp. 109–115, 2015.

Sanjay Chaudhary, M. Bhise, A. Banerjee, A. Goyal, C. Moradiya, "Agro Advisory System for Cotton Crop," *Workshop on Networks and Systems for Agriculture AGRINETS 2015, collocated with 7th International Conference on Communication Systems and Networks COMSNETS,* pp. 1–6, 2015.

Sherine M. Abd El-kader, Basma M. Mohammad El-Basioni, "Precision Farming Solution in Egypt Using the Wireless Sensor Network Technology," *Journal of Egyptian Informatics,* vol. 14, Issue 3, pp. 221–233, November 2013.

X. Jiang, G. Zhou, Y. Liu, Y. Wang, "Wireless Sensor Networks for Forest Environmental Monitoring," Innovations and trends in environmental and agricultural informatics, pp. 2–5, 2010.

Xie Nengfu, Wang Wensheng, "Ontology and Acquiring of Agriculture Knowledge," *Journal of Agriculture Network Information,* vol. 8, pp. 13–14, 2007.

Y. Zhang, "An excellent web content management system," *International Conference on Multimedia Technology,* Hangzhou, pp. 3305–3307, June 26–28, 2011.

Y. Zhu, J. Song, F. Dong, "Applications of Wireless Sensor Network in the Agriculture Environment Monitoring," *The Procedia Engineering,* vol. 16, pp. 608–614, January 2011.

11

Application of Drones with Variable Area Nozzles for Effective Smart Farming Activities

J. Bruce Ralphin Rose
Anna University Regional Campus
Tirunelveli, India

V. Saravana Kumar and V.T. Gopinathan
Hindusthan College of Engineering and Technology
Coimbatore, India

CONTENTS

11.1 Introduction 164
    11.1.1 Roles of Drones in Agriculture 165
        11.1.1.1 Soil Analysis for Field Planning 165
        11.1.1.2 Seed Pod Planting 166
        11.1.1.3 Crop Monitoring 166
        11.1.1.4 Pesticide Spraying 167
        11.1.1.5 Irrigation Planning 167
        11.1.1.6 Crop Health Assessment 167
        11.1.1.7 Controlling Weed, Insect, Pest, and Diseases 167
        11.1.1.8 Tree/Crop Biomass Estimation 167
        11.1.1.9 Scaring Birds 168
        11.1.1.10 Man–Animal Conflict 168
11.2 Existing Spraying Methods 168
11.3 Proposed VAN Integrated Quadcopter Sprayer 169
    11.3.1 Components 171
        11.3.1.1 BLDC Motor 172
        11.3.1.2 Propellers 172
        11.3.1.3 Electronic Speed Controller 172
        11.3.1.4 Frames 172
        11.3.1.5 Li-Po Battery 172
        11.3.1.6 Transmitter and Receiver 173
        11.3.1.7 Flight Controller 173
    11.3.2 Drone Components with Specification 174
11.4 Variable Area Nozzle Sprayer 175
    11.4.1 Type of Nozzles Used in the Spraying System 175
    11.4.2 Working Principle of Sprayer System 176
    11.4.3 Performance of VAN System 177
    11.4.4 Challenges for Using VAN in Agriculture 179

DOI: 10.1201/9781003185413-11

11.5   Smart Farming through Internet of Things.. 179
     11.5.1   Arduino Board 180
     11.5.2   Sensors.. 180
     11.5.3   Advantages of IoT-Enabled Spraying System 181
11.6   Agriculture Drone Usage Statistics for Smart Farming........................ 181
     11.6.1   Smart Farming in Asian Countries.. 182
     11.6.2   Smart Farming in European Countries 182
11.7   Current Status of Regulations and Provisions for Agriculture Drones.................. 183
11.8   Conclusions.. 183
     11.8.1   Future Scope for Drones in Agriculture....................................... 184
References... 184

11.1 Introduction

Drones have several applications in agriculture and they are widely used for extended farming activities in many countries across the globe. Agriculture is the backbone of every country, and the transformation from conventional farming to smart farming is very much essential to cater to the food requirements of the masses. Furthermore, approximately 2 billion people (26.7% of the world population) around the globe depend on agriculture for their livelihood. It also contributes to 4% of global Gross Domestic Product (GDP) and it accounts for more than 25% in developing countries like India [1, 2]. Specifically, the rural population mainly depends on agriculture, where the conventional farming methodologies are still in practice because of the lack of awareness, farm policies, and capital investments. Recently, the use of drones for various agriculture activities has been proved to be very effective in the case of pesticide spraying, Crop monitoring, Crop Damage Assessment (CDA), Surveillance, Irrigation, Planting, replacing labor-intensive and hazardous conventional methods [3].

Among the various applications of drones for smart agriculture activities, the handling of pesticide/fertilizer is a preferred task because of the hazardous nature of the work and it is well-connected with the health of the plants. Hence, Internet of Things (IoT)-enabled systems play a vital role in performing crop health monitoring activities to decide the amount of fertilizer needed at different phases of crop growth [4]. In actual agriculture production, Pests and illnesses of harvests are the central points that influence the yield and nature of yields, and the use of synthetic pesticides is the primary method for their avoidance and control. Additionally, the seriousness of plant illness and insect pests varies with respect to different locations according to the weather and soil conditions. The broad utilization of pesticides straightforwardly jeopardizes the environment and human well-being. Subsequently, disproportionate pesticide usage decreases productivity and it has been acknowledged around the world through various studies. Drones spray the pesticide/fertilizer through a nozzle which is fixed along with the sprayer system. Though agricultural drones have many benefits, the wastage of pesticides or fertilizer is considered to be a major disadvantage because of the absence of sophisticated remotely controlled mechanisms.

In the field of plant security, the variable shower innovation can be applied based on the field of interest and other crop specifications. It has distinct possibilities for improving the usage pace of pesticides and lessening the pesticide deposits. To reduce the wastage of pesticides, it should be applied according to the severity of pests, insects, and weeds with complete field data [5]. To achieve this, variable area nozzles (VANs) are designed and are synchronized with the IoT-enabled sprayer system. Here, the drone camera observes the crop images to determine its health through the image processing software (IPS) deployed in the cloud hub. The IPS computes the degree of infection and the proportionate quantity of pesticide to be sprayed. Subsequently, the required quantity will be communicated to the IoT sensors to trigger the spraying event with appropriate flow rates. Here, the spray system consists of a reservoir, flow lines, throttle valves, and nozzle. Typically, this VAN has three different flow ranges such as 0.16 L/min, 0.32 L/min, and 0.54 L/min, respectively.

As the drone is operated over agricultural land, different areas can be captured according to the field of interest using a multispectral camera. Multispectral camera remote sensing (RS) imaging technology uses green, red, red-edge, and near-infrared wavebands to capture both visible and invisible images of crops and vegetation [6]. With the captured images and subsequent output received from the IPS, the drone operator could plan the flow range required at different locations of the land and use the sprayer system effectively with VANs. It reduces the amount of pesticides being wasted and also reduces the operational cost involved. Hence, this chapter is focused on the design of an IoT-enabled VAN sprayer system (Figure 11.1) and its applications to perform effective smart farming activities.

11.1.1 Roles of Drones in Agriculture

The smart agriculture activities revolve around the IoT systems and drones in terms of several potential applications as listed below:

11.1.1.1 Soil Analysis for Field Planning

Drones can be effectively utilized for the soil and field examination to identify the water system, planting arrangement, and nitrogen (N_2) levels in the soil, as shown in Figure 11.2(a).

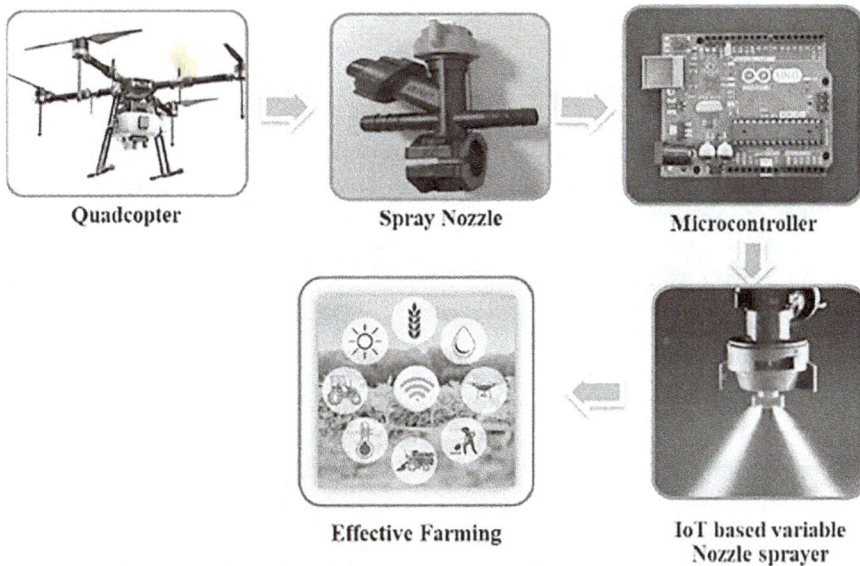

| Quadcopter | Spray Nozzle | Microcontroller |

| Effective Farming | IoT based variable Nozzle sprayer |

FIGURE 11.1
Schematic diagram for IoT-enabled spraying system.

FIGURE 11.2
Roles of drones in agriculture: (a) soil analysis for field planning, (b) seed pod planting, (c) crop monitoring, (d) pesticide spraying, (e) irrigation planning, (f) crop health assessment, (g) controlling weed, insect, pest, and diseases, (h) tree/crop biomass estimation, (i) scaring birds, and (j) man–animal conflict.

In addition, drones are useful to deliver the precise 3D guides that can be employed for direct soil investigation in terms of soil properties, dampness substance, and soil disintegration [7].

11.1.1.2 Seed Pod Planting

In recent times, many kinds of drones have been developed with additional attachments according to various field applications below the flight controller system. As an example, the shoot pod containing seed and plant nutrients is capable of planting seeds into the already-prepared soil with precise spacing at different terrains (Figure 11.2(b)). It helps to reduce the manpower and planting costs with minimum time consumption [7].

11.1.1.3 Crop Monitoring

Crop monitoring is the biggest challenge not only for farmers but also for various stakeholders associated with the agriculture business [8]. This challenge becomes worse in the event of unpredictable weather patterns, which lead to rising crop losses, risks, and maintenance costs. Agronomists and agricultural engineers prefer to use drones with smart crop tracker tools set for large-scale precision farming operations [4]. It monitors the leaves and routes by gathering multispectral geospatial and temporal datasets at predefined scales that are related to crop development and health, as shown in Figure 11.2(c). Drones prepared with special imaging equipment known as the normalized difference vegetation index (NDVI) utilize comprehensive color statistics to show the health of the plants. This advanced information helps to monitor the crop wellbeing much before it is being investigated by manual field observations [7].

11.1.1.4 Pesticide Spraying

The modern drones are optimized to carry the pesticide spray systems of different capacities for proper spraying of pesticides at various phases. Spraying pesticides is a very crucial activity to maintain crop health and these drones are more efficient than any other manual spraying method. (Figure 11.2(d)). Most importantly, the concept of precision agriculture (PA) can be fulfilled only when the time and labor costs are saved significantly. Alternatively, PA also protects the farmers from coming into contact with the various toxic chemicals [9].

11.1.1.5 Irrigation Planning

Drones are equipped with thermal, multispectral, or hyperspectral sensors which can distinguish the pieces of the field with dampness shortages by utilizing multispectral records. The irrigation planning should be aligned with moisture maps and ground reality to overcome the drought and climate change issues. The microwave sensing technology on drones offers perfect images that are better than optical mapping and the picture of the soil will not be affected by vegetation [10]. This system helps to arrange the appropriate water resources to the distinguished regions with exactness on moisture mapping (Figure 11.2(e)).

11.1.1.6 Crop Health Assessment

The GIS plays a vital role in the PA in terms of soil mapping and relevant crop health assessments. Crops reflect visible and clear infrared light, the power level of which changes with wellbeing status and stress concentration experienced by crops. By using the NDVI data, it is much easier to classify the dead leaf, stressed leaf, and healthy leaf of a plant. Drones fitted with health monitoring sensors are suitable for examining the crops by utilizing the infrared light data, and they can also be utilized to follow the wellbeing of the crop throughout a given timeframe (Figure 11.2(f)) [11].

11.1.1.7 Controlling Weed, Insect, Pest, and Diseases

Apart from the evaluation of soil conditions, drones could identify and educate the farmers about field regions perpetrated by weeds, diseases, and insect pests through detailed aerial surveying, as displayed in Figure 11.2(g). With the help of these data, farmers can improve the utilization of synthetics expected to battle against various diseases; henceforth the abrupt losses are reduced by increasing the crops' wellbeing [12].

11.1.1.8 Tree/Crop Biomass Estimation

Crop/tree canopy density and distance from the ground surface can be measured using ultra-compact light detection and ranging (Li-DAR) sensors mounted on the drones. This helps in the estimation of the tree/crop biomass change by studying the variations in height measurements that form a basis for estimating timber production in forest and production estimates in the crops like sugarcanes (Figure 11.2(h)) [1].

11.1.1.9 Scaring Birds

Birds are considered to be a serious issue subsequent to planting seeds on numerous yields. It increases the requirement of manpower to secure the fields, especially in the remote areas, where it takes hours of walking to roam around, for which drones with speakers can be used to produce annoying sounds to frighten the birds off from the fields in alignment with the regulations prescribed by the concerned authorities (Figure 11.2(i)) [13]. Hence, farm management against the scaring birds can now be completed in a few minutes by bird-chasing drones like ProHawk UAV.

11.1.1.10 Man–Animal Conflict

Human–wildlife conflict (HWC) results in crop destruction, loss of human life, and injuries to humans/animals. Recent HWC data shows that the wildlife conflict is continuously soaring over the years across the globe because of the encroachments and negative interactions. Farmers living adjacent to the forests have such major issues during their harvesting times due to monoculture fields. The improved artificial intelligence (AI)-based drones could identify and recognize the increased crop diversity to minimize the HWC. Further, the animals are migrating to society in big numbers in search of drinking water, and food and their migrating path can be well-defined through drone mapping [14]. By drones, one can avoid man–animal conflict by releasing the smoke or making a relevant frequency of sound against the animals through onboard speakers (Figure 11.2(j)).

From the various roles of drones stated above, it is inferred that more than 20 day-to-day applications exist for the agriculture drones in the PA domain. Specifically, pesticide spraying and handling is a key application because of its ill-effects on the farmer's health. Hence, the present chapter is motivated to investigate the challenges associated with the drone spraying mechanisms and the rectification measures as explained in the subsequent sections.

11.2 Existing Spraying Methods

The primary motivation behind any pesticide application method (spraying, dusting, etc.) is to cover the objective with the extreme effectiveness and least activities to monitor the vermin attacks. All the pesticides contain toxic substances which would cause harmful effects to every living organism. Thus, the utilization of pesticides should be extremely prudent with the least pollution to non-targets. Conventional sprayers are available in a variety of sizes and specifications, depending on the requirements of a plant or crop cultivated.

An overview of conventional sprayers is discussed herein to emphasize the comparative advantages of IoT-enabled VAN sprayers in drones. Following are the main types of sprayers used for insecticide or pesticide sprays: (a) low-pressure sprayer (tractor-mounted, high-clearance sprayer, trailer-mounted sprayers, and truck-mounted sprayers), (b) air carrier sprayer, (c) high-pressure sprayer, (d) fogger (mist blowers), and (e) hand-operated sprayer [15].

The widely used method to shower the pesticide in the field by the Tractor mounted low-pressure sprayers are presented in Figures 11.3(a) and (b). It is an expensive method in which the wastage and excessive spraying of pesticides occur very often that eliminates

FIGURE 11.3
Types of sprayers: (a) tractor-mounted low-pressure sprayer, (b) air carrier sprayer, (c) high-pressure sprayer, (d) fogger – mist blowers, (e) hand-operated sprayer.

a great variety of living organisms in agriculture. The high-pressure sprayer splashes the pesticides on the field by revolving around 360°, as highlighted in Figure 11.3(c). For small and medium area land-based agriculture, the farmers carry large shower bottles with them to spray the pesticides that cause numerous medical issues, as displayed in Figures 11.3(d) and (e). A detailed comparison of different kinds of sprayers used for various agricultural activities is summarized in Table 11.1.

As a result of the widespread diffusion of pesticides, a great part of the people involved in agribusiness may get exposed to pesticides due to their occupation [17]. The World Health Organization (WHO) estimates there are more than 1 million occupational exposure pesticide cases reported every year. Among the reported cases, more than one lakh deaths occur every year, especially in the developing countries where pesticides are handled exclusively by human beings [18]. It should be noted that the adverse health effects of pesticides include asthma, allergies, hypersensitivity, cancer, hormone disruption, and problems with reproduction. Hence, the PA and farm management is the need of the hour to create a rapid evolution with smart agriculture activities. The major advantage of drones for crop monitoring and spraying is that it completes the entire process over a large area within a limited time period as compared with other traditional methods.

11.3 Proposed VAN Integrated Quadcopter Sprayer

To overcome the issues related to manual spraying, a novel VAN system integrated Quadcopter sprayer is proposed in this chapter. The component-wise features and functions of the Quadcopter to achieve the objectives of PA are also presented with the schematic

TABLE 11.1

Performance Comparison of Different Spraying Mechanisms [16]

Parameters	Drone Sprayer	Hand-Operated Sprayer	Tractor-Mounted Sprayer	Truck-Mounted Sprayer
Application/day	20–25 hectares/day	0.8–1.3 hectares/day	6.7–20 hectares/day	66.7–80 hectares/day
Pesticide utilization Efficiency	85%–90%	30%–40%	30%–40%	30%–40%
Adaptability	Can be operated at mountains, hilly terrains, and paddy fields.	Few crops, flower, and fruit are easily damaged, trampled or dropped by human-operated sprayer.	Cannot be used in the mountain or hilly terrains and narrow access routes.	Cannot be used in mountains or hilly terrains.
Water consumption per hectare	VAN sprayer ensures uniformity, low dilution rate, and even with the use of highly concentrated liquid pesticide, the water can be saved up to 90%.	Traditional jet spraying, resulting in waste of water, and most of the pesticides lost into the soil along with water.	Traditional jet spraying, resulting in wastage of water, and most of the pesticides lost into the soil along with water.	Traditional jet spraying, resulting in wastage of water, and most of the pesticides lost into the soil along with water.
Safety	Operated away from the field during spraying to avoid pesticide poisoning.	Pesticides enter into the human body by mouth that easily leads to pesticide poisoning.	Applying pesticide from a close range easily leads to pesticide poisoning.	Pesticide is applied from short range and vulnerable to poisoning.

circuits. The nozzle spray nomenclature is illustrated in Figure 11.4(b). Basically, drones are remote-controlled UAVs that use Information and Communications Technology (ICT) to increase agricultural productivity through PA. These UAVs have a huge potential in agriculture in supporting evidence-based planning and in spatial data collection. Despite some inherent limitations, these tools and technologies can provide valuable data that can then be used to influence policies and decisions [8]. The proposed UAV is a combination of a Quadcopter and a spraying system (Figure 11.4(a)) with the following components.

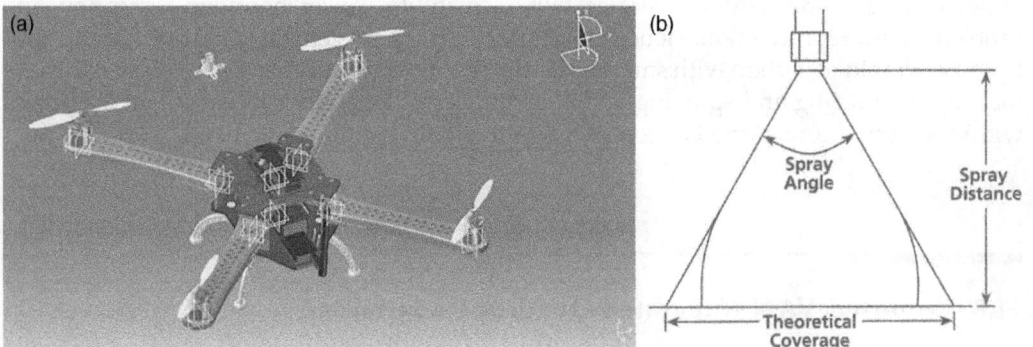

FIGURE 11.4
(a) Quadcopter design prepared using CATIA V5. (b) Nozzle spray nomenclature.

11.3.1 Components

The essential components of the VAN integrated Quadcopter sprayer are as follows: (i) brushless DC motor (BLDC) motors (4 nos.), (ii) propellers (4 nos.), (iii) electronic speed controller (ESC), (iv) Flight control Board (Pix hawk), (v) Li-Po battery, (vi) transmitter with receiver (Fly Sky), (vii) carbon frame, (viii) sprayer (variable nozzle). The actual view of the key components of the Quadcopter sprayer is presented in Figures 11.5(a–h). The size and geometric specifications are not disclosed herein because of the patents involved with each component. The essential functions of basic components of a Quadcopter are discussed briefly herein to offer the key insights about the robustness and reliable working of the system.

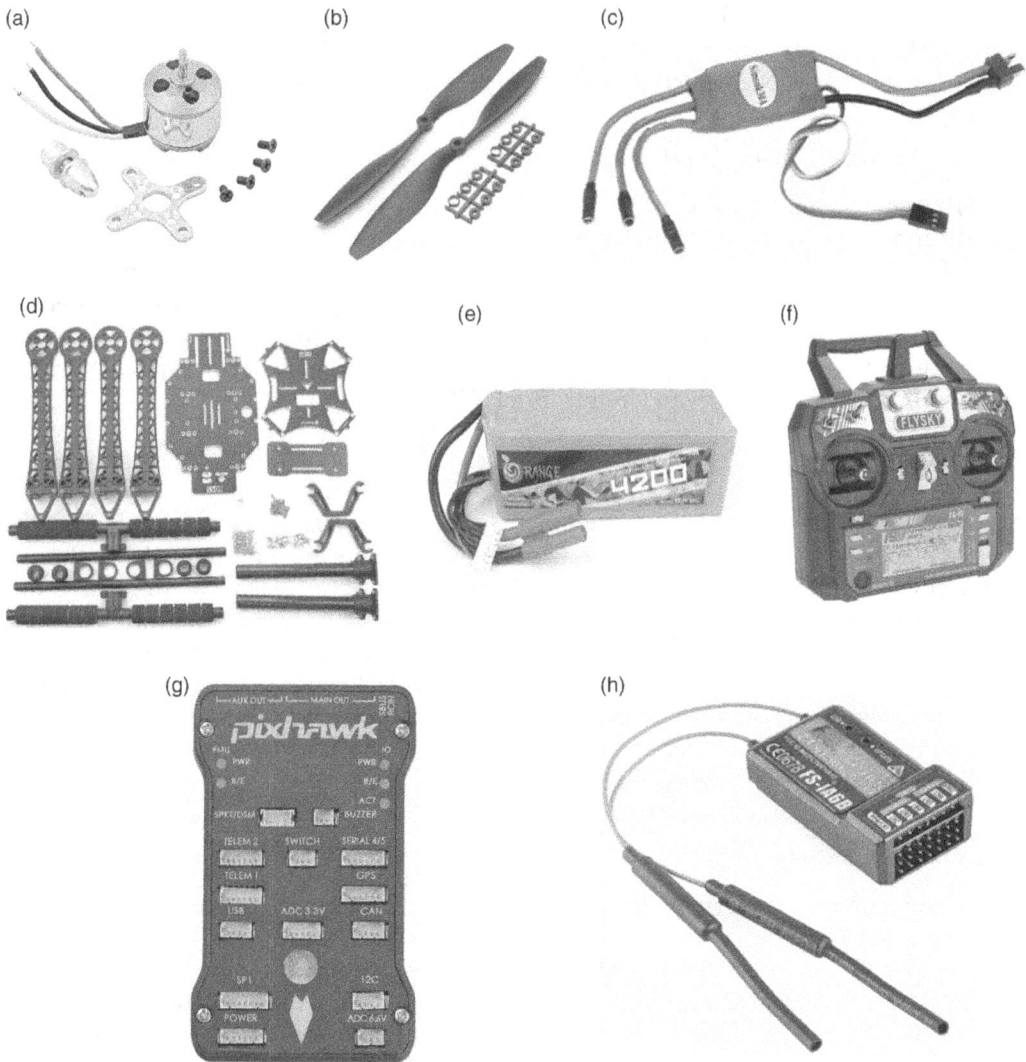

FIGURE 11.5
Basic components of a Quadcopter: (a) BLDC motor, (b) propeller, (c) ESC – red brick, (d) frames, (e) battery, (f) transmitter, (g) flight controller, (h) receiver.

11.3.1.1 BLDC Motor

A BLDC is a DC electric motor that uses an electronically controlled commutation system instead of a mechanical commutation system to deliver the required amount of power to the drone components (Figure 11.5(a)). Generally, BLDC motors run at very high RPM and it is available in both clockwise (CW) and counter-clockwise (CCW) configurations. The power to weight ratio of 2:1 to 4:1 is typically fixed based on the payload and auxiliary system requirements of the spraying drones.

11.3.1.2 Propellers

The purpose of Quadcopter propellers is to generate the required thrust and torque to keep the drone flying and maneuvering. The upward thrust force generated by the propellers is usually measured in pounds or grams. To keep the drone flying at a hover, the upward thrust needs to be equal to the weight of the drone (Figure 11.5(b)). Propellers can be made from wood, plastics, composites, and metallic materials.

11.3.1.3 Electronic Speed Controller

An Electronic Speed Controller (ESCs) is a device that allows a drone flight controller to control and adjust the speed of the BLDC motors. A signal from the flight controller causes the ESC to raise or lower the supply voltage to the motor as required, thus changing the speed of the propellers via motor. Different types of speed controllers are required for brushed DC motors and brushless DC motors. The brake setting, low voltage protection, and start-up modes can be programmed in the ESCs that are available in different varieties like Red Brick (Figure 11.5(c)) and Yellow Brick.

11.3.1.4 Frames

The diagonal distance between the motors determines the size of the frames and propellers for a specific Quadcopter configuration. However, there is no thumb rule for the size of the frames, and the most commonly used dimensions vary from 180 mm to 800 mm according to the various types and functions of the drones. The frame is the part that holds all the parts of the drone and mainly it supports the motors and electronic accessories and prevents them from excessive vibrations. Frames can be made of composites (carbon/glass fiber, high-density polyethylene, etc.), and aluminum alloys as well (Figure 11.5(d)). The selection of frame material and size should be made precisely as per the strength-to-weight ratio requirements of agriculture drones [11].

11.3.1.5 Li-Po Battery

A lithium polymer battery or lithium-ion polymer battery is a rechargeable battery of lithium-ion technology using a polymer electrolyte instead of a liquid electrolyte (Figure 11.5(e)). Based on the power-to-weight (P/W) ratio to be produced, range, and endurance of the drones, a specific type of battery with maximum C-rating should be selected for the sprayer drones [19]. The autonomy of the drones increases with the addition of batteries that helps to achieve rapid response during high power demands. Li-Po batteries are competent to deliver the Specific power of 2800 W/kg and the Energy density of 300 Wh/m^2 to accomplish the given agriculture mission of UAVs. A block diagram of Quadcopter arrangement is shown (Figure 11.6) with all necessary components.

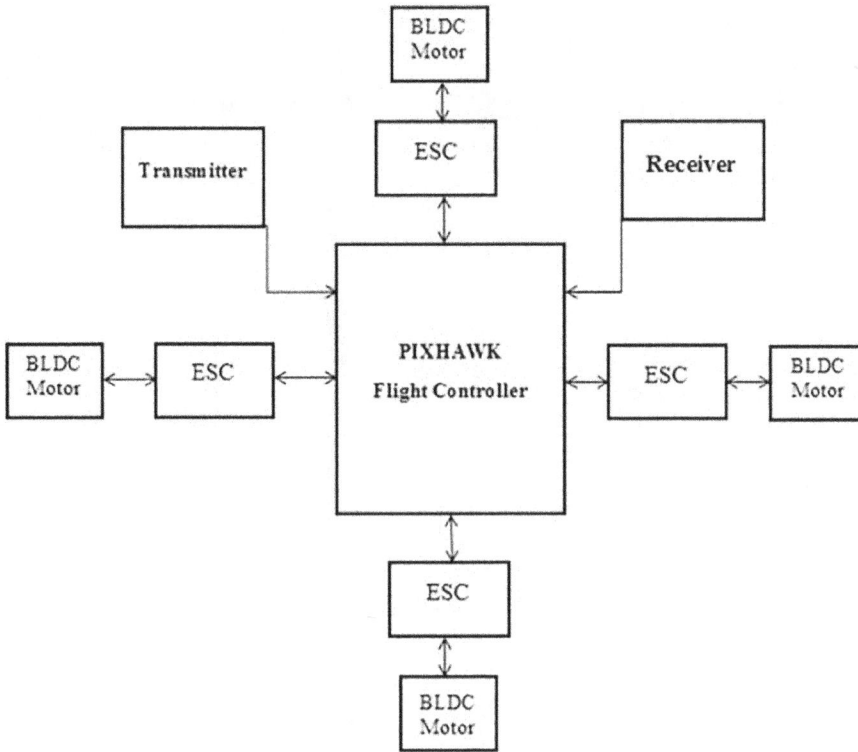

FIGURE 11.6
Block diagram of Quadcopter arrangement.

11.3.1.6 Transmitter and Receiver

The radio control system in the drones is made up of two elements, namely the transmitter and receiver. The transmitter reads the control stick inputs given by the operator or pilot and transmits the commands through a set radio frequency to the radio receiver, which is remotely controlled and fixed inside the drone. Once the receiver obtains the information, it will be passed to the drone's flight controller unit, which controls the stability and performance characteristics of the drone. Generally, a radio control system will have four separate channels for each direction on the sticks along with a few extra ones for any auxiliary switches. Transmitters are of different variants like 6 channel (6CH), 4 channel (4 CH), etc. (Figure 11.5(f)).

11.3.1.7 Flight Controller

A flight controller (FC) is a small circuit board with a range of sensors to perceive the user commands of varying complexity. Further, its function is to optimize and control the RPM of each motor in response to the input commands. A command from the pilot for the multi-rotor system moves forward and is fed into the flight controller that determines how to manipulate the motor's performance for various flight conditions. Simple gyroscopes are used for retaining the orientation and GPS systems are employed for auto-pilot or failsafe purposes. Many modern FCs allow for different flight modes that are selectable

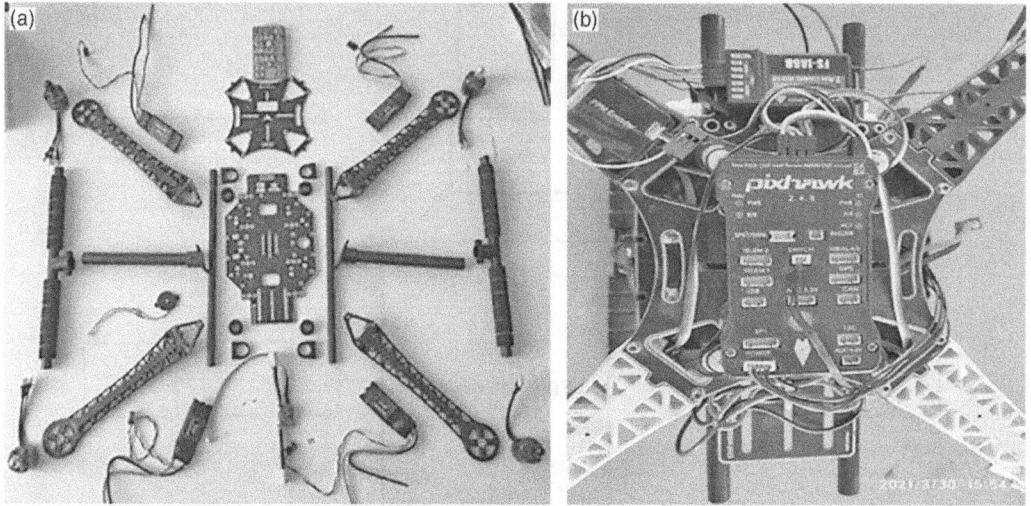

FIGURE 11.7
Quadcopter component views: (a) before assembly, (b) after assembly.

using a transmitter switch. An example of a typical three-position setup is the GPS lock mode, a self-leveling mode, and a manual mode. Different settings can be applied through intelligent electronics and software interface to each profile for achieving required flight characteristics [20]. Quadcopter components view is shown (Figure 11.7) in two phases before assembly and after assembly.

11.3.2 Drone Components with Specification

As discussed in the introduction part of this chapter, there are several types of drones available for agriculture purposes, such as spraying drones, NVDI drones, seeding drones, and surveillance drones. These drones can be made either fully or partially autonomous depending on the field mapping and other requirements in addition to the capital investment constraints to improve productivity. The presented Quadcopter sprayer system is a basic type multirotor model used for agriculture purposes, and high-performance UAVs can be selected for large-scale applications [19]. Table 11.2 provides the specifications of

TABLE 11.2

Quadcopter Components with Specifications

Sl. No.	Component Names	Specifications
1	Four BLDC motors	980 kV
2	Four propellers	10:4.5
3	Electronic speed controller (ESC)	Red brick – 30 A
4	Li-Po battery	4200 mAh
5	Flight control board	Pixhawk 4
6	Transmitter with receiver	Fly Sky 6 CH
7	Frame	Material – carbon fiber
8	Spraying system	Self-developed

different components used in the sprayer drones to offer some basic insights about the proposed design.

11.4 Variable Area Nozzle Sprayer

Agriculture spraying is the primary takeaway regarding the use of drone technology for effective farming and PA. Sprayer and plant-assurance drones are the new devices accessible to farmers which can be utilized to apply pesticides to small land zones and acreages. This sort of drone sprayer could get into the lands that are either excessively wet or, in any case, difficult to reach by the farmers. In addition, the people involved in such activities are removed from the spraying operations, which can greatly help to reduce the toxic effects on humans. The disadvantages involved in the sprayer drones are that they can carry only 8 to 10 liters of liquid with a flight endurance of 20 to 30 minutes. However, these problems have turned into opportunities for small and medium enterprises (SME) in agriculture to deliver sustainable agriculture through sprayer drones. Particularly, if the crops are cultivated adjacent to hilly areas where the complex varying terrains exist, then the shower of pesticides with adequate plant coverage is extremely difficult by manual spraying operations.

The sprayer drones use ultrasonic echoing devices and lasers to adjust their altitude with respect to the changes in topography and geography [21]. Their ability to scan and modulate their distance from the ground enables them to spray the correct amount of the desired liquid evenly in real-time. This results in increased efficiency since the amount of water infiltrating into groundwater is also limited. Drone spraying has also been proved to be a much more effective mechanism by recent studies than any other existing method. The spraying system with a VAN consists of a tank, spray nozzle, actuator, and pump. This auxiliary system can be installed along with the Quadcopter for spraying pesticides or fertilizers on agricultural lands without any additional regulatory requirements.

The raw data composed by the drones would be translated into convenient and coherent information to reduce the cost of spraying by one-third as compared with existing methods. It is well-known that the nozzle is a critical part of any sprayer and it performs the following functions [16]: (i) regulate flow, (ii) atomize the mixture into droplets, and (iii) disperse the spray in a desirable pattern. In any spraying system, Nozzles determine the rate of pesticide distribution at a particular pressure, forward speed, and nozzle spacing [11]. Nozzles are made from different types of materials. The most commonly used materials are brass, plastic, nylon, stainless steel, hardened stainless steel, and ceramics.

11.4.1 Type of Nozzles Used in the Spraying System

Spraying system performance mainly depends on the type of nozzle used [4]. There are four different variations of nozzles used for agriculture applications: (i) flat fan, (ii) hollow cone, (iii) full cone, and (iv) flooding fan. Different nozzles and their droplet size is shown (Figure 11.8)

The performance of the sprayer can be studied or evaluated using various parameters such as spray rate, droplet size, adhesion rate of spray particles, and scattering characteristics. Here, the spray rate helps to determine the area that can be covered per hour and the variation of

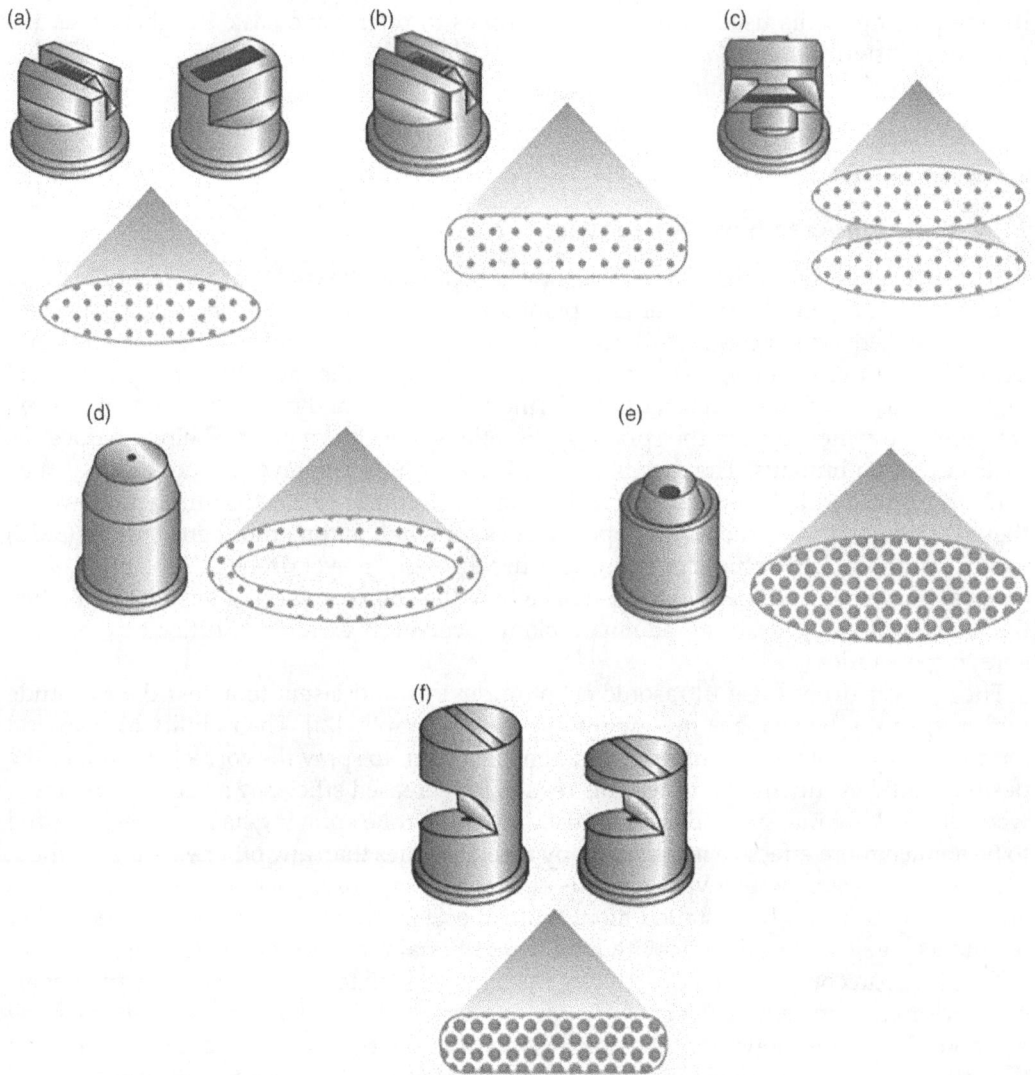

FIGURE 11.8
Different nozzles and their droplet size: (a) flat fan, (b) even flat fan, (c) twin orifice flat fan, (d) hollow cone, (e) full cone, (f) flood fan.

flow pattern as per requirement is fixed according to the droplet size (in microns). Adhesion rate is the strength of the bond between a spraying element and the application surface. Further, the efficiency of the sprayer is measured by the scattering characteristics.

11.4.2 Working Principle of Sprayer System

As the drone is operated over the agricultural lands, it captures the data about the zones of interest in the field, which require additional fertilizers or pesticides using sensors, infrared cameras, multispectral cameras, etc. The data obtained through these systems will be processed by the IPS to enable the IoT sensors to customize the amount of fertilizers or pesticides to be sprayed over the affected areas. The required variation in flow rate can be achieved by a VAN spraying system. It consists of a storage tank, water pump, flow lines, regulator, splitter, BLDC motor, ESC, and nozzles. The water pump is operated by the BLDC motor with ESC instead of DC motors to reduce the weight of the entire system. The VAN has three different flow ranges about 0.16 L/min, 0.32 L/min, and 0.54 L/min, respectively. Then, according to the data obtained from the different sensors, it can be customized by the operator to achieve a balanced distribution of pesticides or fertilizers for different crops based on their growth and requirement. Block diagram with various components of a sprayer system is shown (Figure 11.9).

11.4.3 Performance of VAN System

Spray-drop size is one of the most important factors which affect the drift and it varies from fine to extremely coarse. Spraying with minimal drift and adequate coverage are essential factors that influence herbicide performance. Spray drift is the physical movement of spray particles through the air from the application site to an area where the treatment is not intended, usually resulting in non-target plant injury. Fine-to-medium size droplets are desirable when applying insecticides and fungicides, because they usually provide better coverage. However, the fine droplets are difficult to make deposition on the target, and hence it remains airborne and drifts long distances because of their small, lightweight size. Therefore, coarse to extremely coarse are the preferred droplet size categories, which are idle to be used with VAN [4].

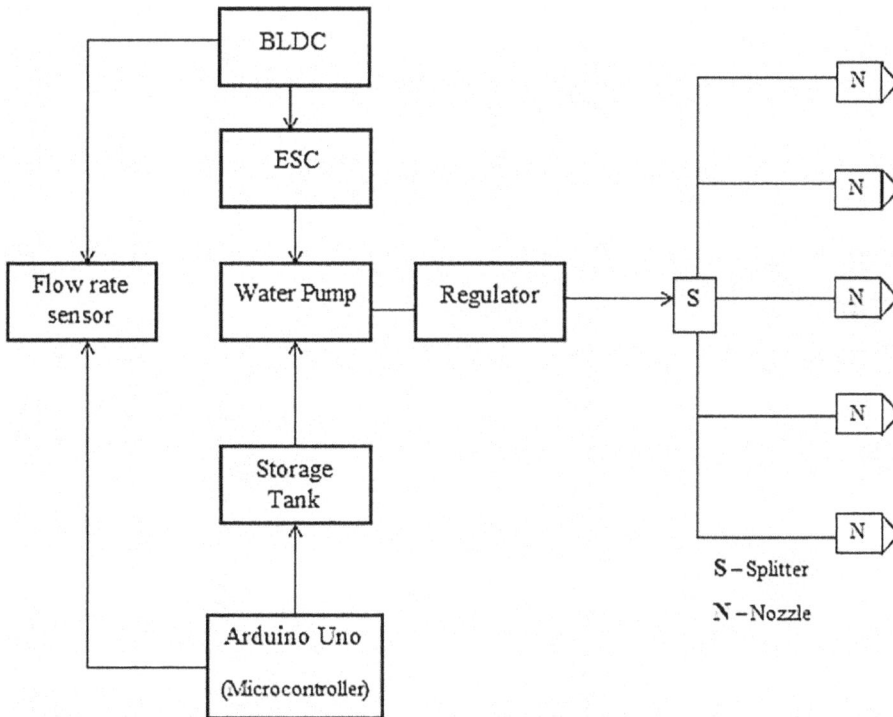

FIGURE 11.9
Block diagram of the sprayer system.

The different factors which would affect the size of a droplet in the nozzle flow are as follows: (i) pressure, (ii) spray pattern type, (iii) spray angle, (iv) nozzle type, (v) specific gravity of the fluid, and (vi) viscosity and surface tension. If the size of the spray droplet is greater than 225 microns, then the drift is less. From the above graph, the droplet size of the full cone and VAN is greater, which will reduce spray particle drift and increase the spraying efficiency (Figure 11.10(a)). Similarly, spray angles have an inverse effect

(a)

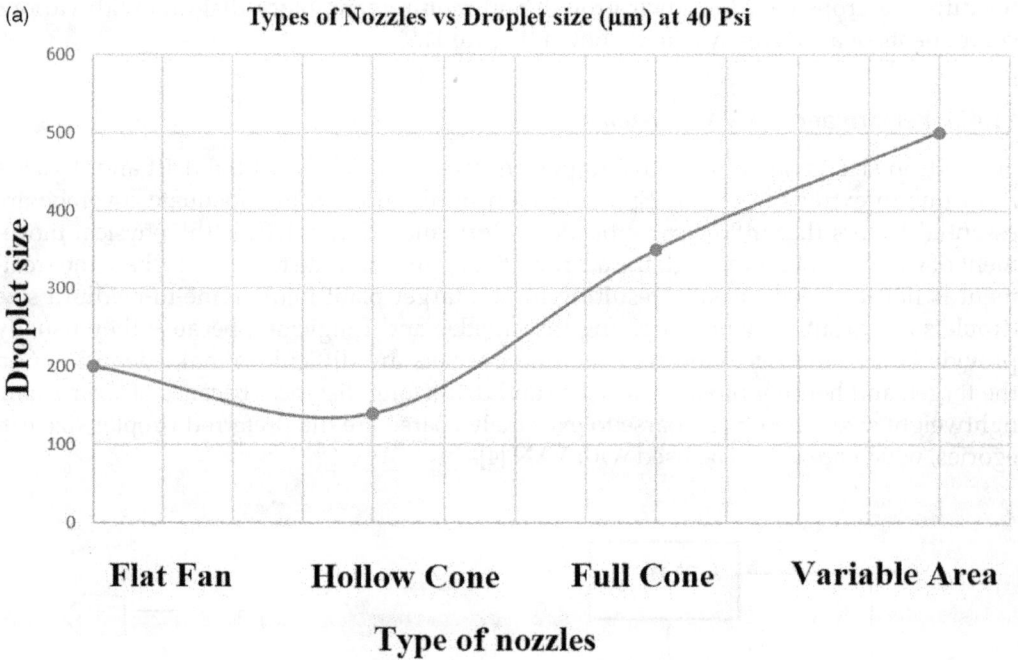

(b)

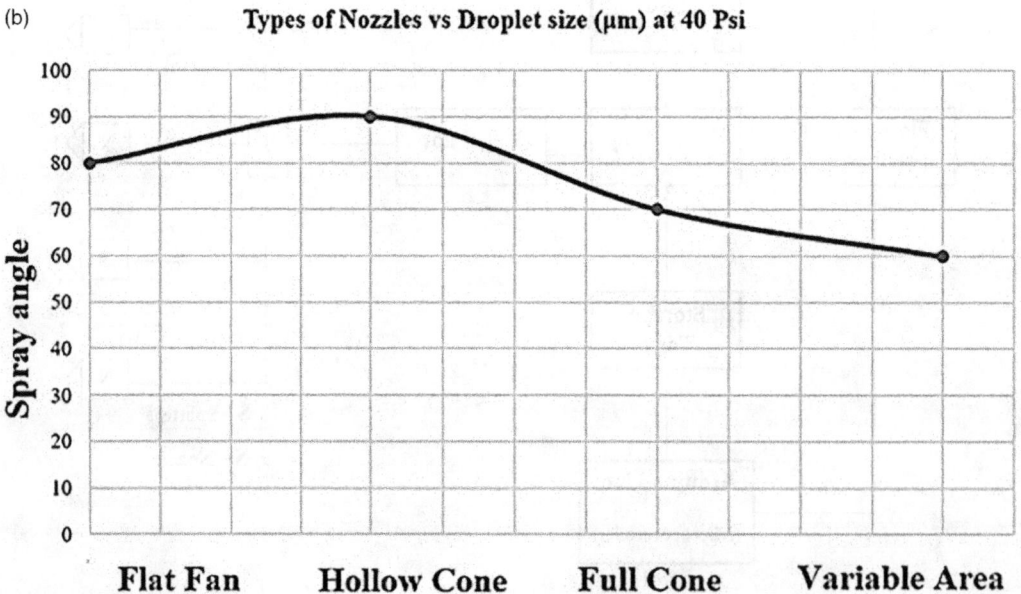

FIGURE 11.10
Performance of VAN Sprayer. (a) Types of nozzles Vs droplet size and (b) spray angle.

on the droplet size. In the spraying drones, an increase in the spray angle reduces the droplet size and vice versa. From the above Figure 11.10(b), when the spray angle for full cone and VAN is minimum, it helps to obtain a large droplet size for better spraying performance.

11.4.4 Challenges for Using VAN in Agriculture

The VAN usage for pesticide spraying to achieve the PA targets has several challenges. Particularly, the height at which the droplets are released, droplet size and the local field wind patterns would affect the process through air dispersion, thus resulting in the wastage of spray chemicals. The crops may appear alike from a specific height but their pesticide requirement could be different and it will affect the efficiency of the system incorporated. Hence, IoT-enabled sensors should be integrated to monitor and communicate the crop health data to the master computer in addition to the live weather statistics. Moreover, to customize the area ratio of the VAN according to the crop requirements, skilled operators are needed. Hence, more autonomous operations are needed with AI-enabled modules to reduce the manpower heads and more profit per operational cycle.

11.5 Smart Farming through Internet of Things

IoT is a promising innovation that provides an efficient and dependable solution toward the modernization of works related to farming. IoT-based solutions are being developed to maintain and monitor agricultural lands with minimal human involvement [15]. IoT sensors are capable of providing information about agriculture fields in alignment with AI, which are essential for farmers to plan their activities to increase production [16]. Five inspirational strategies of IoT that enable PA in various aspects are summarized below.

1. *Data collected by smart agriculture sensors*: The data collected by smart IoT sensors (weather conditions, soil quality, crop growth progress, etc.) can be used to track the state of the Agribusiness in terms of staff performance and equipment efficiency, etc.

2. *Better control over the internal processes*: The ability to forecast the output of production based on the crop health data allows the farmers to plan for better product distribution. If a farmer knows the exact amount of crops going to be harvested, then it is easy to deal with different stakeholders to promote the product with a minimum guaranteed price.

3. *Cost management and waste reduction*: The UAV applications in agriculture have been increasing day by day because of the autonomy delivered by IoT and fifth Generation (5G) technologies [22]. Hence, the IPS could easily identify anomalies in crop growth or livestock health to mitigate the risks of losing their yield.

4. *Increased business efficiency through process automation*: By using smart IoT-enabled devices, farmers can automate multiple PA processes across the production cycle (e.g., irrigation, fertilizing, or pest control and drone pollination) [23].

5. *Enhanced product quality and volumes*: Better control over the production process is possible through PA and higher standards of crop quality, and growth capacity can also be maintained through automation.

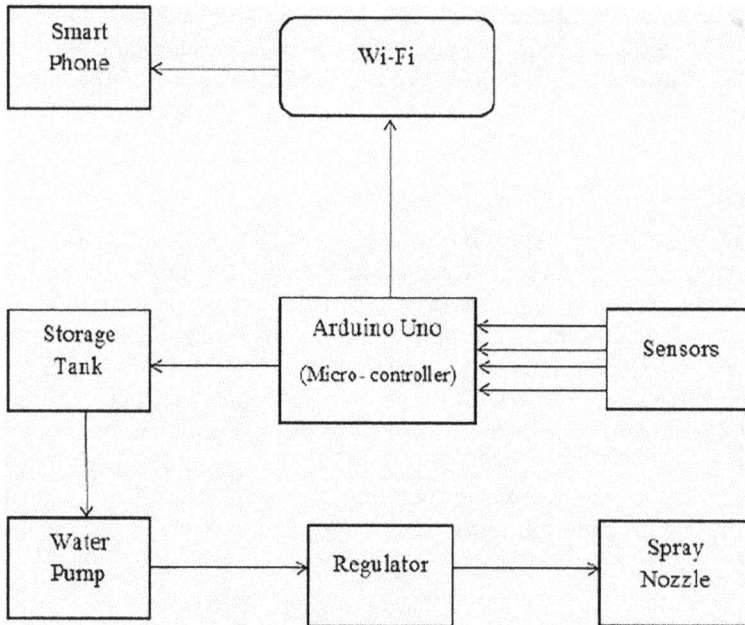

FIGURE 11.11
Block diagram of IoT-based VAN sprayer.

An IoT-based VAN sprayer consists of several varieties of sensors to actuate the micro-controllers based on the field of application. Basically, the medium-weight UAV segment (25 kg ≤ W ≤ 150 kg) is widely used for agricultural activities that works based on the vigor maps developed by NDVI data. The spatial and temporal characteristics of soil data and the corresponding fertilizer usage are the significant information obtained through the sensor networks. The primary components of IoT-based VAN sprayers are illustrated in the block diagram presented in Figure 11.11.

11.5.1 Arduino Board

Arduino is an open-source programmable circuit board. This board contains a micro-controller that can be programmed to sense and control the objects in the physical world. By responding to sensors and inputs, the Arduino can interact with a large array of outputs such as LEDs, motors, and displays [7]. The inputs for the microcontroller are received from different sensors like temperature, moisture sensors, multispectral camera images, etc.

11.5.2 Sensors

There are a variety of sensors used to provide the necessary information/input to the microcontrollers. They are: (i) location-based Sensors, (ii) electrochemical sensors, (iii) temperature or humidity sensor, (iv) optical sensor, and (v) multispectral sensors. Here, location-based sensors are used for locating the different areas and spots in the agriculture fields [8, 24]. Normally, GPS receivers are used for finding the longitude and latitude of a particular point on the earth's surface with the help of a GPS satellite network. These smart location sensors play an important role in PA by pointing out the location in the fields for monitoring the growing crops toward effective watering, fertilization, and treatment of weeds.

Electrochemical sensors are used to extract a specific composition from a particular biological sample such as plants, soil, etc. [17]. In smart agriculture, these sensors are generally used to detect the pH levels and soil nutrient levels where the sensor electrodes detect specific ions within a soil. Moreover, temperature and humidity are the most important weather factors which directly affect the health and growth of all types of crops. Correct measurement of these environmental factors would be helping the farmers to adjust the quantity of fertilizer and water [21]. Temperature and humidity sensors are available in wireless-enabled and battery-operated types according to the level of autonomy required.

Optical sensors work on the principle of converting light rays into an electrical signal [24]. Several types of optical sensors (such as RGB camera, converted near-infrared camera, six-band multispectral camera, high spectral resolution spectrometer) have been used in UAVs for PA-related applications [25]. Similarly, Multi-spectral sensors are extremely appropriate for UAV-based agricultural analytics. These sensors capture images with exceptional spatial resolution and also possess the capability to determine reflectance in near-infrared [13]. The collection of multispectral data is an absolute necessity for performing analysis of crop health. The multiple bands of light enable the researchers to conduct precision analytic studies and offer precise insights on plant vigor, canopy cover, leaf, and various other parts. The absence of such multispectral data would make the early detection of plant diseases, weeds, pests, and calculation of vegetative biomass almost impossible.

11.5.3 Advantages of IoT-Enabled Spraying System

a. *Easy Control*: Intelligent flight with autopilot according to real-time environmental data [26].

b. *Adaptability*: It can be operated over mountains, hilly terrains, and highly wetted regions.

c. *Foldable*: Foldable frames, easy to transport and versatile to handle by the farmers.

d. *Maintenance*: Modularized design, easy to carry out disassembly and inspection.

e. *Cost Control*: Optimal usage of fertilizers/pesticides and reduced labor costs.

f. *Saves Water*: Less water requirement for spraying activities.

g. *Productivity*: Increase in production/yield through detailed coverage of spraying.

11.6 Agriculture Drone Usage Statistics for Smart Farming

The IoT-based agriculture policies play a crucial role in the growth of smart PA activities across the globe. A substantial revision in the drone usage regulations for the PA activities is the need of the hour in many countries to achieve the goals in food production. Specifically, Internet-of-Ag-Things (IoAT) is an emerging project in countries like the United States of America (USA) with state-of-the-art sensing technologies. However, the investment toward expensive instrumentation is a key deciding factor for the SMEs to move forward with PA with drone applications. Hence, the trade-off studies and advantages of potential applications should be well understood before deploying drones for crop monitoring and control activities. In countries like Brazil, South Africa, and Australia, during 2014 itself several flight trials were conducted with UAVs to ascertain the plant health before and after the application of organic nutrition. As the agriculture business is a labor-intensive industry, the outdated legislative systems present in many countries are the real barrier to deploying spraying and monitoring drones on a large scale. The current usage statistics of agriculture spraying drones in different countries are discussed below, with some key insights to achieve the best results.

11.6.1 Smart Farming in Asian Countries

China's agriculture drone revolution in the Agriculture Ecosystem makes up the largest supply of civilian drones in the world. Globally, it was estimated that approximately 70% of all civilian drones were sold by Chinese manufacturers in the year 2017. Among the commercial drones operated in China, it is estimated that about 13% to 17% of drones are deployed for agricultural purposes [27]. Narrow Band-IoT (NB-IoT) enabled with Low Power Wide Area (LPWA) technology has transformed the PA culture of China through wide coverage and a large number of IoT devices. Meanwhile, in India, 40 drone start-ups are engaged in enhancing the technological standards and reducing the prices of agriculture drones via the "Make in India" initiative to make them affordable and prevalent among educated and uneducated farmers. Farmers of Dahanu-Palghar tribal villages of Maharashtra state have learned to use drones for organic farming, fish farming, crop rotation, bio-control, hydroponics, and biowaste management, on their orchards and farms [9].

Israel has a strong history of high-tech agriculture innovations and it invented the concept of drip irrigation during the 1970s [28]. It is a hot spot for agriculture technology start-ups and PA research and the current trends are summarized as follows:

- Overall, about 30% of farmers use Global Navigation Satellite System (GNSS) auto steer and 75%–85% of large companies use the sprayer drone technology.
- About 10% of existing sprayers and most new sprayers use GNSS boom control.
- About 50% of farmers use satellite images and drone imagery is used at more than 30 farms.
- Variable-rate N_2 fertilizer application is used by 5% of farmers. A few farmers use site-specific weed management and precision-guided cultivation.

Drones now serve to fulfill the objectives of PA in Israel by a variety of methods: monitoring and pesticide use, soil surveys, monitoring irrigation problems, identifying lack of uniformity in the field, as well as fruit picking. But overall adoption of PA technology, except for soil moisture sensors for irrigation management, seems similar to that of Europe [28].

11.6.2 Smart Farming in European Countries

Most of the farmers and SMEs in Germany prefer to use agriculture drones for effective farming if the area of vegetation exceeds 100 hectares. Interestingly, one-tenth of German farmers use drones for their agricultural operations, in which 4% fly their own drones and 6% depend on external drone service providers (DSPs). Similarly, Switzerland operates at the global forefront in the development of drone technologies. It applies in particular to the fields of sensor technology, drone control systems, and data processing, in which the country is setting new standards. Even though the use of drones in Swiss farming is currently still limited, the technology holds significant potential for the country's diverse and highly structured agriculture [29].

11.7 Current Status of Regulations and Provisions for Agriculture Drones

Rules and regulations for the use of drones vary across the countries and four major elements are considered in this regard. They are (i) drone registration, (ii) airspace, (iii) insurance, and (iv) licensed operator (pilot). The requirements of these four elements vary based on the drone mass, altitude, application, and level of pilot license. Considering the variation in these four components of regulations across the countries, six broad approaches to national commercial drone regulations become apparent as follows: [1]

a. *Outright ban:* Countries do not allow drones at all for commercial use without license.

b. *Effective ban:* Countries should have a formal process for commercial drone licensing, but licenses do not appear to have been approved before they comply with all requirements.

c. *Requirement for constant Visual Line of Sight (VLOS):* A drone must be operated within the pilot's VLOS, thus limiting potential range.

d. *Experimental uses of beyond visual line of sight (BVLOS):* With certain restrictions and pilot ratings, certain exceptions to the constant VLOS requirement are possible.

e. *Permissive:* Countries have enacted relatively unrestricted legislation in commercial drone use. These countries have a body of regulations that may give operational guidelines or require licensing, registration, and insurance upon following proper straightforward procedures.

f. *Wait-and-see:* Countries have enacted little, if any, drone-related legislation and monitor the outcomes of other countries' regulations.

Hence, it is perceived that certain unique approaches and provisions are needed from different stakeholders around the globe to accelerate the usage of agriculture drones to ensure smart PA at all levels.

11.8 Conclusions

Quadcopters with IoT-enabled systems would massively transform the farmer's current way of doing agricultural activities in terms of crop health monitoring, soil moisture data, fertilizer usage, crop damage assessment, etc. It is inferred that the payload capacity of Agriculture drones is continually increasing through state-of-the-art propulsive system designs such as with Hexacopter, Octo-rotor designs to expand their usage. Multi-rotor configuration also provides more space for the higher capacity nozzle sprayer systems, which will be very helpful for covering large farmlands in a limited time period as compared with the Quadcopter designs. However, the selection of appropriate dimensions of the drone is determined based on the budget, regulatory provisions, and wireless communication modes available in the region.

Farmers can utilize the VAN spraying system effectively to shower the pesticides for various crops at different stages of their growth. As the splashing can be done from variable heights, ecological contamination is diminished by achieving PA farming with minimum wastage. The objective of such an IoT-enabled spraying system could also reduce the exposure of profoundly harmful pesticides to the people around. One can adapt the hovering mode to customize the droplet size of the pesticides and spraying angle to cover large regions in a short time, according to the crop monitoring data. It is more significant for the farmers as well as consumers to ensure the optimal wellbeing of the yields to gain more ideal opportunities for the promotion of their quality products across the nations.

11.8.1 Future Scope for Drones in Agriculture

Many autonomous drones are still in the testing and development phase that are to be deployed for agriculture activities in the near future. One of the most publicized uses is the pollinating drone technology. Researchers in the Netherlands and Japan are developing small drones that are capable of pollinating plants without causing any damage to them. The next step is to create autonomous pollinating drones that will work and monitor crop health without any constant instructions from the operators [25].

Another remarkable development in drone technology involves machine learning techniques with advanced IPS. Improved AI in drones is important to be able to make them more useful to small farmers in developing nations. Current drone technologies are more effective for monitoring the well-known crops which are planted in the large field patterns in different continents. Drone monitoring programs will have a hard time while recognizing the areas with increased crop diversity, less well-known products, and grains that look similar throughout their growth stages and so are less effective in monitoring the crop growth and health. More intensive research and studies are needed to train AI systems with large datasets to recognize the less common crops and more diverse planting patterns [27].

References

1. Daniel Frona, Janos Szenderak and Monika Harangi-Rakos, "The Challenge of Feeding the World," Sustainability, vol. 11, pp. 1–18, 2009. https://doi.org/10.3390/su11205816
2. The Future of Food and Agriculture: Trends and Challenges. http://www.fao.org/3/i6583e/i6583e.pdf
3. Nadia Delavarpour, Cengiz Koparan, John Nowatzki, Sreekala Bajwa and Xin Sun, "Technical Study on UAV Characteristics for Precision Agriculture Applications and Associated Practical Challenges," Remote sensing, vol. 13, pp. 1–25, 2021. https://doi.org/10.3390/rs13061204
4. Arnab Kumar Saha, Jayeeta Saha, Radhika Ray, Sachet Sircar, Subhojit Dutta, Soummyo Priyo Chattopadhya and Himadri Nath Saha, "IOT-Based Drone for Improvement of Crop Quality in Agricultural Field," IEEE 8th Annual Computing and Communication Workshop and Conference (CCWC), 2018.

5. Hongbin Dou, Chengliang Zhang, Lei Li, Guangfa Hao, Bofeng Ding, Weike Gong and Panlin Huang, "Application of Variable Spray Technology in Agriculture," IOP Conf. Series: Earth and Environmental Science, 2018. https://doi.org/10.1088/1755-1315/186/5/012007

6. Vijai Singh, Namita Sharma and Shikha Singh, "A Review of Imaging Techniques for Plant Disease Detection," Artificial intelligence in agriculture, science direct publications, vol. 4, pp. 229–242, 2020. https://doi.org/10.1016/j.aiia.2020.10.002

7. H. Pathak, G.A.K. Kumar, S.D. Mohapatra, B.B. Gaikwad and J. Rane, (2020), "Use of Drones in Agriculture: Potentials, Problems and Policy Needs", Publication no. 300, ICAR-NIASM, pp 13+iv.

8. E-Agriculture in action: Drones for Agriculture, Food and Agriculture Organization of the United Nations and International Telecommunication Union Bangkok, 2018.

9. Raghav Pimplapure and Yash Kshirsagar et al. "Design and Analysis of Pesticides and Insecticide Spraying Machine," International journal of innovations in engineering and science, vol. 2, no. 5, pp. 33–35, 2017.

10. Giovanni Ludeno, Ilaria Catapano, Alfredo Renga, Amedeo Rodi Vetrella, Giancarmine Fasano and Francesco Soldovieri, "Assessment of a Micro-UAV System for Microwave Tomography Radar Imaging," Remote sensing of environment, vol. 212, pp. 90–102, 2018.

11. Panagiotis Radoglou-Grammatikis, Panagiotis Sarigiannidis, Thomas Lagkas and Ioannis Moscholios, "A Compilation of UAV Applications for Precision Agriculture," Computer networks, vol. 172, pp. 107148, 2020.

12. John W. Slocombe and Ajay Sharda, "Agricultural Spray Nozzles: Selection and Sizing," K-State Research and Extension, Kansas State University, March 2015.

13. https://www.dailydot.com/debug/prohawk-uav-bird-chasing-drone/, Accessed on 25 March 2021.

14. Muhammad Shoaib Farooq, Shamyla Riaz, Adnan Abid, Kamran Abid and Muhammad Azharnaeem, "A Survey on the Role of IoT in Agriculture for the Implementation of Smart Farming," Special section on new technologies for smart farming 4.0: Research challenges and opportunities, IEE access, vol. 7, pp. 15637–15671, 2019.

15. Drones for Precision Agriculture: Case Study Brazil Drone Project July 2018, v1r4: https://www.qualcomm.com/, Accessed on 31 March 2021.

16. Muhammed Enes Bayrakdar, "Employing Sensor Network Based Opportunistic Spectrum Utilization for Agricultural Monitoring," Sustainable computing: Informatics and systems, vol. 27, pp 1–10, 2020.

17. https://www.makerspaces.com/wp-content/uploads/2017/02/Arduino-For-Beginners-REV2.pdf, Accessed on 25 March 2021.

18. https://www.farmpractices.com/agricultural-sprayers, Accessed on 15 April 2021.

19. Mohamed Nadir Boukoberinea, Zhibin Zhoub and Mohamed Benbouzid, "A Critical Review on Unmanned Aerial Vehicles Power Supply and Energy Management: Solutions, Strategies, and Prospects," Applied energy, vol. 255, pp. 1–22, 113823, 2019.

20. Vemema Kangunde, Rodrigo S. Jamisola Jr. and Emmanuel K. Theophilus, "A Review on Drones Controlled in Real-time," International journal of dynamics and control, 2021. https://doi.org/10.1007/s40435-020-00737-5

21. Seon-Woo Lee and K. Mase, "Activity and location recognition using wearable sensors," in IEEE Pervasive Computing, vol. 1, no. 3, pp. 24–32, July-Sept. 2002, doi: 10.1109/MPRV.2002.1037719.

22. Abdul Salam, Mehmet C. Vuran, Rigoberto Wong and Suat Irmak, "Internet of things in agricultural innovation and security," Internet of Things for Sustainable Community Development. Springer, 2020, pp. 71–112.

23. Ritesh Kumar Singh, Michiel Aernouts, Mats De Meyer, Maarten Weyn and Rafael Berkvens, "Leveraging LoRaWAN Technology for Precision Agriculture in Greenhouses," Sensors, vol. 20, no. 7, p. 1827, 2020.

24. Jorge Alvar-Beltrán, Carolina Fabbri, Leonardo Verdi, Stefania Truschi, Anna Dalla Marta and Simone Orlandini, "Testing Proximal Optical Sensors on Quinoa Growth and Development," Remote sensing, vol. 12, no. 12, p. 1958, 2020.

25. Stefanie K Von Bueren, Andreas Burkart, Andreas Hueni, Uwe Rascher, Mike P Tuohy and Ian J yule, "Deploying four optical UAV-based sensors over grassland: challenges and limitations," Biogeosciences, vol. 12, no. 1, pp. 163–175, 2015.
26. Fu Bing, "Research on the agriculture intelligent system based on IOT", In Proceedings of the 2012 International Conference on Image Analysis and Signal Processing, Zhejiang, China, 9–11 November; pp. 1–4, 2012.
27. Muhammad Shoaib Farooq, Shamyla Riaz, Adnan Abid, Tariq Umer and Yousaf Bin Zikria, "Role of IoT Technology in Agriculture: A Systematic Literature Review," Electronics, vol. 9, p. 319, 2020. https://doi.org/10.3390/electronics9020319
28. https://www.israelagri.com/?CategoryID=403&ArticleID=1861, Accessed on 10 April 2021.
29. https://www.croptracker.com/blog/drone-technology-in-agriculture.html, Accessed on 11 April 2021.

12

Standards and Protocols for Agro-IoT

S. Mythili
Bannari Amman Institute of Technology
Sathyamangalam, India

K. Nithya
Kongu Engineering College
Perundurai, India

M. Krishnamoorthi
Dr. N.G.P. Institute of Technology
Coimbatore, India

M. Kalamani
KPR Institute of Engineering and Technology
Coimbatore, India

CONTENTS

12.1 Introduction to Standards and Protocols in Agro-IoT 188
12.2 Agro-IoT Sensor Network 188
12.3 International Standards Support for Agro-IoT Communication 189
12.4 Communication Technology for Agro-IoT 191
    12.4.1 Bluetooth 191
    12.4.2 Zigbee 191
    12.4.3 Wi-Fi 193
    12.4.4 WiMAX 193
    12.4.5 LoRaWAN 194
    12.4.6 RFID 196
12.5 Performance-Measuring Factors 196
    12.5.1 Packet Loss Percentage 196
    12.5.2 Node Connectivity 197
    12.5.3 Transmission Throughput 197
    12.5.4 Scalability 197
    12.5.5 Interoperability 197
12.6 Comprehensive Overview of Communication Protocols 197
    12.6.1 MQTT 197
    12.6.2 DDS 199
    12.6.3 CoAP 200
    12.6.4 AMQP 201

DOI: 10.1201/9781003185413-12

12.7   Security Protocols Used to Secure the Network 203
      12.7.1   DTLS 203
      12.7.2   TCP 203
      12.7.3   UDP 203
12.8   IoT-Based Solution for Leaf Disease Detection Using Machine Learning
      Classifiers for Precision Agriculture 203
12.9   Implementation of Machine Learning Classifiers and Its Performance Study
      for Leaf Disease Detection in Agriculture Field 204
12.10   Real-Time Processing of Sensor Data Using IoT 206
12.11   Conclusion 207
References 207

12.1 Introduction to Standards and Protocols in Agro-IoT

In the agriculture industry the impact of the Internet of Things (IoT) plays a vital role in smartness in all spheres like surveillance, automation, and precision farming even without major human intervention. The realization of sensory architecture is implicitly defined by its inherent nature. The sensor-acquired data has to be transferred to different nodes connected to the network to reach the recipient side irrespective of the distance. This leads to a focus on standardization. Standard is defined as the specifications or rules to be followed to enable the effectiveness of communication systems that are deployed for diverse applications. The method of transmitting information in a huge network and interacting among the devices, even though it depends on different vendors and different communication technology, is possible with the help of a defined term "standard." It leads to enhanced characteristic natures like interoperability, security, accessibility, and scalability [1]. To perform the designed task the set of protocols are essential along with its standards for having control over information and signal level between the sender and receiver.

There are different aspects of technologies looking for minimal human intervention by maintaining appropriate standards. One such field developed for maintaining and monitoring the agriculture sector is called Agro–IoT [2]. It is purely based on different sensors and its configuration applied for smart farming [3]. The structure of protocols and standards leads to a systematic process of Agro-IoT network that ensures the attainment of the desired output. Figure 12.1 depicts the flow of standards and protocols and their necessity. The network setting and its process flow are initialized by the defined standard. The sustainability of the network depends on the subset functionalities like potentially executed tasks and communication flow involved in the network. The specific network message formatting is to be set between the communication devices at each point to validate the transmission flow for measuring the system accuracy.

12.2 Agro-IoT Sensor Network

The revolution of machine-to-machine communication plays a central role for the different kinds of features in application scenarios such as field management, crop monitoring, soil monitoring, wild animals attack monitoring, etc. Figure 12.2 illustrates that the primary objective of the sensor is to collect all the physical information of the desired application from the environment and convert it into digitized signal for further transmission process.

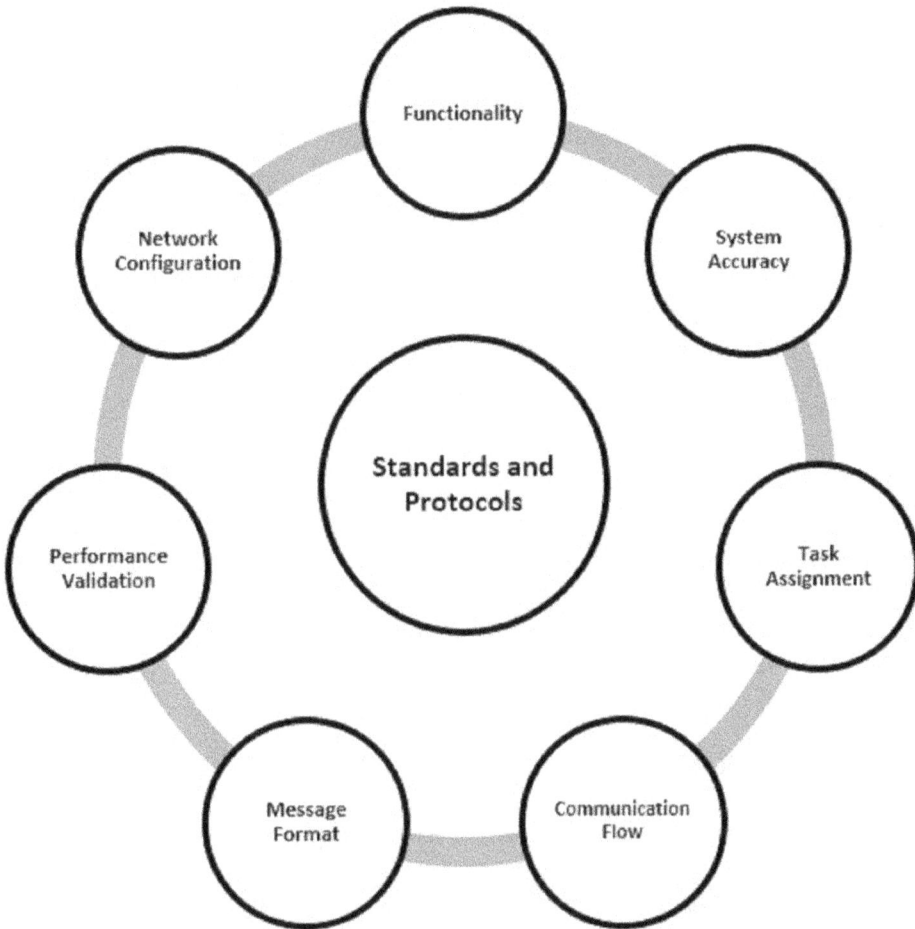

FIGURE 12.1
Importance of standards and protocols.

The digitized output signal will be strengthened with the help of a signal conditioning unit which becomes easier to detect the output. There are different types of sensors used for precision farming such as temperature sensor, humidity sensor, soil sensor, leaf wetness sensor, rainfall sensor, wind direction, and atmospheric pressure sensor.

12.3 International Standards Support for Agro-IoT Communication

The advancement of communication is by many professional societies embodied together to set a standard for communication through various protocol structures in the IoT network, as shown in Figure 12.3 and detailed in Table 12.1. The framework and adoption of various characteristics for carrying out the transmission and sharing the information between any devices is defined by various standards such as IEEE (Institute of Electrical and Electronics Engineers), IETE (Institution of Electronics and Telecommunication Engineers),

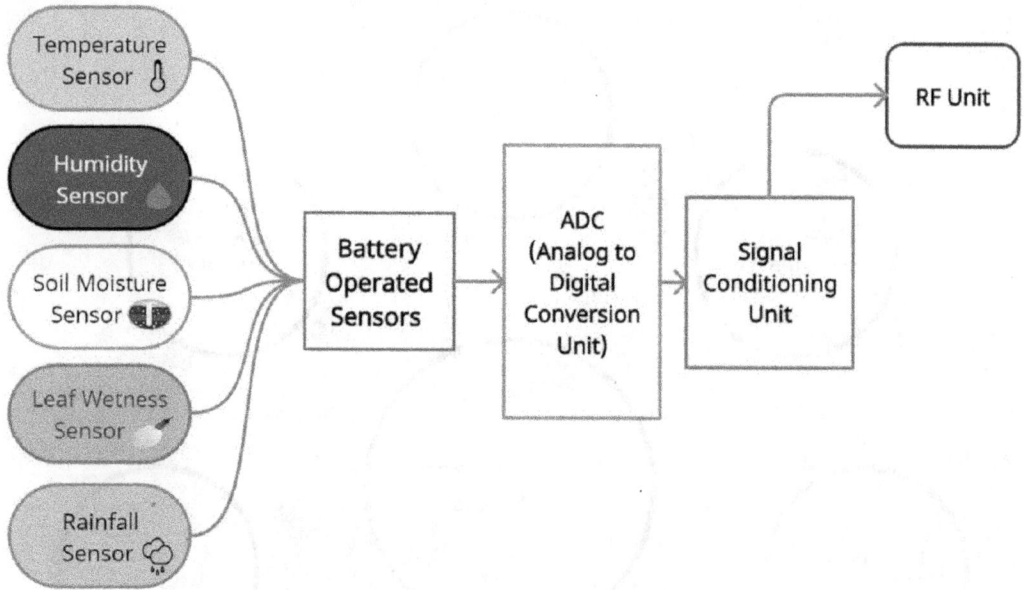

FIGURE 12.2
Block diagram of sensor module working.

IETF (Internet Engineering Task Force) and ISO (International Organization for Standardization). The regulatory parameters which follow the defined standards are Bandwidth allocation, Network range, Affordability, Device power consumption, Network latency, and Security. There are two categories in utilizing the spectrum resources one is licensed and the other is unlicensed. If the radio network is opting for the licensed spectrum, then there is no constraint for security provision and uninterrupted data transmission is possible even though there is a huge amount of data. It is applicable for Wide Area Network communication, whereas the unlicensed one is a bit controversial and useful for the Local Area Networks. The ISM band is free for Industrial, Scientific, and Medical applications, so it is useful for IoT-based Agricultural purposes.

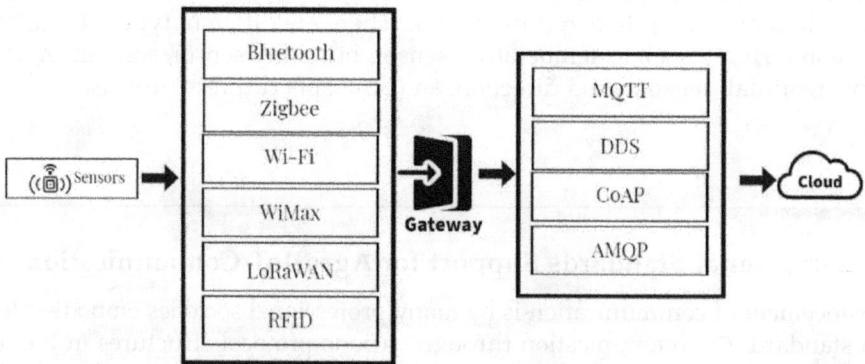

FIGURE 12.3
Overview of standards and protocols in IoT-based communication network.

12.4 Communication Technology for Agro-IoT

12.4.1 Bluetooth

There are many Agro-IoT sensors like Soil moisture sensors and RGB sensors to make the correct decision with the sensed data irrespective of the situation. It is possible with the help of Bluetooth communication module IEEE 802.15.1 for short-distance communication with the unlicensed ISM frequency band of 2.4 GHz [4]. To establish the connection among the nodes the inquiry paging procedure is being done. The communication range is about 10 m. The asymmetric data is being transferred with the data rate of 721 Kbps through the total bandwidth of 1 MB/s [5]. The transmitting method between transmitting end to receiving end followed in Bluetooth is Frequency hopping [6]. It is supportive of eight channels and provides the functional component of interoperability. For the fine-tuning of higher data rate, advanced security and low energy consumption, various standards have been developed such as 1.0,1.0 a, 1.0 b, 1.0 b + ce, 1.1, 1.2, 2.0 + edr, 2.1, 3.0 + hs, 4.0, 5. The autonomous ubiquitous communication using Bluetooth-enabled sensor network deployed for intelligent farming [7] and it is shown in Figure 12.4

12.4.2 Zigbee

For the necessity of low power consumption, the IEEE 802.15.4 protocol named Zigbee comes into the picture that supports the distance of 10 to 20 m in a communication network [8]. The Zigbee operates at three different frequencies such as 868 MHz, 902–928 MHz, and

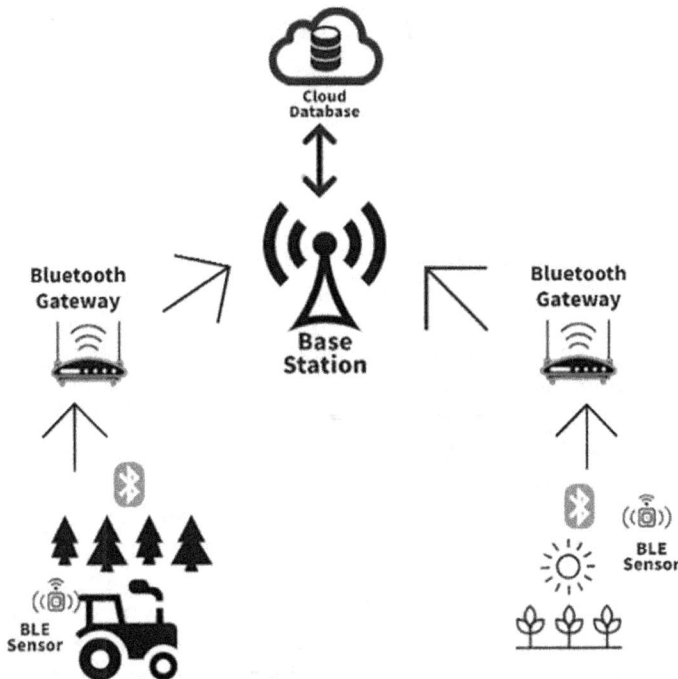

FIGURE 12.4
Bluetooth-enabled sensor communication.

FIGURE 12.5
Zigbee-enabled sensor communication.

2.45 GHz. The periodic data transmission takes place at the low data rate range of 40 kbps to 250 kbps. As shown in Figure 12.5, there are three different types of Zigbee devices: router, coordinator, and end device, which helps for precision farming. The Zigbee routers act as intermediate nodes which permit the data to pass over from one node to the other node [9]. The coordinator node handles the data storing and its transmission operation within the network. The end device has limited functions on the coordinator and router nodes which directly reduces the power consumption. All three ensure the reliable operation of the network functionality by concentrating on the power consumption of each node specifically. The sensor used here is the Soil sensor in which the air, water, salt, and soil minerals (potassium, magnesium, iron, calcium, sulfur) are the major constituents of soil. The volume of water content present in the soil is measured by a soil moisture sensor. Based on varying factors such as temperature, soil type, and geographical area, the measured soil moisture content will vary. It is useful for soil health monitoring, plant ecology, and plant disease forecasting and for increasing the crop quality in different plant growth stages.

12.4.3 Wi-Fi

In addition to these technologies, IEEE 802.11 Wi-Fi standard technology operates at 2.4 and 5 GHz frequency bands with the data rate of 1 Mbps to 6.75 Gbps. It helps to distribute the access points load to avoid the interruptions between the individually connected devices supported for 20–100 meters. Wi-Fi is one of the best supporting wireless protocols for Agro-IoT. Real-time weather forecast data update to the farmers is possible with the help of various wireless sensor nodes communicating through Wi-Fi will reach the end-user using the intermediate gateways as shown in Figure 12.6. There are various Wi-Fi standards such as 802.11 a, b, g, n, ac used for diverse applications. The varying characteristics of different standards are data rate, signal interference, cost, etc. The selection is made based on the data transfer requirements. The sensor used are: (i) Temperature sensor in which the periodical temperature variance is observed. It is helpful for remote ambient condition monitoring for harvesting. (ii) Measurement of moisture content and air temperature all these sensors will be helpful for the remote ambient condition monitoring for harvesting. The applications like drip irrigation are in need of accurate measurement of moisture content, which supports the growth of plants and indoor vegetation.

Wi-Fi HaLow (802.11ah) is the unlicensed band focused on the extension of network coverage by using relays and optimization in power consumption using the predefined busy/idle period. The contention is also getting reduced via station grouping. The Wi-Fi wavebands highly facilitate the interoperability feature.

12.4.4 WiMAX

IEEE 802.16 is the standard named WiMAX (Worldwide Interoperability for Microwave Access), which helps for long-distance communication up to 50 km [10]. Orthogonal

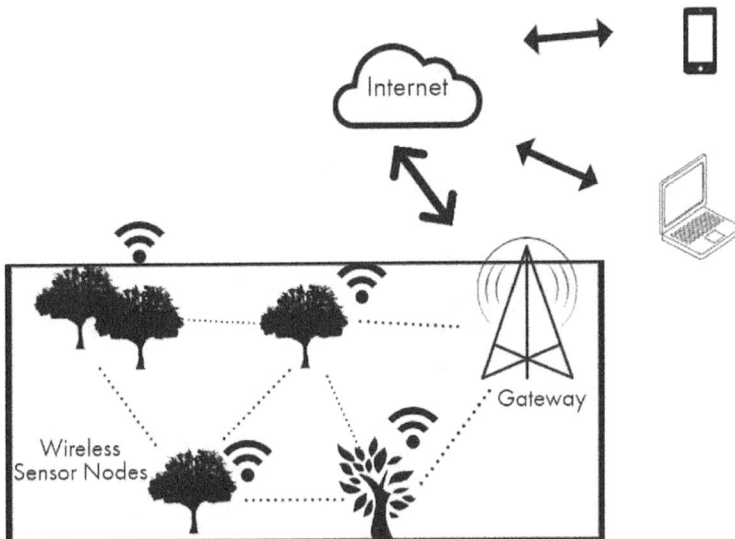

FIGURE 12.6
Wi-Fi-enabled sensor communication.

Frequency Division Multiplexing (OFDM) is the modulation concept involved in WiMAX, which enables multicarrier parallel transmission at a high data rate with better connectivity to ensure efficiency. The range of data rates from 1.5 Mbps to 1 Gbps is scalable under varying channel bandwidth. For the frequency of 10 MHz the data rate will be 25 Mbps in downlink and 6.7 Mbps in uplink using Time Division Duplexing (TDD) scheme. It also supports a very high data rate of 74 Mbps when operating at 20 MHz. This kind of wireless broadband will be a better alternative to achieve long-distance communication; thus it is a better suitable technology for crop area monitoring, as detailed in Figure 12.7.

12.4.5 LoRaWAN

For better interoperability features the Long-Range Wide Area Network (LoRaWAN R1.0) is used, which helps to achieve the minimum energy consumption for the multi-layer architecture of IoT scenarios [11]. It is a technology that supports a long-distance communication range of 10 to 20 km. LoRaWAN has a data rate from 0.3 kb/s to 50 kb/s with an operating frequency of 868 MHz and 900 MHz [12–14]. The sustained battery limit is up to 10 years which is very adaptable for IoT enabling networks. The end-to-end security provision using the AES algorithm is the best feature of LoRaWAN. The information of the leaf wetness sensor is being communicated among the nodes and the collected information reaches the receiver end through gateway collector communication using LoRAWAN as given in Figure 12.8. The sensors used are: (i) to measure the minute water

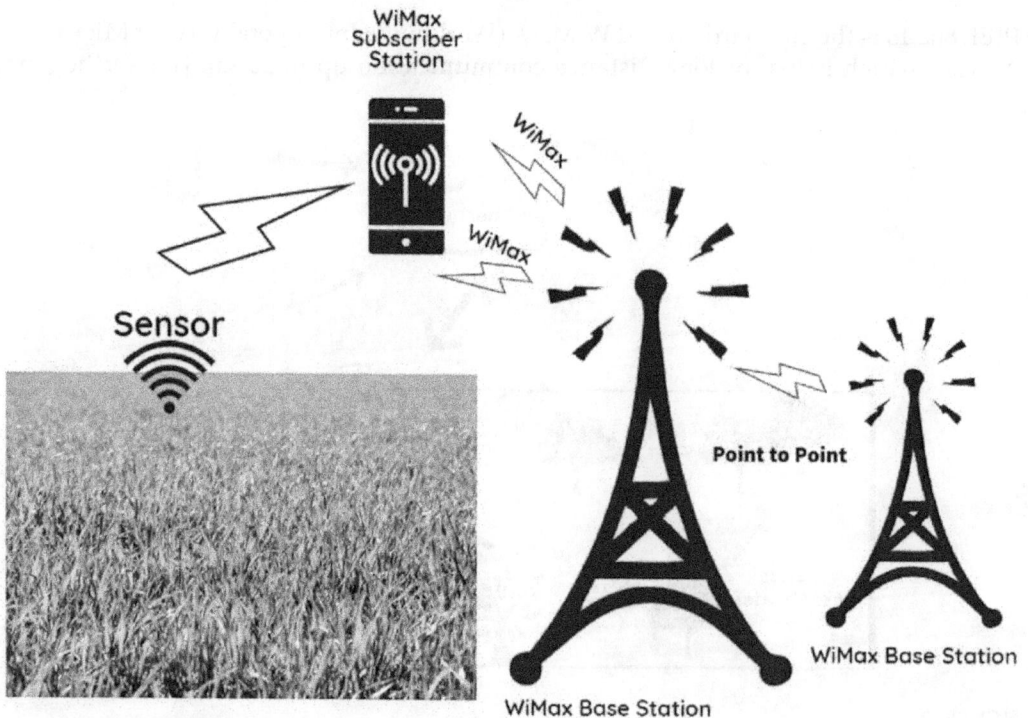

FIGURE 12.7
WiMAX-enabled sensor communication.

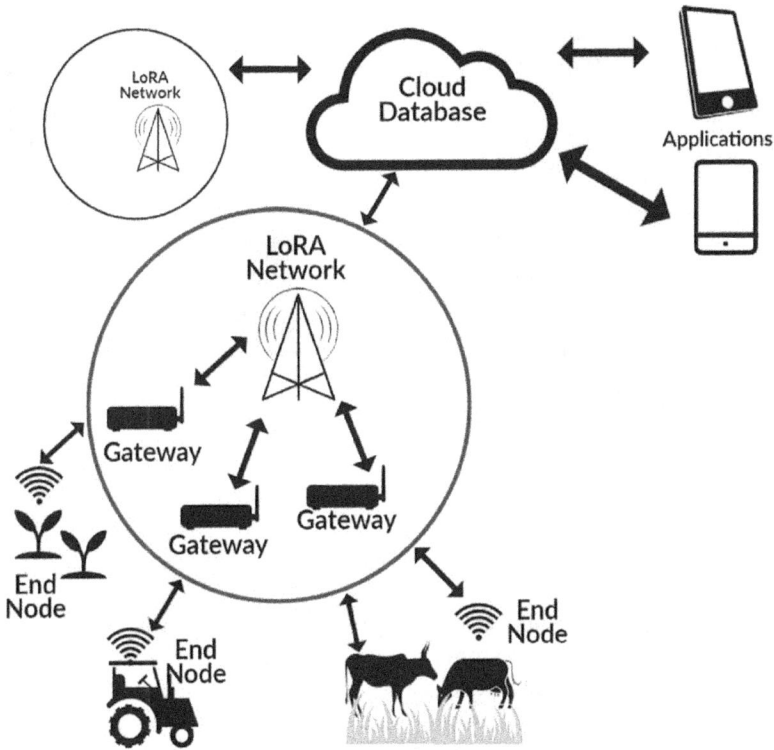

FIGURE 12.8
LoRaWAN enabled sensor communication.

TABLE 12.1

Agro-IoT Applications for Different Communication Standards

Standard	Application
Bluetooth (IEEE 802.15.1)	• Water pump and gate valve controllers • Bluetooth-enabled boundary tracking like intruder alert system in agri-land
Zigbee (IEEE 802.15.4)	• Field maintenance sensor communication such as the soil moisture content, nutrient level, and detection of viruses in the plants
Wi-Fi (IEEE 802.11)	• Monitor the environmental conditions • Drone operations – seed, fertilizer, water, and pesticides
LoRaWAN (LoRaWAN R1.0)	• For capturing and storing water • Irrigation scheduling • Organic farming • Increasing crop yield
WiMAX (IEEE 802.16)	• Remote diagnosis of the farming system-productivity tracking • Fault diagnosis in agri-machines and fields
RFID (ISO18000-6C)	• Greenhouse regulation – water consumption, fruits and vegetables nutrient level identification, and harvest time detection • Animal detection using unique ID

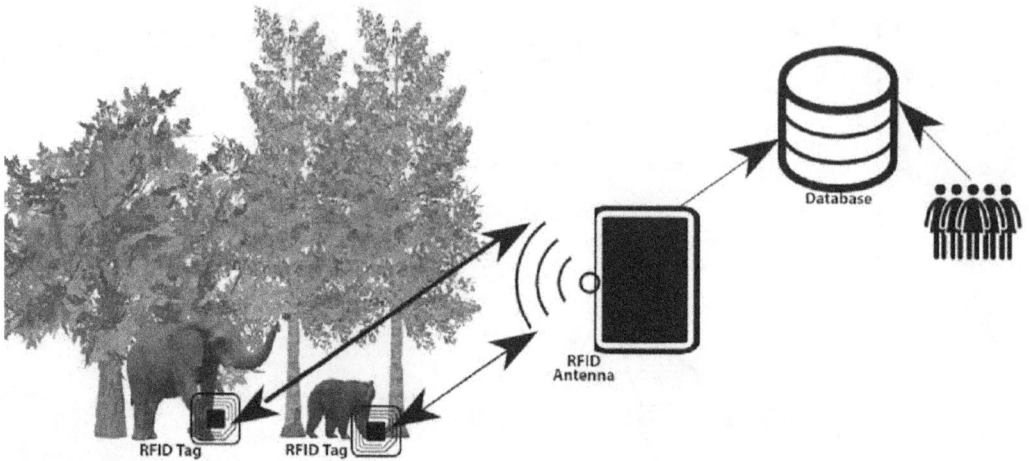

FIGURE 12.9
RFID-enabled sensor communication.

drop content to quantify the water level and predict when to spray the crop; (ii) the water conservation and preservation for smart irrigation systems and weather condition monitoring is required to determine the excessive rain condition which causes floods carried by the rain sensors.

12.4.6 RFID

Object detection will be quite efficient in the case of Radio Frequency Identification (RFID) (ISO18000-6C) usage with unique IDs to identify and track the object or event [15, 16]. It supports calibration if the sensed data is inaccurate. For example, the sensors that have been buried into the soil have to measure the designed parameters like temperature, humidity to maintain the quality of the crop for production and animal tracking [17]. The sensed information will be forwarded to the end-user database with the help of an RFID tag and signal transceiver, as shown in Figure 12.9. If anything, what the user needs then that will be directed to RFID as a command to the sensor. It helps the farmers abundantly with various factors such as time and money for better crop yield.

12.5 Performance-Measuring Factors

12.5.1 Packet Loss Percentage

Due to poor network connection, there may be a loss in transmission of information; it is measured by the following formula:

$$LP\% = \left\{ 1 - \left(\frac{TRP}{TTP} \right) 100 \right\}\%$$

where TPR is total received packet and TTR is total transmitted packet.

Packet loss will generally reduce the speed or throughput of the given packet.

12.5.2 Node Connectivity

Node connectivity is used to transmit and receive the data between two nodes. It quantifies the direct connectivity between sensor nodes on the basis of used communication standards like Bluetooth and Zigbee.

12.5.3 Transmission Throughput

Throughput is a performance measure to indicate the successful transmission from the source node to the destination node.

12.5.4 Scalability

Scalability refers to the number of nodes involved in the transmission range of the network for communication with respect to distance. It is based on communication standards involved in the network. It is a capability of optimal working even if the traffic increases; thus poor scalability leads to poor network performance [18].

12.5.5 Interoperability

The network integration challenges are overcome with the help of the characteristic nature of interoperability. It helps to enhance the Quality of Service (QoS) with respect to the network configuration.

The comparison of various supportive communication technologies (Bluetooth, Zigbee, Wi-Fi, WiMAX, LoRaWAN, RFID) for Agro-IoT is discussed in Table 12.2.

12.6 Comprehensive Overview of Communication Protocols

More consideration is given to the protocols listed in Table 12.3 thus helps to select it for the intended applications of Agro-IoT.

12.6.1 MQTT

The IoT network is in need of bandwidth optimization that can be achieved by using MQTT, a message transferring protocol. It is expanded as Message Queuing Telemetry Transport protocol. It follows the publish/subscribe [19] conceptual scheme as illustrated in Figure 12.10. The node which has the information starts publishing the topic – message to the intermediate nodes called brokers. If any node is interested to get that information it will subscribe through the broker. Then that subscribed node will receive the message from the broker node. MQTT brokers will receive all the messages from the MQTT clients, then filter and forward them to the interested MQTT clients, which will be considered as the appropriate receiver. It supports bidirectional communication to broadcast the message from one point to another point. The control information such as header, trailer, and

TABLE 12.2

Comparison of Various Supportive Communication Technologies for Agro-IoT

Protocol	Loss of Packets (%)	Node Connectivity	Transmission Throughput	Scalability	Inter-operability
Bluetooth	2%	Star topology and mesh topology	0.7–2.1 Mbps	Eight nodes can be involved in the network basis to carry out the Bluetooth communication. The network covers 10-meter distance	Yes
Zigbee	7.80%	Mesh topology	10–115.2 kbps	Due to the mesh topology the network can be extended to several hundreds of nodes for communications with a distance of 10–100 meters	Yes
Wi-Fi	5%	Mesh topology	>100 kbps	Using the mesh topology 32 nodes can be connected with a 100-meter distance in standard Wi-Fi. With the help of wireless extender, it is possible for several thousand nodes to be in connection	Yes
WiMAX	4.3%	Mesh topology	10–15.84 Mbps	Since the nodes are mobile in nature and multi-hop relays ensure the transmission of about 50 km for broadband complementary network	Supports WiMAX equipment. Does not support other vendors' equipment
LoRaWAN	2–25%	Star topology	7.80 bytes per second	High density of 120 nodes can be in a range to carry out the communication of 10–20 km	Yes
RFID	5–10 %	Star topology and point-to-point topology	180 bytes to 800 bytes per second	Depends on the implementation of Agro-IoT platform	Yes

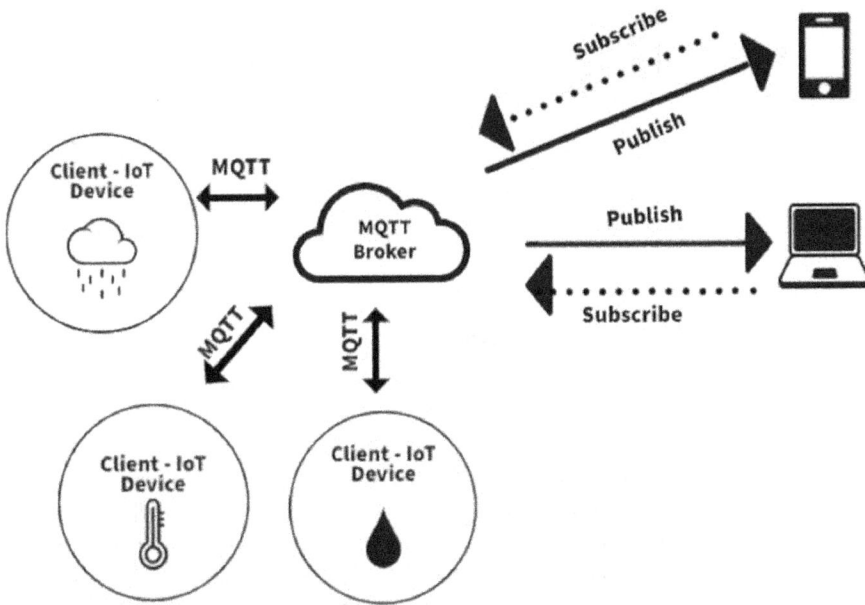

FIGURE 12.10
Architecture of the publish/subscribe mechanism and transportation of messages between devices in MQTT.

other information needed to make the actual message reach the destination is considered to be less; thus MQTT has less overhead. Whenever there is a necessity of control over the system then this protocol paves the way for it.

There are two use cases in IoT that are reliable and unreliable in nature. Reliability is the major factor that is considered to be essential in Agro-IoT networks if the nodes are connected in a wireless manner. There is a chance for the node to be affected by various environmental factors. In that case the network can be recovered if the node has the feature of reliability. MQTT broker serves as an MQTT server, holds the information, and is not dispatched when the receiving MQTT clients are not connected properly, called an unreliable network. After the acknowledgement is received from the broker only the topics (e.g., temperature sensor in the agri-field) will be forwarded to the intended receiver. The essential requirements and new features incorporated in MQTT are categorized into different types: MQTT 3.1.1 and MQTT 5. This protocol is applicable to the temperature, humidity, and soil monitoring sensors. In order to improve the network quality, the QoS is a parameter that has to be maintained in an optimal way. The QoS includes bandwidth, delay, data rate etc. There are three different QoS messages. If the QoS message carries the value of 0 then it indicates at most once; if the value is 1 then the condition is at least once; if it is 2 then it is exactly once. It is considered to be the security-enabled protocol that helps encrypt the actual message using TLS and the authentication process is done by executing a mechanism called OAuth.

12.6.2 DDS

For exchanging the data in real time, the Data Distribution Service (DDS) is required. It is one of the standardized protocols in which peer-to-peer basis communication is being carried out [20]. The discovery mechanism involved here is that the peer will discover and

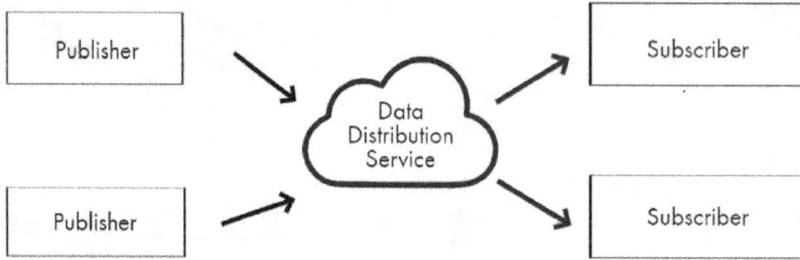

FIGURE 12.11
Interoperable DDS mechanism for machine-to-machine communication.

identify the matching peer with its flexible QoS configuration and topic name. Basically, it applies to a data-centric network. Similar to the MQTT protocol DDS also follows the publish/subscribe mechanism for transferring messages. The discovery mechanism is dynamic and the endpoints are automatically discovered through DDS. The only difference is that there is no broker between the clients since it has the direct data bus between the publishers and subscribers, as shown in Figure 12.11. It leads to less complexity. Adding a new DDS participant is a simple process since it supports scalability [18]. Even if the nodes are not reachable at the receiver side the communication will not be affected because of the multicasting nature and node's reliability in the network. Hence it is best suitable for machine-to-machine communication and uses minimum energy resources and low cost for deployment. It is widely used in mission-critical applications because of the major consideration of security. It supports intelligent farm management in Agro-IoT.

12.6.3 CoAP

One of the specialized application protocols is a Constrained Application Protocol (CoAP). It is discovered by the IETF standard elaborated as Internet Engineering Task Force. Basically there are many constrained devices in the IoT network such as minimum bandwidth, low battery, minimum latency, and less power consumption of the network. A simple node that helps to have communication is a battery-constrained one that can convey its message using the wider internet with CoAP, as represented in Figure 12.12.

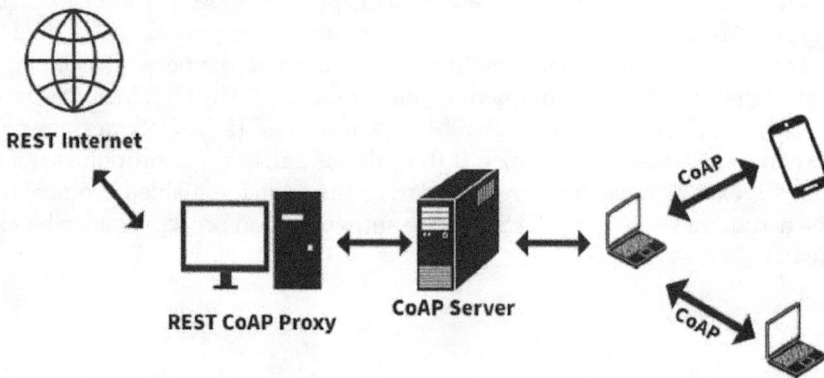

FIGURE 12.12
Generic web transfer mechanism of CoAP.

It is an application layer protocol famous for machine-to-machine communication runs on User Datagram Protocol (UDP). It is similar to HTTP, which is based on Transmission Control Protocol (TCP). CoAP logically deals with asynchronous message transfer through request/response mechanism. The client sends one or more requests to the server and services the requests via the response message by the server. The sequence of numbers from 0 to 8 is considered to be a token to match the requests with the response. It supports both unicast and multicast data transmission. It has four bytes of header addresses along with the data, which provides the minimum overhead feature in the network. Each message holds a message ID that helps detect duplicate messages that are being transmitted to the receiver. Reliability is an optional header added to make the message a confirmable (CON) message; it supports having retransmissions in case of time-outs, and it automatically resends the message until the acknowledgment is being received by the transmitting node from the receiver. The acknowledgment is an ID which is the same message ID used while initiating the transmission. Even if the message does not want reliable transmission, then without acknowledgment the transmission takes place with the message type as a non-confirmable (NON) message called unreliable message transmission. Here also the message ID is used for identifying the duplication of the same message. If a NON message is not getting processed then the Reset (RST) option will be preferred. It has control over simultaneous data transmission. Datagram Transport Layer Security (DTLS) is the security model designed for CoAP communication. Proxying is an intermediary concerned with forwarding the request and responses, caching operations, and protocol translation, useful in resource-constrained networks in order to minimize network traffic, improve the network efficiency, utilize the resources that are under sleeping condition, etc.

12.6.4 AMQP

AMQP is a location-based protocol, expanded as Advanced Message Queuing Protocol, supports the compact messaging application and communication pattern. The message transmission between sender and receiver in a distribution system is by a hardware or software architect infrastructure support called message-oriented middleware, simply known as MOM, used at the API level for standardization in AMQP [21]. The messaging pattern used here is the publish/subscribe mechanism where the publishers are called senders who will not send the message directly to the receiver, called the subscriber, who will receive only the interested message subscribed.

If the messages are passing continuously among the nodes, then there is a chance of getting overloaded by the nodes, so the message used for later purposes is stored in a message queue through the channel, as shown in Figure 12.13. The storing is based on data exchange type and binding. The binding is the set of rules to route the message to a queue based on the exchange type. The message attribute is declared as at least one for transmission. There are four exchange message types for AMQP 0.9.1 broker: direct, fanout, topic, and header. Using the application interface the nodes in different networks can communicate easily. This is known as interoperability. Here binary protocol and wire level protocol are used to transmit data from one point to another point and perform the data operation in a distributed system.

AMQP efficiently supports flow control over transmitted messages and ensures the message delivery via floating the acknowledgment message. The message delivery guarantee from one point to another point is by using three verticals: at most once – the message gets delivered once or never, at least once – minimum one message has to be delivered, or it

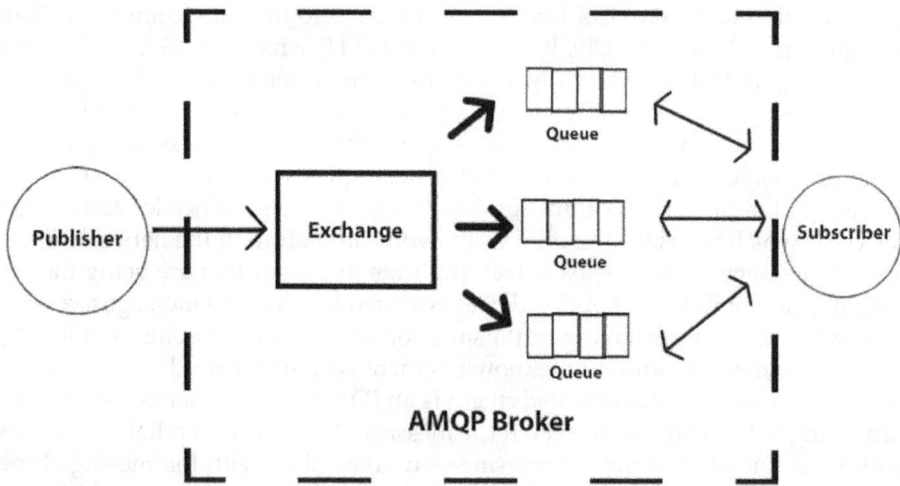

FIGURE 12.13
Transmission perspective of AMQP protocol.

can be multiple times, exactly once – only once. TLS and STLS are the encryption methods used in AMQP to secure the message while transmitting. The newer version of AMQP 1.0 supports to direct the message with little less intervention of brokers through the mentioned address containing routing key as a field. Reliability and extendibility and security are the important features and the major advantages of AMQP.

TABLE 12.3

Pros and Cons of IoT Protocols

Protocol	Advantages	Disadvantages
MQTT	• Ensures message delivery with minimal network traffic • Consumes low power • Works well even with unreliable internet connection	• High latency, which affects the speed • Lack of security while transferring messages from one end to another end
DDS	• Supports interoperability • Holds the communication architecture with low latency • Secured connection	• Consumes twice the bandwidth as MQTT • Web service interface not supported
CoAP	• Consumes less power • Synchronous communication is not needed for data transmission • Reliable communication using the acknowledgment message	• The cost for additional security provision is high since it is an unencrypted protocol • CoAP messages get lost due to the use of UDP
AMQP	• Interoperability ensured via the use of wire-level protocol • Simple peer-to-peer communication • Secured connection using SSL protocol	• High bandwidth required • Unsupported discovery of network attributes

12.7 Security Protocols Used to Secure the Network

12.7.1 DTLS

DTLS helps to secure the Datagram-based network. There are various possible attacks to steal private information. Eavesdropping is one of the common attacks which will listen to private information without the knowledge of the users in the transmitting and receiving end.

12.7.2 TCP

It is a standard connection-oriented protocol used to establish a connection between the nodes and maintain the internet-based communication between sender and receiver. It helps the application program to exchange the data. It is a set of rules defining how the application data is segmented as packets, routing of packets to the destined nodes. It has the responsibility of flow control to control the flow of traffic with the help of exchanging control information like RTS (request to send) and CTS (clear to send) in case of additional device connection and overflow of data. It further ensures the uniformity of data being carried out. And error control is a functionality identification of error location and retransmission process at which the packet gets missed. TCP is a very important model that helps to exchange the data over the network securely with the help of SSH (Secure Shell).

12.7.3 UDP

It is a connectionless message-oriented transport layer protocol in which the entire data is divided into several data units, each addressed with a unique ID and forward toward the receiver by taking multipath. Since there is no end-to-end connection between sender and receiver the UDP feels tedious to order the forwarded packets at the receiver end. It is considered to be an unreliable network since it has no prior communication to set the data path for the transmission. UDP is applicable for the time-sensitive applications where there is no necessity for error checking and data handling during data transmission and even for the error checking case where the packet dropping is appropriate rather than retransmission of an errored data unit. It is supportive of broadcasting the data and multicasting is also possible in this packet switching network.

12.8 IoT-Based Solution for Leaf Disease Detection Using Machine Learning Classifiers for Precision Agriculture

The farm's output reflects the country's economy. It is necessary to provide the facility of better yield through automated processes. Prior knowledge about the quality of plant and its strength through observing the condition of leaves helps to achieve the optimum yield. Figure 12.14 describes the leaf disease detection process; it comprises pre-processing where the filtering is done. The filtering used here is a non-linear filtering method known as median filtering. It is a technique to remove the unwanted noise from the image by

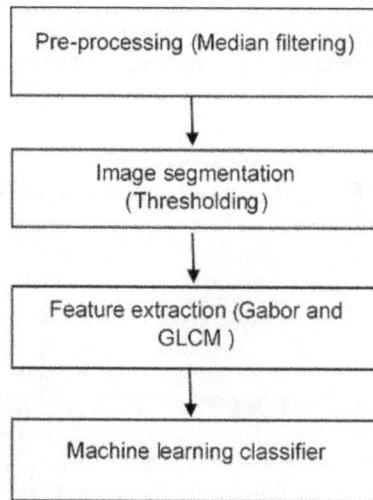

FIGURE 12.14
Flowchart of leaf disease detection and classification using machine learning algorithm.

considering the nearby pixels as the pixel value, then its own pixel as the median. So that the surrounding blurs are removed, and thus the edge points are sharpened through it.

After filtering the region of interest from the image has to be extracted through a technique known as segmentation. In segmentation, the entire image is divided into many portions, which leads to analyzing the required image in a clear way, which thus helps create the exact map of the analyzing part. The simplest thresholding method is used where the grayscale image is converted into a binary scale image and based on the average value the threshold is set and the analysis is done. The next process is a feature extraction where the desired attributes of the data are identified and driven with the non-redundant data for further process. The enhanced predictability is achieved through Gabor filter and Gray Level Co-occurrence Matrix (GLCM) statistical features. The Gabor filter is a band-pass filter in which it provides the frequency and orientation representations of an image. Gabor features are best and outperforms when compared to GLCM. The distinctive features are classified by using classification algorithms such as KNN, Naive Bayes, Random Forest, Decision tree, and SVM.

12.9 Implementation of Machine Learning Classifiers and Its Performance Study for Leaf Disease Detection in Agriculture Field

For the implementation, the tomato leaf disease detection dataset is opted and used for analyzing the quality of the leaf and its strength. The dataset comprises 10,000 trained images of healthy and disease-affected tomato leaves, in which 9 disease classes and 1 healthy class are present [22]. The images from each category are extracted and considered to validate the performance of the classifiers. The dataset contains 1000 images in each category that are extracted from the dataset. In this experimental setup, 70% from

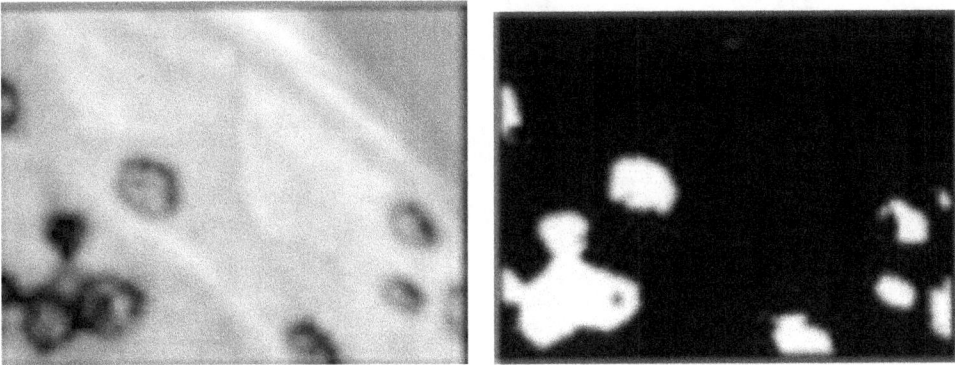

FIGURE 12.15
Segmented bacterial spot in tomato leaf image using thresholding.

each category is used for training and the remaining 30% is used for testing. The affected area of each image is segmented out through thresholding.

Figure 12.15 shows the segmented bacterial spot in tomato leaf image using thresholding. GLCM and Gabor features are extracted for each segmented test and train tomato leaf image and it is used as a feature vector for all the classifiers. The machine learning classifiers with GLCM and Gabor features are simulated for the abovementioned dataset and its performance measures such as classification accuracy, F-measure, precision, and recall are computed and compared. The performance comparison of machine learning classifiers with GLCM and Gabor features are shown in Tables 12.4 and 12.5, respectively. From the

TABLE 12.4

Performance Comparison of Machine Learning Classifiers with GLCM Features

Classifiers/ Performance Measures	Precision	Recall	F-Measure
Decision Tree	0.11	0.63	0.19
Naïve Bayes	0.15	0.72	0.25
KNN	0.18	0.79	0.29
Random Forest	0.20	0.84	0.32
SVM	0.22	0.91	0.35

TABLE 12.5

Performance Comparison of Machine Learning Classifiers with Gabor Features

Classifiers/ Performance Measures	Precision	Recall	F-Measure
Decision Tree	0.19	0.75	0.30
Naïve Bayes	0.21	0.81	0.33
KNN	0.24	0.86	0.38
Random Forest	0.26	0.92	0.41
SVM	0.29	0.97	0.45

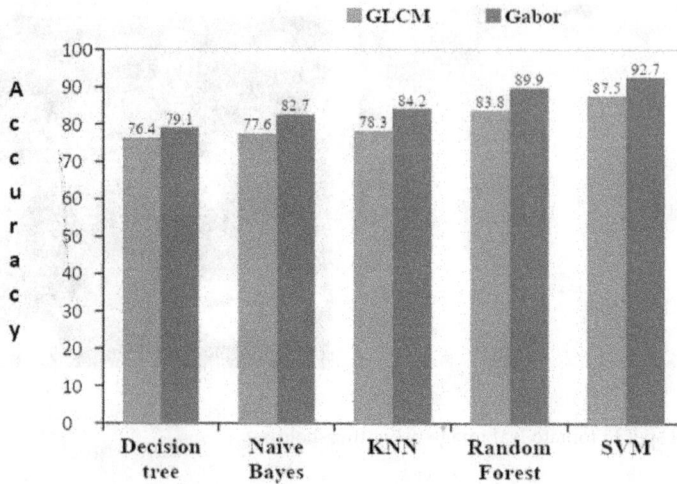

FIGURE 12.16
Performance comparison of classification accuracy (%) for various machine learning classifiers with different features.

simulated results, it is observed that SVM classifiers with Gabor features provide significant improvement in terms of precision, recall, and F-measure when compared to other classifiers with GLCM and Gabor features.

The performance comparison of classification accuracy (%) for various machine learning classifiers with different features is shown in Figure 12.16. From the results, it is observed that SVM classifiers with Gabor features efficiently classify tomato leaf disease when compared to other classifiers with GLCM and Gabor features.

12.10 Real-Time Processing of Sensor Data Using IoT

The top-level components of IoT are hardware, gateway, and cloud. At the initial stage the optical sensors capture the leaves' images, which are analyzed using machine learning algorithms to classify the healthy and unhealthy nature of captured leaves. IoT protocols such as MQTT and CoAP make sense for the data stored in the cloud and again respond to the farmer side with appropriate standards to make decisions further for the better way of crop yield. The scalability features of the protocols are tested with the help of the IoT Simulator. In which 20 sensors are considered to be active for publishing the messages using MQTT protocol with the public broker of IoT.eclispse.org [23]. The considered subscriber client is publicly available in Hive MQ. And the sensed data is generated every 10 seconds in the simulation and it is categorized using an SVM classifier to update the status of the plant to the farmer then and there for optimum yield with the help of IoT network. The overall analytics process is possible with the help of the salient feature known as interoperability which provides the sensor structured data integration [24].

12.11 Conclusion

This study is aimed at providing insights into various standards and protocols used for Agro-IoT. The findings are based on three verticals. First is about the technologies that pose its advances and its beneficial operations, which are useful for deploying an efficient Agro-IoT network. The dilemma on determining the standards and technologies to be deployed is considered and provided the solution based on applications. The communication technologies such as Bluetooth, Wi-Fi, Zigbee, WiMAX, LoRaWAN, and RFID developed by the standards like IEEE, ISO, and IETE are highlighted with their significant factors for use cases of Agro-IoT. Along with this the oversight on the scope and responsibilities of various sensors such as temperature, soil, humidity, leaf wetness, rain sensors, optical sensors that are useful for remote sensing of agriculture fields. Second vertical is about the most demanding phase in networks known as communication protocols. A detailed view of various Protocols such as MQTT, AMQP, DDS, CoAP, and its major constraints are examined and analyzed. The complications arose due to the fact that wireless networks in terms of security are controlled with the help of layered protocols such as DTLS, TCP, and UDP, which support secured communication from the sensor nodes to the end-user. Third vertical is about the comparison of various algorithms for analyzing the sensor captured image through classifier algorithms like KNN, Naive Bayes, Decision Tree, Random Forest, and SVM. The analysis is done based on the features like precision, recall, and F-measures. Through these measures the efficient classifier SVM holds accuracy values of 82.5% and 92.7% based on the GLCM and Gabor feature extraction methods applied for the process of exact diseased leaf detection. Based on the classified data the accurate result is obtained and passed to the farmer with the appropriate communication standards and protocols of IoT network for the suggestion on the composition of pesticide application on the affected plants in the field. As a whole the Agro-IoT is an area that has a great deal for field management and farm management with proper attribute setups through the deployment of sensors along with its communication standards and protocols.

References

1. O. Elijah, T. A. Rahman, I. Orikumhi et al., "An Overview of Internet of Things (IoT) and Data Analytics in Agriculture: Benefits and Challenges," in IEEE Internet of Things Journal, vol. 5, no. 5, pp. 3758–3773, 2018, doi: 10.1109/JIOT.2018.2844296.
2. V. P. Kour and S. Arora, "Recent Developments of the Internet of Things in Agriculture: A Survey," in IEEE Access, vol. 8, pp. 129924–129957, 2020, doi: 10.1109/ACCESS.2020.3009298.
3. M. S. Farooq, S. Riaz, A. Abid, K et al., "A Survey on the Role of IoT in Agriculture for the Implementation of Smart Farming," in IEEE Access, vol. 7, pp. 156237–156271, 2019, doi: 10.1109/ACCESS.2019.2949703.
4. H. Fornazier et al., "Wireless Communication: Wi-Fi, Bluetooth, IEEE 802.15.4, DASH7," ROSE 2012 ELECINF344/ELECINF381, Télécom ParisTech, web site: http://rose.eu.org/2012/category/admin
5. O. Javeri and A. Jeyakumar, "Wireless Sensor Network Using Bluetooth." In: Unnikrishnan S., Surve S., Bhoir D. (eds) Advances in Computing, Communication and Control. ICAC3 2011. Communications in Computer and Information Science, vol. 125. Springer, Berlin, Heidelberg, 2011. https://doi.org/10.1007/978-3-642-18440-6_54

6. C. G. Naik, A. Krishna, Srinitha et al., "Integrated Farming System Using IoT and Bluetooth," in International Research Journal of Engineering and Technology, vol. 07 no. 04, pp. 6459–6464, 2020.

7. K. Chang, "Bluetooth: A Viable Solution for IoT? [Industry Perspectives]," in IEEE Wireless Communications, vol. 21, no. 6, pp. 6–7, 2014, doi: 10.1109/MWC.2014.7000963.

8. C. M. Ramya, M. Shanmugaraj and R. Prabakaran, Study on ZigBee technology. 2011 3rd International Conference on Electronics Computer Technology, 2011, pp. 297–301, doi: 10.1109/ICECTECH.2011.5942102.

9. A. Kalra, R. Chechi, and Khanna R. et al., Role of Zigbee Technology in Agriculture Sector. NCCI 2010 – National Conference on Computational Instrumentation CSIO Chandigarh, India, 19–20 March 2010.

10. V. Daravath and A. Daravath, WIMAX (IEEE 802.16) broad band technology for smart grid applications. 2015 International Conference on Communications and Signal Processing (ICCSP), 2015, pp. 1273–1275, doi: 10.1109/ICCSP.2015.7322712.

11. D. Davcev, K. Mitreski, S. Trajkovic et al., IoT agriculture system based on LoRaWAN. 2018 14th IEEE International Workshop on Factory Communication Systems (WFCS), Imperia, Italy, 13-15 June 2018. doi:10.1109/wfcs.2018.8402368

12. G. Hornero, J. E. Gaitán-Pitre, E. Serrano-Finetti et al., "A Novel Low-Cost Smart Leaf Wetness Sensor," in Computers and Electronics in Agriculture, vol 143, pp. 286–292, 2017, ISSN: 0168-1699, https://doi.org/10.1016/j.compag.2017.11.001

13. P. P. Ray, "Internet of Things for Smart Agriculture: Technologies, Practices and Future Direction," in Journal of Ambient Intelligence and Smart Environments, vol. 9, pp. 395–420, 2017, doi: 10.3233/AIS-170440.

14. J. Haxhibeqiri, E. De Poorter, I. Moerman and J. Hoebeke, "A Survey of LoRaWAN for IoT: From Technology to Application," Sensors, vol. 18, no. 11, p. 3995, 2018. https://doi.org/10.3390/s18113995

15. R. Want, "An Introduction to RFID Technology," in IEEE Pervasive Computing, vol. 5, no. 1, pp. 25–33, 2006, doi: 10.1109/MPRV.2006.2.

16. Dong-Liang Wu, W. W. Y. Ng, D. S. Yeung and H-L Ding, A brief survey on current RFID applications. 2009 International Conference on Machine Learning and Cybernetics, 2009, pp. 2330–2335, doi: 10.1109/ICMLC.2009.5212147.

17. T. Wasson, T. Choudhury, S. Sharma et al., Integration of RFID and sensor in agriculture using IOT. 2017 International Conference On Smart Technologies For Smart Nation (SmartTechCon), REVA University, Bengaluru, India, 17 – 19 August 2017.

18. André B. Bondi Characteristics of Scalability and Their Impact on Performance. WOSP '00: Proceedings of the 2nd International Workshop on Software and Performance, https://doi.org/10.1145/350391.350432

19. B. Mishra and A. Kertesz, "The Use of MQTT in M2M and IoT Systems: A Survey," in IEEE Access, vol. 8, pp. 201071–201086, 2020, doi: 10.1109/ACCESS.2020.3035849.

20. A. Corsaro and D. C. Schmidt, "The Data Distribution Service." The Communication Middleware Fabric for Scalable and Extensible Systems-of-Systems. PrismTech, Vanderbilt University, Nashville, TN, March 2012.

21. J. E. Luzuriaga, M. Perez, P. Boronat, J. C. Cano, C. Calafate and P. Manzoni, A comparative evaluation of AMQP and MQTT protocols over unstable and mobile networks. 2015 12th Annual IEEE Consumer Communications and Networking Conference (CCNC), 2015, pp. 931–936, doi: 10.1109/CCNC.2015.7158101.

22. https://www.kaggle.com/kaustubhb999/tomatoleaf

23. https://mqttlab.iotsim.io/

24. J. del Rio Fernandez, D. M. Toma, E. Martinez, S. Jirka, "Tom O'Reilly3 From Sensor to User—Interoperability of Sensors and Data Systems." In: Challenges and Innovations in Ocean in Situ Sensors, Elsevier, 2019, pp. 289–337, ISBN 9780128098868, https://doi.org/10.1016/B978-0-12-809886-8.00006-5

13

Research Issues and Solutions in Agro-IoT

N. Vijaya

K. Ramakrishnan College of Technology
Tiruchirappalli, India

T. Vigneswari

Sri Manakula Vinayagar Engineering College
Puducherry, India

CONTENTS

13.1 Introduction...209
     13.1.1 Issues and Challenges in a Nutshell 210
13.2 Integration/Communication Issues and Solutions..................................... 211
     13.2.1 LP-WAN (Low-Power Wide Area Network) 212
          13.2.1.1 LoRa (Long Range) 213
          13.2.1.2 NB-IoT 213
          13.2.1.3 Sigfox... 214
13.3 Failure Issues and Solutions.. 215
          13.2.2.1 Predictive Fault-Tolerant Systems................................... 215
13.4 Data Management Issues and Solutions... 217
13.5 Security Issues and Solutions 218
13.6 Summary.. 219
References.. 220

13.1 Introduction

World population growth is increasing tremendously and is expected to reach 9.7 billion in 2050, according to the reports of the United Nations in 2019 [1]. The population growth coupled with global threatening news such as water shortage, land shortage, climate change, etc., will put additional pressure on agriculture to feed the population. Till now, farmers have been struggling a lot in monitoring and protecting their livestock and farms against natural disasters, damage, theft, etc. Adopting agro-based IoT technologies increases yields in farms, reduces theft of cattle, reduces human effort, and many more. As per Juniper Research, the number of IoT devices in agriculture is 38.5 billion in 2020, which is a major rise of 285% when compared with 13.4 billion devices in 2015 [2]. Even though Agro-IoT products offer numerous benefits to farmers, agriculturists, and industrial food producers; yet many challenges are associated with selecting, instrumenting, and integrating IoT devices [3]. As these IoT devices have the ability to connect and communicate with each other in mobile and web-based application platforms, the interoperability of these devices is still a challenging issue.

When instrumenting livestock with sensors [4], the battery life of wireless devices is a major issue. The battery power and coverage range are interrelated. Because the selection of wireless devices highly depends on the distance the data needs to travel between the communicating devices. Figuring out sensors with enough battery life, choosing the range of wireless network, i.e., whether short distance RFID/NFC (100 meters), Bluetooth low energy (within 10 meters) or WAN (1000 meters), and positioning the sensors for effective communication remains challenging. Even though IoT resolved many traditional farming problems such as drought response, yield optimization, land selection, irrigation, pest control, etc., the installation, maintenance, and protection of electronic circuits during natural calamities is really problematic and also requires high cost. Since the farms are located in remote areas, internet connectivity, storing data in the cloud, and accessing data from the cloud also remains challenging [5, 6]. Providing Quality of Service (QoS), reliability, scalability, and efficiency in smart agriculture is still an unfolded issue. This chapter highlights the following issues and discusses the respective solutions in detail.

13.1.1 Issues and Challenges in a Nutshell

A smart agriculture system is framed by four main parameters: IoT devices, communication technology, internet, and data processing units [7–9]. Figure 13.1 specifies the collaboration of these parameters for the successful deployment of Agro-IoT technology. Here, sensors and actuators are used to collect information about the crops, and the communication technology transfers information to the cloud, then the cloud stores and processes information, and results are sent to the agriculturist for further action.

1. Various tools in smart farming do not follow the same set of standards; hence more and more gateways are needed to transform these smart tools into farmer-friendly platforms.

2. Big data centers and gateways consume enormous energy; hence smart agricultural tools should additionally focus on the energy depletion risks. Improper maintenance of these resources can also cause failure and damages.

3. Lack of IoT knowledge among farmers can also be alleviated by giving adequate training on tools/devices such as sensors, drones, and other technologies since the failure of irrigation sensors may lead to overwatering or under watering of crops which leads to huge irrecoverable loss.

4. Implementation of cloud services in farmlands with hilly terrains restricts data transmission and data storage; hence the solution lies in improving network bandwidth and speed.

5. Analyzing big data collected from smart farms every day requires skilled persons and smart agriculture demands the correct production function in order to optimize the output since the incorrect application of inputs will lead to sub-optimal results.

6. As the resources and tools provided by big original equipment manufacturer (OEM) may not be compatible with smaller OEM, hence migrating from an older platform to a new platform leads to compatible problems. Providing cross-platform migration solutions will address this issue.

The next section discusses each and every problem in detail and also provides respective solutions.

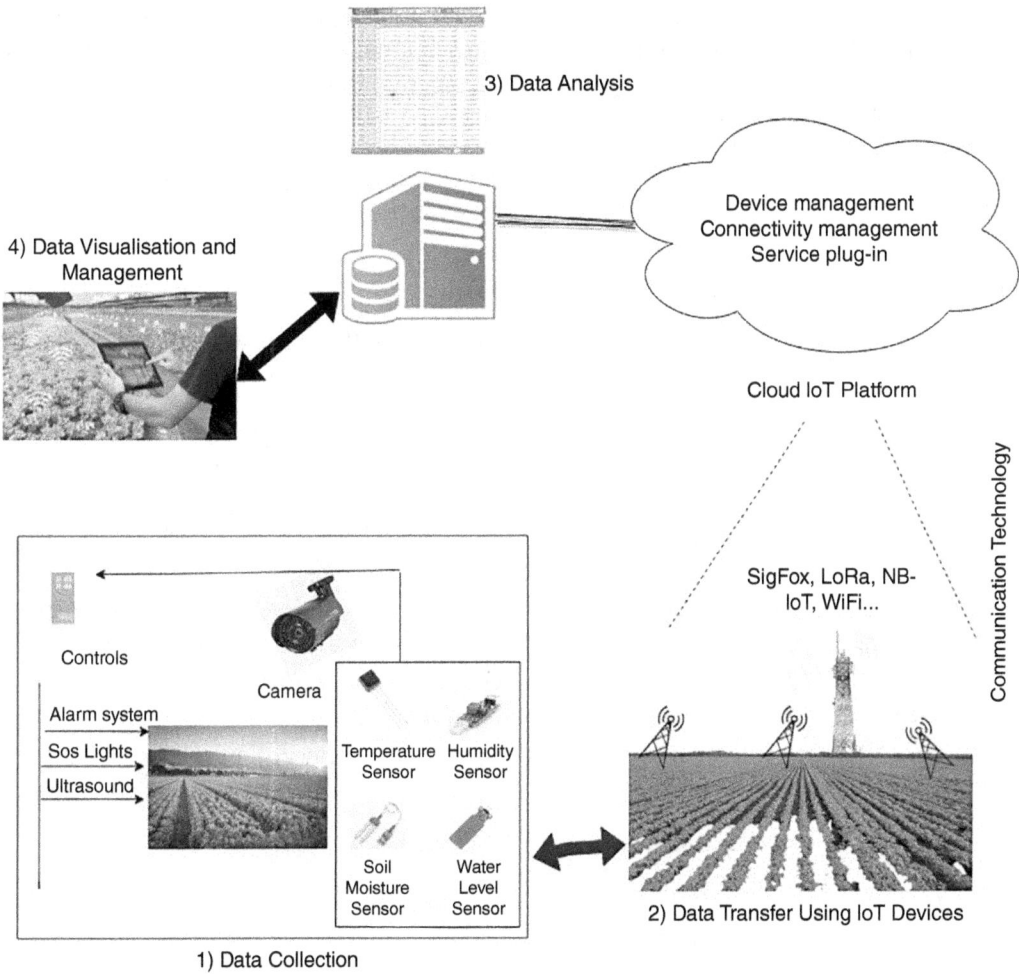

FIGURE 13.1
Smart agriculture system.

13.2 Integration/Communication Issues and Solutions

The new common agricultural policy demands farmers to reduce the use of herbicides and fertilizers under the new term called Greening. The introduction of new communication technologies in farming should reduce the use of chemicals and focus on traditional methods of agriculture in the growth of plants. The ultimate objective is to produce quality and healthy food products for the human community. Nowadays, the usage of modern technologies such as aerial drones, IoT agrosensors, drools, farm controller networks, cloud computing, multimedia view browsers, etc., are very common in agriculture [10]. But integrating these technologies remains challenging since each wireless protocol has its own communication paradigms and network mechanisms. In order to enable communication among these devices, a smart wireless network topology is needed.

Table 13.1 gives the various IoT technologies and their benefits in agriculture. Traditional wired and wireless networks options such as mesh networks, Bluetooth, cellular networks, Wi-Fi, etc., are not suitable for IoT applications as they cannot meet the cost, coverage and power requirements. In order to ensure reliable and scalable connectivity across heterogeneous networks, many new protocols have been proposed [11]. The protocols discussed in the next section serve as compatible products in gathering data about agricultural conditions at a lower cost, thereby increasing farming efficiency.

13.2.1 LP-WAN (Low-Power Wide Area Network)

This technology provides solutions for smart agriculture developments by providing the following promising services [12].

a. LPWAN offers high data reception rate by operating in licensed or license-free spectrum. LPWAN ensures high Quality of Service and reliability.

b. Day by day usage of IoT devices is getting increased; hence network expansion and addition of IoT devices are more essential. LPWAN ensures high scalability by providing a large network capacity.

c. LPWAN offers the integration of low-cost battery constraint sensor devices which consume low power to achieve more environmentally sustainable architecture.

d. LPWAN offers mobility by providing high-speed data transmission between the edge nodes and sink nodes in precision farming.

e. LPWAN offers high security by providing multilayer encryption solutions. It also provides secure authentication, end-user identification, secure data transmission, and high-level data integrity.

TABLE 13.1

IoT Technologies for Smart Agriculture System

	Impact in Agriculture	
IoT Technologies	**Support to Agriculture**	**Applications in Agriculture**
Embedded systems (devices that consist of both hardware and software)	Remote monitoring and controlling of equipment increases profitability and sustainability. It also decreases production costs.	Load shedding (solves electricity problems), water management and usage of drones.
Wireless sensor networks (sensor nodes with radio communication capabilities)	Tools to integrate different sensors. Facilitate collection and management of data from different sensors.	Livestock tracking, equipment tracking, soil moisture monitoring, and temperature monitoring.
Communication Protocols (backbone of IoT to establish connectivity)	By using various data exchange formats, these protocols are used to exchange data over different networks.	Supports long-range communication.
Big data analytics (process of examining and analyzing large data sets)	Providing schemas, dashboards, and analytical reports to farmers to take decisions.	Optimize usage of pesticides, farm equipment management, prediction of yield, and managing the problems in supply chain.
Cloud computing (a type of internet-based computing)	Services are available on demand. It provides real-time computation and data access to shared resources and also provides enormous storage capacity.	Crop-related information, soil information, farmers' data, and e-commerce.

The most promising LPWAN technologies that provide high-quality Agro-IoT solutions across the world are: (i) LoRa (Long Range), (ii) Sigfox, (iii) Narrowband IoT (NB-IoT), and (iv) myThings.

13.2.1.1 LoRa (Long Range)

This technology was mainly introduced for long-range data transmission using low power consumption. Its geolocation capabilities facilitate devices to exchange their operating locations with the help of gateways. It operates under a license-free sub-gigahertz radio frequency band and operates in different frequency bands in different regions such as 868 MHz (Europe), 915 MHz (Australia and North America), 865 MHz to 867 MHz (India), and 923 MHz (Asia). It provides data rates in the range of 0.3 kbits/s and 27 kbits/s. It operates at a range of 50 km [13]. Even though Wi-Fi/Bluetooth Low Energy (BLE) technology-based networks serve as an optimal solution for smart agriculture, it requires high bandwidth and high power, covers a limited range, and also demands line-of-sight proximity [14]. Its inability to penetrate deep environments as the interference may lead to the adsorption of RF signals.

Lora in precision farming:

1. Suitable for large farms where even 5G technology is difficult to penetrate the large physical structures in the environment. Moreover, the communication range of the LoRa WAN protocol is up to 30 miles. Hence it is more optimal for applications such as water and gas metering, asset tracking, supply chain and logistics, smart homes and buildings, and smart agriculture.

2. It uses unlicensed spectrum and ISM (Industrial, Scientific, and Medical) frequency bands for defining its architecture and communication protocol. It is used for quick deployment of private or public IoT networks using hardware or software anywhere. The physical layer in the protocol provides a long-range communication medium between different sensors and gateways in the precision farming environment [15].

3. It provides a good solution for security by providing end-to-end AES-128 encryption and solutions for mutual authentication, data confidentiality, and data integrity.

4. It provides quality of service by providing seamless handoff while maintaining communication between devices in motion.

5. It offers interoperability between different IoT devices and facilitates the quick deployment of IoT applications anywhere, anytime.

6. Minimizes battery replacement cost by offering low power consumption of IoT devices hence devices equipped with a lithium battery will long last up to 10 years.

13.2.1.2 NB-IoT

NB-IoT is a fast-growing and leading LPWAN technology developed and standardized by 3GPP (3rd Generation Partnership Project) in 2016. It is classified under 5G technology and can co-exist with 2G, 3G, and 4G technologies. NB-IoT has 26 frequency bands, and out of 26, about 18 to 19 frequency bands are in the sub-GHz frequency range and about 7 frequency bands are above 1800 MHz [16].

NB-IoT in precision farming:

- NB-IoT is highly robust in nature and can connect a large fleet of about 50,000 IoT devices, with low power consumption and increased network coverage. Since it satisfies the industry-based standards, it has a wide range of professional applications such as smartmetering, Object tracking, smart cities, fire alarms, and connecting industrial appliances.
- NB-IoT is highly resource efficient i.e., by minimizing the power consumption, it extends the battery life of IoT devices to about 10 years. It also serves as a low-cost communication protocol.
- NB-IoT is less sophisticated and easy for OEMs to design, develop and deploy compared to cellular networks. When compared with LTE M1 (Long-Term Evolution machine-type communications (MTC)), NB-IoT can have good communication among devices with deep penetration in indoor, underground and rural areas.
- NB-IoT offers similar security features as that of LTE networks such as user-based authentication, user confidentiality, device identification, and data integrity.
- NB-IoT is characterized by its optimal network architecture and it eliminates the need for gateway and serves as cost-efficient structure with spatial diversity of +20 dB.

13.2.1.3 Sigfox

Sigfox enables low power connectivity of inexpensive objects to communicate with the cloud infrastructure. Sigfox uses a co-operative reception strategy, where the devices are not attached to any base station. But it can transfer information through nearby base stations. Hence Sigfox enables coverage over large areas using the minimum number of base stations. It uses an unlicensed ISM radio frequency band and the range highly depends on location: In Europe the band used is 868 MHz, in Asia the band used is 433 MHz and in the United States the band used is 915 MHz, which is restricted to the national regulations. Sigfox provides coverage of 30–50 km, whereas in areas with a lot of obstructions it covers up to 3–10 km. Hence it is suitable for applications such as asset tracking, health applications, automotive communication in transport, and precision agriculture.

Sigfox in Precision Farming:

- Sigfox allows devices to communicate with a minimal number of base stations. It also uses the cellular network basics for the remote devices to communicate with base stations using internet. Hence if there is internet connectivity the devices can transfer, access, and control over the specified region.
- Sigfox facilitates devices to effectively communicate using low power consumption, i.e., 10 mA to 50 mA. It doesn't demand synchronization between devices for sending or receiving data, i.e., no pairing is recommended. This facility allows devices to run for a long time with minimal battery charge.
- Sigfox is highly resilient to interference. It utilizes 192 KHz Ultra Narrow Band (UNB), which allows transmission of signals under the presence of jamming signals. This anti-jamming feature is achieved when UNB is coupled with the base station's spatial diversity of +20 dB.

- Sigfox has an inbuilt security mechanism and provides secure authentication, data integrity, confidentiality, and anti-replay mechanisms for data exchanged over the network. Its in-built firewall facility secures devices from internet-related attacks.
- Sigfox provides a software-based computing solution where all data are computed and managed in the cloud creating an energy-efficient infrastructure.

13.3 Failure Issues and Solutions

Agro-IoT enables different components to communicate with each other by sending observed data to selected nodes, thereby instructing the actuators to perform the required change in the farming environment. Here, a diverse array of sensor devices are used to measure soil moisture, light intensity, water level, temperature, humidity, CO_2 rates, etc., and these devices are interconnected to monitor the system in real-time. Improper functioning of these devices leads to incorrect interpretation. Moreover, the lifetime of the entire network also depends on the energy level of these sensor nodes. The energy gets depleted based on the location of sensor nodes and the positioning of relay nodes. The poor internet connectivity in the region also creates data loss [17]. Hence in order to meet the challenges, the IoT environment must be resilient to the abovementioned problems. One of the approaches suggests that multiple redundant sensors can be used for failure detection, but the cost of deployment in large farms leads to a major concern [18].

13.2.2.1 Predictive Fault-Tolerant Systems

The main problem with sensors in the context of agriculture is the humidity and temperature. Because the temperature rise causes a huge impact on signal strength when there is a rise of above 50°C. Moreover, the radio wave propagation is affected by humidity, since the sensor nodes are deployed in open farms and are exposed to rain and irrigation. Additionally, many factors such as distance between nodes, the height of the antenna, and frequency range must be taken into consideration while deploying the transceiver in large farms [19]. Figure 13.2 specifies the list of smart sensors used in smart Agriculture.

It is highly challenging to manually detect faulty devices due to time constraints, and also these devices are easily subject to human errors. The Agro-IoT environment should have an automated mechanism to predict the faulty devices and also failures in service provision. Many fault prediction approaches are proposed to mitigate the failures before they occur. In these approaches, machine learning capabilities along with real-time complex event processing mechanisms are used for the prediction of device faults. But, a thorough knowledge of the following parameters is essential to construct the machine learning model.

- What kind of faults will occur in the system?
- How will the faults get classified?
- What is the severity of the fault?
- What will be the result of the fault?
- Which device is highly prone to risk or fault?

Here the machine learning framework will collect data in real-time and observe the patterns in them. The objective is to maintain an acceptable level of service even if faults occur in the system.

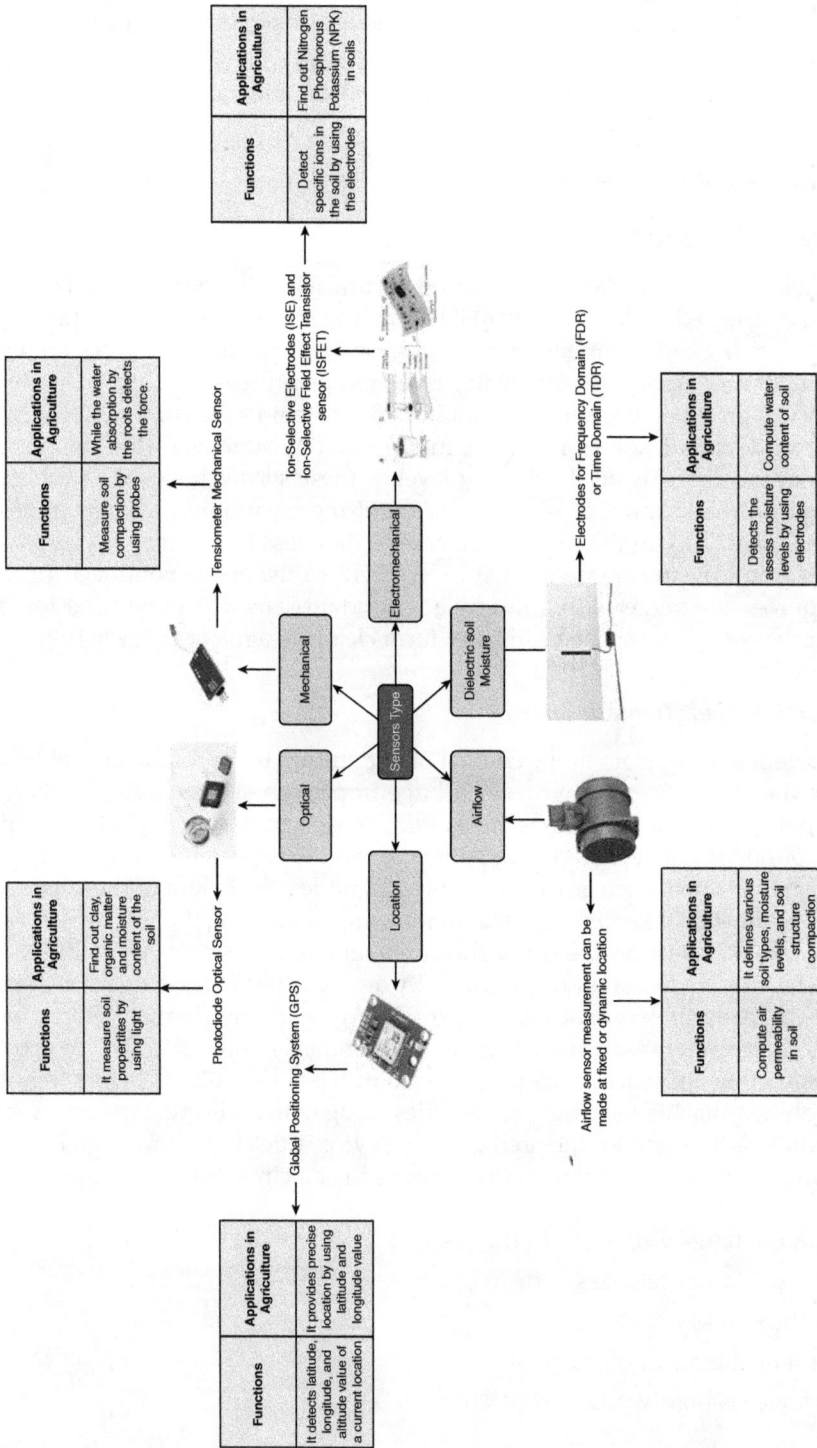

FIGURE 13.2
Smart sensors for smart agriculture.

13.2.2.1.1 Fault Anticipation System

Once the historical data related to the faults are gathered, a supervised learning classifier can use this training data to identify and classify the faulty devices by observing fault patterns in the real-time data collected from the system. As the classifier is trained on known faults in the system, the probability of device failure can be anticipated in good time so that the propagation of fault with the rest of the system is prevented. Many different classifiers are used to select the relevant set of attributes for failure prediction. (1) Perceptron – a more sophisticated kind of neural network classifier exclusively used by supervised learning model for binary classification and they are applicable only to linearly separable objects. (2) Decision tree – a non-parametric supervised learning classifier used for classification and the regression models in which a learning model predicts the value of the target variable based on simple decision rules inferred from the attributes of the dataset. (3) Logistic regression – a linear regression model which uses logistic function particularly to classify list of observations to the discrete set of classes. (4) K-nearest neighbor – an on-demand or lazy classifier that belongs to the category of supervised learning algorithm, which predicts the class of a particular object based on K nearest neighbors to the selected object. (5) Support Vector Machines – a vector space model in which the objects are represented in N-dimensional space and a decision hyperplane is formulated which separates the objects in two different classes; here the prediction is based on the relative position of the object in the hyperplane or decision surface.

13.4 Data Management Issues and Solutions

As agriculture is experiencing the digital revolution with the use of IoT devices, a huge amount of information is passed as data [20]. As data is transmitted wirelessly across several miles, data management involves gathering data from several sources, processing the data, figuring out the information such as receiving mobile alerts, remote monitoring, remote servicing, tracking the device failure, etc. [21]. Even though it is simpler in today's scenario, yet some challenges arise as a result of data sharing and transmission. These challenges need to be addressed. This section presents the challenges and solutions associated with managing IoT data.

- Data security
  When data is interpreted at the control center, there may be a possibility that data gets tampered which leads to incorrect interpretation. This may lead to making wrong decisions in field monitoring. Additionally, there may be a situation where the entire control of the farm can be allegedly handled by an intruder and he can do anything that leads to collapsing the farm. Hence addressing this issue requires employing high data encryption standards and providing strong authorization and access control mechanisms. Today's advanced artificial intelligence techniques can be deployed to perform security audits over the farm regularly.

- Scalability
  Every day enormous amount of data is collected from IoT farms and this is expected to grow exponentially; hence before installing the infrastructure, an agriculturist should have answers to the following questions:
  i.   Whether the infrastructure will support the growth of data for the next five years?
  ii. Will the infrastructure handle large volumes of data in terms of zettabytes?

   iii. Which platform is suitable for analyzing large volumes of data?

   iv. Who is allowed to access and manipulate the data from the cloud?

It is more important to educate and train agriculturists about IoT solutions, data management, IoT success stories, etc.; hence, hiring tech-savvy personnel to maintain data across cloud infrastructure will ensure effective IoT adoption and execution in large farms.

- Reliability
When there is a power failure or any problem in the local ISP, then the entire IoT system will go down. During that time data collection, monitoring, and reporting processes will get interrupted. Natural calamities also make devices unusable. In order to ensure reliability in this adverse condition, two important features like low power support and offline support should be addressed by the IoT market. This facilitates to achieve reliability to utilize sensors and devices under unexpected power or weather conditions.

- Energy Consumption
Agro-IoT devices are intended to operate 24/7 with no exception; hence all electronic devices require energy to operate continuously. These devices should actively participate in data transmission to and from the network adapters, gateways, etc. Apart from cloud storage, all devices require minimum physical storage capacity to handle data. Also remote servers located in large farms need to house the digital content and require excessive energy to operate continuously. Under heavy loads these data centers require a large-scale cooling system that in turn leads to the demand for huge electrical power. In order to meet out these challenges, solar energy source is the best alternative solution. Nowadays, the cost of installing solar is almost declined.

13.5 Security Issues and Solutions

Even though IoT-based agrosystems benefit the agriculture industry, it is highly prone to security breaches and cyber-attacks. For example, while using aerial drones the intruders can take control of drones in spraying pesticides and fertilizers in large amounts leading to the damage of the crops resulting in poor productivity [22]. The intruders can also control the underground sensors and surface sensors to maliciously update the data during transmission. This will affect the growth of crops resulting in the exploitation of the country's economy. When these types of attacks occur on a large scale over smart agricultural system may lead to agroterrorism [23]. Security vulnerabilities are characterized under the following major categories:

   i. The most common attacks targeting smart farming are (1) Data injection attack, where the data collected from several sensor devices are modified to take wrong decisions. (2) Malware injection attack, where the software installed in the smart devices can get infected, leading to improper functioning of devices.

   ii. The next category of attacks that occur on the farmer's side when accessing web portal at the entry level includes weak authentication mechanisms, open ports, and poor update policies.

   iii. Another category of attacks that targets WSNs such as node compromise, open ports, eves dropping, and location tracking are common threats for the destruction of the functionality of IoT devices.

The following solutions may be adopted to address the security issues:

- Lightweight privacy-preserving data aggregation methodology can be deployed in smart agriculture systems that will integrate cryptographic techniques such as homomorphicPaillier encryption Chinese remainder theorem and one-way hash chain. The mean and variance of data collected from edge nodes are calculated using the Chinese theorem. The report received from IoT devices gives huge information about sensing data. This data is encrypted using homomorphicPaillier encryption. Additionally, IoT devices achieve Lightweight authentication through a one-way hash chain technique. Hence false data injection can be avoided.
- Anonymous and privacy-preserving data aggregation protocol (APPA) provides anonymity and unforgeability. This APPA protocol uses the Paillier cryptosystem along with signature-of-knowledge to prevent false data injection attacks and eavesdropping attacks.
- The Agro-IoT enables the collection of data from different sensor sources, which leads to issues concerned with farmers' data privacy. Dynamic privacy protection (DPP) model provides privacy by means of classifying the privacy protection levels. Blockchain-based PKI is also available to provide privacy.
- User anonymity is achieved by deploying hybrid linear combination encryption among IoT devices during communication in order to avoid DoS and impersonation attacks.
- Colluding and inference attacks can be prevented by providing dummy locations that are identified using a greedy algorithm.
- Delegated authentication is required as data is forwarded through many untrusted networks. It is provided by Semi-outsourcing privacy-preserving (SOPP) policy, which incorporates applied elliptic curve cryptography to acquire one-way authentication. SOPP avoids invalid data access and imparts a high level of data integrity.
- The sensor data of Agro-IoT should be equipped with privacy-preserving protocol and also provisioned with authentication codes (Message Authentication Codes [MAC]). This solution ensures that the data received by the farmers is not tampered with during communication.
- Lightweight label-based access control (LACS) provide authentication to Agro-IoT user by substantiating the data integrity while using 5G network. The label-based authentication is used against two attacks, namely, disturbing attack and ignoring attack.
- Secured access to data is enabled by lightweight integrity verification (LIVE) architecture. Merkle Hash Tree algorithm is used to generate tokens that provide access to data through various security levels.
- The Fair Access framework deploys blockchain technology to provide access control such as grant, get, revoke and delegate permissions to use data.

13.6 Summary

To summarize, this chapter provides various issues and solutions in adopting Agro-IoT technologies. Agro-IoT-based technologies provide continuous transformation in the agriculture sector worldwide. If the abovementioned challenges are treated effectively, then the benefits will be more evident and more sustainable. The IoT devices protect the ecosystem and nature of soil by ensuring the proper usage of fertilizers and pesticides. They serve as a win-win solution model by feeding the growing population and preserving nature.

References

1. United Nations. "World Population Prospects 2019: Data Booklet." Statistical Papers – United Nations, Population and Vital Statistics Report (2019). https://population.un.org/wpp/Publications/Files/WPP2019_Highlights.pdf.
2. Rothmuller, Markus, and Barker, Sam. "IOT – the internet of transformation 2020." Juniper Research, April 2020. https://www.juniperresearch.com/document-library/white-papers/iot-the-internet-of- transformation-2020.
3. Anand, Pooja, et al. "IoT vulnerability assessment for sustainable computing: Threats, current solutions, and open challenges." IEEE Access 8 (2020): 168825–168853.
4. Saravanan, K., and S. Saraniya. "Cloud IOT based novel livestock monitoring and identification system using UID." Sensor Review 38 (2018): 21–33.
5. Antony, Anish Paul, et al. "A review of practice and implementation of the Internet of Things (IoT) for smallholder agriculture." Sustainability 12.9 (2020): 3750.
6. Ayaz, Muhammad, et al. "Internet-of-Things (IoT)-based smart agriculture: Toward making the fields talk." IEEE Access 7 (2019): 129551–129583.
7. Soumyalatha, Shruti G. Hegde. "Study of IoT: Understanding IoT architecture, applications, issues and challenges." 1st International Conference on Innovations in Computing & Net-working (ICICN16), CSE, RRCE. International Journal of Advanced Networking & Applications. No. 478. 2016.
8. Talavera, Jesús Martín, et al. "Review of IoT applications in agro-industrial and environmental fields." Computers and Electronics in Agriculture 142 (2017): 283–297.
9. Yadav, Er Pooja, Er Ankur Mittal, and Hemant Yadav. "IoT: Challenges and issues in Indian perspective." 2018 3rd International Conference on Internet of Things: Smart Innovation and Usages (IoT-SIU). IEEE, 2018.
10. Boursianis, Achilles D., et al. "Internet of things (IoT) and agricultural unmanned aerial vehicles (UAVs) in smart farming: a comprehensive review." Internet of Things (2020): 100187.
11. Cambra, Carlos, et al. "An IoT service-oriented system for agriculture monitoring." 2017 IEEE International Conference on Communications (ICC). IEEE, 2017.
12. Mekki, Kais, et al. "Overview of cellular LPWAN technologies for IoT deployment: Sigfox, LoRaWAN, and NB-IoT." 2018 IEEE International Conference on Pervasive Computing and Communications Workshops (Percom Workshops). IEEE, 2018.
13. Codeluppi, Gaia, et al. "LoRaFarM: A LoRaWAN-based smart farming modular IoT architecture." Sensors 20.7 (2020): 2028.
14. Elijah, Olakunle, et al. "An overview of Internet of Things (IoT) and data analytics in agriculture: Benefits and challenges." IEEE Internet of Things Journal 5.5 (2018): 3758–3773.
15. Torroglosa-Garcia, Elena M., et al. "Enabling roaming across heterogeneous IoT wireless networks: LoRaWAN MEETS 5G." IEEE Access 8 (2020): 103164–103180.
16. Petrenko, Alexey S., et al."The IIoT/IoT device control model based on narrow-band IoT (NB-IoT)." 2018 IEEE Conference of Russian Young Researchers in Electrical and Electronic Engineering (EIConRus). IEEE, 2018.
17. Lin, Yi-Bing, et al. "SensorTalk: An IoT device failure detection and calibration mechanism for smart farming." Sensors 19.21 (2019): 4788.
18. Swain, Rakesh Ranjan, Tirtharaj Dash, and Pabitra Mohan Khilar. "A complete diagnosis of faulty sensor modules in a wireless sensor network." Ad Hoc Networks 93 (2019): 101924.

19. Power, Alexander. "A predictive fault-tolerance framework for IoT systems." Diss. Lancaster University, 2020.
20. Shu, Lei, et al. "Challenges and research issues of data management in IoT for large-scale petrochemical plants." IEEE Systems Journal 12.3 (2017): 2509–2523.
21. Ma, Meng, Ping Wang, and Chao-Hsien Chu. "Data management for internet of things: Challenges, approaches and opportunities." 2013 IEEE International Conference on Green Computing and Communications and IEEE Internet of Things and IEEE Cyber, Physical and Social Computing. IEEE, 2013.
22. Ferrag, Mohamed Amine, et al. "Security and privacy for green IoT-based agriculture: Review, blockchain solutions, and challenges." IEEE access 8 (2020): 32031–32053.
23. Görmüş, Sedat, Hakan Aydin, and Güzin Ulutaş. "Security for the internet of things: a survey of existing mechanisms, protocols and open research issues." Journal of the Faculty of Engineering and Architecture of Gazi University 33.4 (2018): 1247–1272.

14

Renewable Energy Sources for Modern Agricultural Trends

S. Arulvel and T. Joshva Devadas

Vellore Institute of Technology
Vellore, India

D. Dsilva Winfred Rufuss

Vellore Institute of Technology
Vellore, India and
University of Birmingham
Birmingham, UK

M. Amutha Prabakar

Vellore Institute of Technology
Vellore, India

CONTENTS

14.1 Introduction 224
14.2 Polyhouse Farming 226
    14.2.1 IoT-Based Polyhouse Farming 227
    14.2.2 Temperature/Humidity/Light Monitoring Sensors 228
    14.2.3 CO_2 Level Monitoring Sensors 228
    14.2.4 Soil Temperature Monitoring and Drip Irrigation 228
    14.2.5 Rodent Monitoring Sensor 228
14.3 Shade Net Farming 228
    14.3.1 IoT-Based Shade-Net Farming 229
14.4 Polybag Farming 231
    14.4.1 IoT-Based Polybag Farming 232
    14.4.2 Soil Moisture Sensor 232
    14.4.3 Lysimeter Sensor 232
14.5 Hydroponic Farming 233
    14.5.1 IoT-Based Hydroponics Farming 234
14.6 Aeroponics Farming 234
    14.6.1 IoT-Based Aeroponics Farming 235
    14.6.2 Humidity Sensor 235
    14.6.3 Temperature Sensor 235
    14.6.4 Pressure Sensor 235
    14.6.5 Water Flow Sensor 236

DOI: 10.1201/9781003185413-14

14.7 Sprinkling-Based Farming 236
     14.7.1 IoT-Based Sprinkling Farming..236
14.8 Conclusions...239
References..241

14.1 Introduction

Agriculture is one of the main sectors, which play a crucial role in the economy of most of the countries. In India, nearly 17% of total GDP was credited to agriculture and also it provides 60% of the employment in the country (Arjun, 2013). Despite the importance of it, the issues faced by the farmers in agriculture were enormous due to the various natural calamities. These natural calamities could cause a great loss to the economy of the individuals as well as to the country. To overcome this, various modern advanced technologies like polyhouse farming (La Notte et al., 2020), shade net farming (Saran et al., 2019), polybag farming (Kasirajan et al., 2012), hydroponics farming (Samreen et al., 2017), aeroponics farming (Tiwari et al., 2020), and sprinkling-based farming (Tarjuelo et al., 1999) were used in the agriculture by the farmers. This has certainly enhanced productivity as well as the barrier for natural calamities. In addition to this, information and Communication Technologies were also integrated in recent years to increase the efficiency, service, and productivity of agriculture products (Nyarko et al., 2021).

Though the technologies are advanced, the sustainability of agriculture is still a problem in terms of quantity with quality aspect. Also, the availability of electricity and water sources could play a vital role in agriculture, which is one of the major drawbacks faced by farmers in recent years. To avoid this, solar energy was integrated with agriculture to provide various functions like electricity (Pascaris et al., 2021) for irrigation, to drive the agriculture machinery (Gorjian et al., 2021; Chadalavada 2021) and to provide a cooling system for storing agriculture-related products (Sadi et al., 2021). Despite the wide usage, there is still greater opportunities for renewable energy sources (solar, geothermal, wind and biomass) in agriculture.

Internet of Things (IoT) is a rapidly emerging technological paradigm over the past decade in which internet plays a significant role to connect the object around the world from anywhere. This is achieved by deploying IoT-associated devices to various domains such as smart homes, wearables, healthcare, and agriculture (Figure 14.1). Rapid deployment of IoT devices becomes easier due to the fast growth in communication technology and decreased cost of sensor devices available in markets.

Sensors were the building blocks of IoT and it can be deployed in agriculture farm fields to improve productivity and to reduce the total cost of farming. It is an undeniable fact that the impact of IoT on connected devices cannot be segregated in this modern world. Today, IoT plays a significant role in almost all domains, especially in agriculture. The deployment of IoT in the domain of agriculture ensures the farmers not to rely on horses and plows. Recent advancements in IoT ensure not to depend on age-old methods and to rely on the idea of IoT-based agriculture and farming. Also, IoT deployment in agriculture not only automates the manual labor but accelerates the farmers to effectively utilize and manage the resources with appropriate precaution in maintaining, monitoring, and predicting the environment. Thus, IoT deployed in agriculture helps the farmers to utilize smart farming gadgets and to have control over the procedure of growing crops. IoT adopted

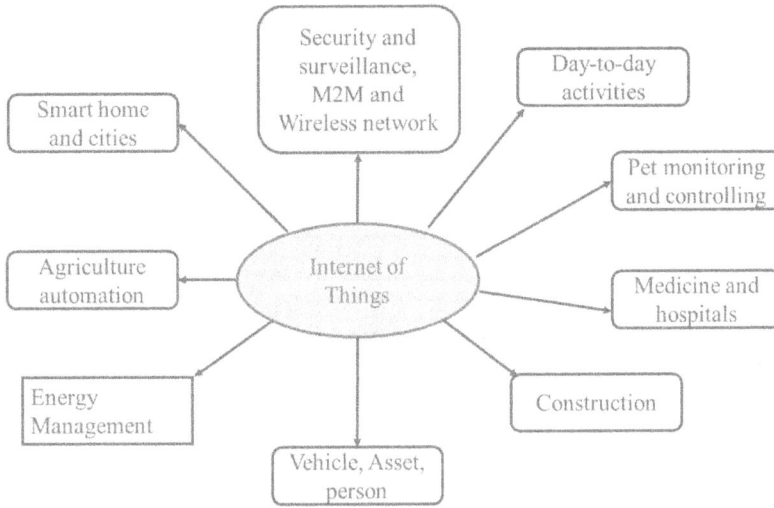

FIGURE 14.1
Domain-specific IoT.

in agriculture predicts the installation of IoT devices may reach 75 million during 2020, which grows by 20% annually and the size may be tripled in 2025. The advancement in IoT technology paves the way for the farmers to cultivate efficiently by suitably deploying appropriate sensors to monitor the field. IoT with its conventional networking standard was enabled suitably using the wireless, Bluetooth, RFID, radio protocols, NFC and Wi-Fi technologies.

Agriculture IoT is explored to focus on water management, soil monitoring, field monitoring, and machines for routine operations. IoT-based smart irrigation system improves crop yields and saves water by determining soil moisture level (Figure 14.2). Also, it determines the amount of water be released when the moisture level goes down using soil sensors and Wi-Fi-enabled computing systems. Moreover, IoT deployed in agriculture

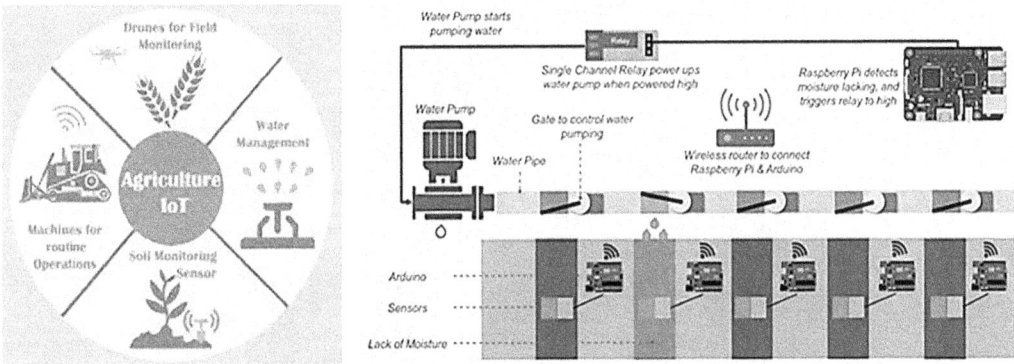

FIGURE 14.2
IoT in agriculture and smart irrigation system flow.

improves productivity by controlling the temperature, humidity, moisture level using actuation devices. Data collected through various sensors are stored, analyzed, and correlated to the yield to improve productivity by optimizing and applying appropriate control strategies.

Also, the effective use of renewable energy like solar, wind, voltaic, solar pumping, and solar space heating improves the yield further and decreases the energy consumption in the modern agriculture system. In addition to this, integrating IoT with modern agriculture enhances productivity, decreases the maintenance cost, and helps to manage the virtual network through online proctoring efficiently. So, the overview of modern agricultural technologies in this chapter could aid the farmers, researchers, and academicians to know and apply IoT and renewable energy in modern agriculture integrated, which could enable a green environment in the near future.

14.2 Polyhouse Farming

Polyhouse technique is one of the modern agricultural greenhouse techniques where different types of covering materials like polythene (polyethene), ethyl vinyl acetate (EVA), and polyvinyl chloride (PVC) are used to provide a safe environment for the crops to grow in a controlled environment (Figure 14.3). This can be done by filtering the harmful radiation from the sun and thus protecting the crops from the adverse effect of natural calamities (Maraveas, 2019). The schematic of the most popular polyhouse structure is depicted in Figure 14.4. Recent research uses various structural modifications to improve the yield, which includes a walk-in tunnel (Figure 14.4(a)), insect-proof net house, gothic roof (Figure 14.4(b)), slant roof, sawtooth (Figure 14.4(c)) and flat roof. Among the various control parameters, thermal environment control and humidity control plays a vital role in improving the crop productivity of polyhouse farming (Von Zabeltitz, 2011).

Renewable energy can be used as an effective choice in polyhouse cultivation by governing the thermal environment and humidity inside the roof structure. Humidity control is

FIGURE 14.3
Actual polyhouse farming (Yelagiri Hills, Tamil Nadu, India).

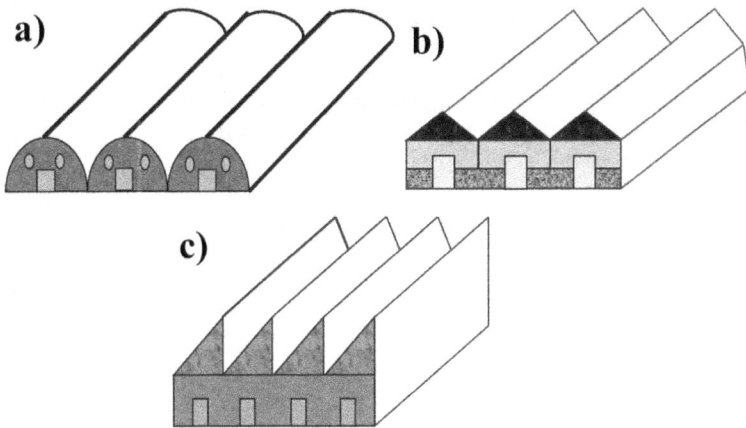

FIGURE 14.4
(a) Walk-in tunnel, (b) gothic roof, and (c) saw tooth.

one of the main aspects of polyhouse farming due to its inevitable control over plant transpiration, photosynthesis, and carbon dioxide exchange between air and leaves (Panwar et al., 2011). This humidity controller can be assisted with the solar photovoltaic (PV) technique for providing the necessary source of power either from a stand-alone PV or grid-connected PV depending on the application and necessity.

Thermal energy storage is a proverbial technique that is getting boomed in the various applications for maintaining the temperature (both hot and cold) during summer and winter seasons. Sensible heat and latent heat are the two techniques used to store the heat energy from the sun during the peak sunny time and release the heat energy during the nocturnal hours. A thermal energy storage tank with latent heat energy storage materials can be integrated into the polyhouse farming structure, which can store both heat energy and cold energy from the sun and atmosphere, respectively (Sarbu et al., 2018). Phase change materials can be used to store the heat energy from the sun and release them during the on-peak period. Depending on the operating environmental parameters such as operating temperature and ambient temperature, the selection of phase change materials can be made for different seasons (summer or winter) to enhance the yield. A sensor can be used to monitor the temperature inside the polyhouse roofing and when the sensor gives a signal, the energy either heat energy or cold energy from the thermal energy storage tank will be released to the enclosure to maintain a controlled environment inside the roof structure.

14.2.1 IoT-Based Polyhouse Farming

This section describes various IoT-based software solutions with highly efficient sensors and internet gateways for storing the track of various data collection points within the environment of poly-house. These data points are connected with the cloud environment or server for analysis. This analysis will identify and provides information about the climate changes inside the polyhouse. If the climate condition exceeds the particular threshold value, an automated alert can be generated and passed to the managerial person. The collected data from sensors can be accessed by the farmers at anytime from anywhere through remote login. The following section explains different kinds of sensors, which are required to enhance polyhouse farming in a better way.

14.2.2 Temperature/Humidity/Light Monitoring Sensors

Temperature, humidity, and light sensors are commonly used sensors for measuring the indoor conditions of polyhouse on a regular interval and this information will be stored as digital data. The collected data is then transferred to IoT applications hosted in a cloud over wireless communication. The climate changes are analyzed based on the threshold value set by the farmers (for the crops). An alarm message will be generated by the cloud server based on the measured data. The immediate change in the Temperature level inside the polyhouse can be maintained by deploying the sprinkler automation. Sprinklers within the polyhouse can be used for temperature control and to improve the ease of maintenance.

14.2.3 CO_2 Level Monitoring Sensors

To improve the growth of plants and to get faster yields, CO_2 enrichments are critical in polyhouse farming. The CO_2 level inside the polyhouse might decrease faster since the crops and plants inhale CO_2 and emit oxygen. Farmers plan to handle this crucial situation and maintain CO_2 levels without jeopardizing worker safety. Periodically, the CO_2 level has been measured through IoT-based CO_2 sensors deployed in the polyhouse environment.

14.2.4 Soil Temperature Monitoring and Drip Irrigation

The soil temperature and humidity are important aspects, which govern the growth of the crops. Drip water irrigation is a globally accepted approach for providing sufficient water supply in farming with a minimal water supply and achieving more profit. The modern polyhouse environment requires an automated drip irrigation approach for maintaining accurate soil conditions. IoT-based soil sensors are deployed in the polyhouse fields farming areas to gather the soil temperature and automatically controls the water supply based on the need.

14.2.5 Rodent Monitoring Sensor

The grasp motion detection can be detected by using proximity sensors in polyhouses to track the people or animal movements at regular intervals. It also checks and alerts in case of rodents entering into the polyhouse structure. IoT technology deployed in polyhouse helps to establish a healthy polyhouse for harvest.

14.3 Shade Net Farming

The shade net farming uses a net house structure (Figure 14.5), which is enclosed with various poly nets to filter the required amount of moisture, sunlight, and air to pass onto the soil. This will also create an appropriate microenvironment for augmenting stable plant growth. Table 14.1 elaborates the various advantages and disadvantages of shade net farming. The major drawback of shade net farming is the presence of high moisture in the soil and plant decay during the winter (Lenka, 2020). Though the soil moisture encourages plant growth, it is important to control its level to avoid the decay of roots of the plants.

FIGURE 14.5
Shade net house farming (Yelagiri Hills, Tamil Nadu, India).

Till now, there is no initiation on overcoming the plant decay during the winter season due to overwatering. Also, overwatering could cause more plant diseases like leaf scorch/burn, edema, and water-soaked spots. This will affect the quality and quantity of the fruits and vegetables during the winter, which leads to the price rise of these products in the market. So, it is important to resist plant decay during the winter to increase the yield of the products. This limitation can be overcome by integrating renewable energy technology with shade net farming. Here, the shade net forming structure will be coupled with the air drier driven by solar energy. The flow path of the heated air from the solar drier will be forced (forced convection) to pass through the way just near the plant to evaporate the required amount of water from the soil to the atmosphere. The rate of hot airflow, the volume of the hot air, the moisture content will be governed using a data acquisition system through which the moisture level will be constantly monitored and the process of the airflow will be stopped once the soil reaches the required moisture. This can be further advanced by employing artificial intelligence with the tri-generation system.

14.3.1 IoT-Based Shade-Net Farming

The invention of emerging technologies paves radical change in the growth and production of crops in the agriculture sector. A protected structure is one among the technology in which shade net is widely used by farmers for cultivation. Initially, the type of the crop,

TABLE 14.1

Advantage and Disadvantages of Shade Net Farming

Advantages	Disadvantages
• Microclimate cultivation.	• High-labor cost.
• Suitable for all types of plants (spices, foliage, flower, vegetables, and medicinal).	• High-production cost.
	• Difficult in drying the soil during the winter.
• Increases moisture content.	• Plant decaying due to overwater during the
• Can be used for nurseries and raising forest species.	winter.
• Quality drying.	
• Prevents pest attacks.	
• Effective graft saplings with low mortality can be produced during summer.	
• Suitable for tissue culture.	

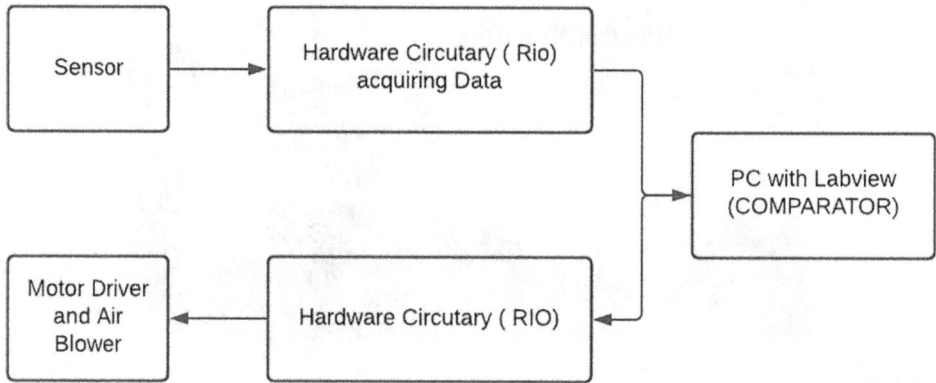

FIGURE 14.6
Functional architecture diagram for IoT-based shade net farming.

climatic condition, and availability of the materials should be analyzed and investigated before the next step i.e., commissioning. Also, the site selection, orientation, and shade-net structure play a significant role before processing the shade-net farming. Varying the percentage of shade factor determines the light intensity, which is chosen based on the crop variety. Choosing appropriate shade-net material is purely relies on the crops being grown such as vegetables, flowers, fruits, and ornamentals. To overcome the deficiencies incurred during winter, IoT-based assistive technologies (Figure 14.6) were deployed in shade-net farming to have better cultivation.

Such integration of shade net farming with IoT devices will monitor soil moisture level (Figure 14.7) and ensures to offer the standard soil moisture level by enabling/disabling the hot airflow with the help of renewable energy resources. To prevent dew formation,

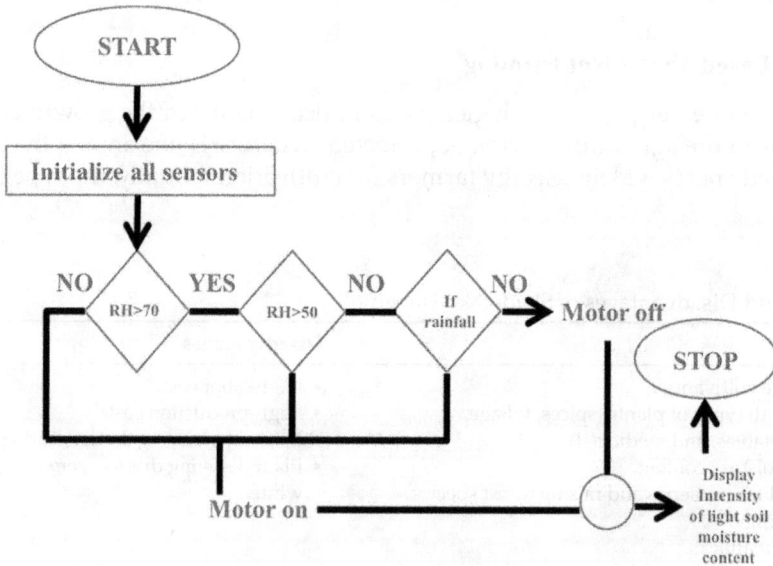

FIGURE 14.7
Flow-chart to control the soil moisture in shade net farming.

the plant leaf may only need 59 degrees Fahrenheit for moisture. So, the soil moisture is recorded in the database by measuring the relative humidity (RH) and temperature through sensors. The ideal humidity for shade net farming is 50–70% RH, but it's tolerable even if it is in between 60% and 80% RH.

This limitation can be overcome by integrating renewable energy technology with shade net farming. Here the shade net forming structure will be coupled with the air drier driven by solar energy. The flow path of the heated air from the solar drier will be forced (forced convection) to pass through the way just near the plant to evaporate the required amount of water from the soil to the atmosphere. The rate of hot airflow, the volume of the hot air, the moisture content will be governed using a data acquisition system through which the moisture level will be constantly monitored and the process of the airflow will be stopped once the soil reaches the required moisture. This can be further advanced by employing artificial intelligence with the tri-generation system.

14.4 Polybag Farming

Polybag farming (Figure 14.8) is one of the effective modern farming techniques that can be implemented in the shorter area space. For example, this type of farming can be used on the terrace of the houses. Polybag farming utilizes polybag materials to store the soil for plant growth. A sufficient amount of soil is mixed with the natural cocopeat and used in the polybags to grow the plants (G Pantuwan et al., 2002). Since cocopeat is mixed with the soil, the moisture along with the nutrition can easily penetrate the roots of the plants and enhances plant growth. Based on the amount of soil and cocopeat is mixed, the growth of the plant can be controlled. Also, the water holding capacity of cocopeat is larger as compared to the normal soil. Hence, it can be very useful in the summer season to maintain the moisture level in the soil. Since individual bags are used for plant growth, the amount of water consumption is greatly reduced for polybag farming as compared to the other modern farming techniques. In addition, if there are any diseases in the plant, they may not transfer to the other plants as they are placed in a separate medium, which is also found to be one of the major advantages of Polybag farming (Al-Shrouf, 2017).

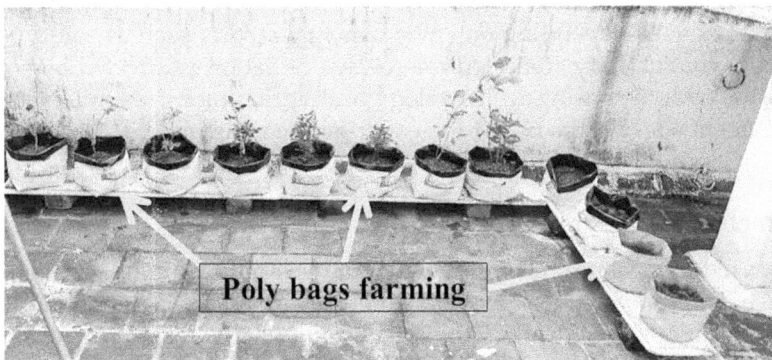

FIGURE 14.8
Polybag farming (Yelagiri Hills, Tamil Nadu, India).

The renewable source of energy like solar photovoltaic or solar thermal can be integrated to govern the temperature of the entire system. Cheaper energy storage materials can also be used in this system to provide the necessary temperature inside the system. Also, when all these systems are integrated, the solar PV or wind turbine technology can act as a primary or secondary source of energy for providing the required amount of power to run the pump thereby the pump work may get reduced which in turn improves the efficiency of the system. Though we have several advantages, polybag farming has few disadvantages like adjusting the nutrient composition, recyclability, soil contaminants, and supporting the weight of the plant. Generally, cocopeat consists of natural salts, which could offer a high cation exchangeability to enhance plant growth (Xiong et al., 2017). However, these salts are not suitable for recyclability and hence, after the utilization, the cocopeat was thrown as waste into the environment. Among the various drawbacks, nutrient composition and soil contaminants play a vital role in plant growth. Hence, it is important to monitor the nutrition composition and soil contaminants regularly to overcome the major drawbacks of polybag farming. For effective integration of the above-said components like cocopeat, solar PV, solar thermal unit, AI plays a vital role in governing the entire circuit and to monitor the same.

14.4.1 IoT-Based Polybag Farming

The role of IoT-based software solutions provides a greater way in polybag farming by applying appropriate sensors along with the internet to collect data from various points from the polybag environment. IoT-based polybag farming is thus improved by monitoring the entire system and improves its performance. To deploy IoT-based polybag farming, the following section elaborates on the various sensors needed to enhance polybag farming efficiently and effectively.

14.4.2 Soil Moisture Sensor

Soil moisture sensors are designed to measure the volumetric water content in soil with indirect measurement. Since the direct gravimetric measurement of free soil moisture requires removing, drying, and weighing of a sample, soil moisture sensors measure the volumetric water content indirectly by using the properties of the soil, such as electrical resistance, dielectric constant, or interaction with neutrons as a proxy for the moisture content. The relationship between the measured property and soil moisture must be calibrated and it may vary depending on environmental factors such as soil type, temperature, or electric conductivity. Reflected microwave radiation is affected by soil moisture and is used for remote sensing in hydrology and agriculture. In some places, portable probe instruments may also be used for measuring the moisture content of the soil.

14.4.3 Lysimeter Sensor

Lysimeter is used to measure the amount of actual evapotranspiration which is released by plants (usually crops or trees). By recording the amount of precipitation that an area receives and the amount lost through the soil, the amount of water lost to evapotranspiration can be calculated. Lysimeter can be classified based on weighing and non-weighing techniques. These techniques can be employed in polybag farming for the effective measurement of evapotranspiration.

14.5 Hydroponic Farming

Hydroponic System is the most advanced farming system used in today's modern agriculture. Hydroponics farming grows the crops without the use of soil and is often referred to as soilless farming (Figure 14.9). The roots of the plant are fully grown inside the liquid nutrient solution.

In the hydroponic system, the plant roots generally grow in a liquid nutrient solution or inside moisture inert materials such as vermiculite and Rockwool. There are five foundational elements like freshwater, oxygen, nutrition, root support, and light which are required for the hydroponic system (Maucieri et al., 2019). Among the five, water is the basic stuff for the hydroponic system. Freshwater with a pH of 6–6.5 is more suitable for the plants to grow healthily. Hence, it is mandatory to maintain the pH level in the hydroponics system to increase the productivity of the crop. Also, the oxygen supply at the roots is required to provide the respiration process. So, it is complicated in designing the hydroponic system to oxygenate the container, where the plants are placed. The lack of nutrition like calcium, phosphorus, and magnesium could also influence the health and productivity of the plants. From the five foundational elements, it is very clear that the process of controlling the pH, oxygen, and nutrition could enhance the productivity of the plants in hydroponics farming (Trejo-Téllez et al., 2012). Solar energy can be used to maintain a certain temperature in the hydroponic system. Also, the oxygen extractor/separator (from air) filter can be used which can operate using the renewable power source (solar/wind) rather than conventional. Thus, the oxygen level, nutrition, and water can be maintained in the system using renewable sources of energy. IoT can also be integrated with hydroponics to increase crop production.

FIGURE 14.9
Hydroponics farming.

14.5.1 IoT-Based Hydroponics Farming

A hydroponic farming system typically contains three water tanks to hold clean freshwater i.e., nutrient-enriched water, pH-controlled water, and wastewater. The nutrient-enriched and pH-controlled water comprise the growth medium for the plants. The pH level and nutrient composition in this water reservoir are controlled using valves and pumps to add nutrients, carbon dioxide, or freshwater. Draining wastewater from the nutrient-enriched water reservoir helps to maintain consistent water levels while adjusting the nutrient concentration and pH.

To set up a hydroponic farming system, conductivity sensors and pH sensors should be installed in the water reservoir that will be used to supply it to the plants' growth medium. The pH controllers and conductivity controllers give signals for the opening and closing of valves or the operation of pumps based on the measurement data from the sensors. For example, if the nutrient concentration in the water reservoir became too high, the conductivity controller could turn on the freshwater pump to dilute the nutrients. Conversely, if nutrient concentration became too low, the conductivity controller could turn on the nutrient pumps. Similarly, if pH became too high, the pH controller could open the solenoid valve, allowing carbon dioxide to flow into the water reservoir. Carbon dioxide reacts with water to form carbonic acid, which lowers the pH of the solution.

- High nutrient concentration > high conductivity reading > add fresh water
- Low nutrient concentration > low conductivity reading > add nutrients
- Overly alkaline solution > high pH reading > add carbon dioxide.

14.6 Aeroponics Farming

Aeroponics farming is the advanced form of hydroponics farming, where the plants are grown completely in the water and nutrition medium (Figure 14.10). The utilization of water in aeroponics farming was nearly 98% lesser than conventional farming and the productivity was 30% higher than traditional farming. Aeroponics farming was suitable for all types of species as in conventional farming (Lakhiar et al., 2018).

Similar to the hydroponics farming system, pH, oxygen, and nutrition are the mandatory requirements for aeroponics farming. However, one of the major drawbacks of aeroponics farming is maintaining the freshness of the water. The water has to be changed continuously to avoid root decay. This type of aeroponics farming gives high quality of vegetables which are mostly leafy crops. Here, the automatic spraying and the closed-loop water circuit play a vital role in the functioning of the entire system. Renewable energy can be employed in automatic spraying and water circulation system by proving the necessary power from solar and wind energy. When this farming is carried out near a system that produces waste heat, the produced waste heat can be utilized to heat or cool the greenhouses, which reduces the energy input to the system and thereby improve the efficiency of the system (González-Briones et al., 2018). The whole system can be integrated with AI for efficient management. Hence, monitoring the freshness of water and providing the actual values of pH, oxygen and nutrition are important in aeroponics farming, which makes IoT come into the picture for ease of use.

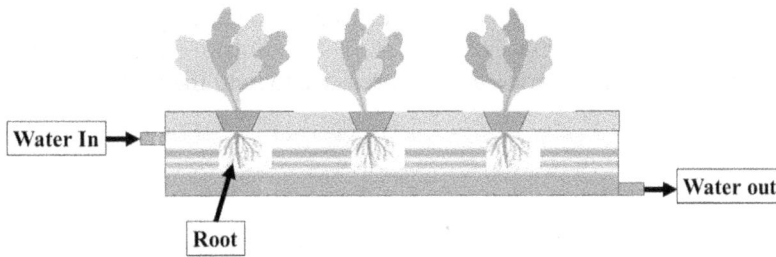

FIGURE 14.10
Aeroponics farming.

14.6.1 IoT-Based Aeroponics Farming

Aeroponics is a culture technique where the plant roots are suspended in the air and are spasmodically sprayed with a nutrient solution. This technique has been used both for research study and to harvests profitable production. Aeroponics presents excessive benefits over traditional agriculture methods. This kind of method reduces the consumption of water and nutrients, increases the growth rate and plant density.

The system should continuously monitor the energy consumption, the level of the nutrient solution and the correct operation of pumps and valves, which can be performed through the application of IoT. The following paragraph describes the methods of monitoring aeroponics using wireless sensors.

14.6.2 Humidity Sensor

Humidity sensor used for measuring the humidity level of water present in the air. The amount of water vapor in the air can affect human comfort as well as many manufacturing processes in industries. The presence of water vapor also influences various physical, chemical, and biological processes. In agriculture, the measurement of humidity is important for plantation protection (dew prevention) and soil moisture monitoring. In agriculture applications, humidity sensors are employed to indicate the moisture levels in the environment.

14.6.3 Temperature Sensor

A temperature sensor is a device, usually a resistance temperature detector (RTD) or a thermocouple that collects the data about temperature from a particular soil and environment. This will be converting the measured data into an understandable form to a device or an observer.

The most common type of temperature sensor is a thermometer, which is used to measure the temperature of solids, liquids, and gases in the agricultural land. The water content of the soil and environment can be changed based on the measurement value. This control can be activated through IoT-enabled devices from the remote or nearby place.

14.6.4 Pressure Sensor

A pressure sensor is a device for pressure measurement in gases or liquids. Pressure is an expression of the force required to stop a fluid from expanding and is usually stated in terms of force per unit area. A pressure sensor usually acts as a transducer; it generates a signal as a function of the pressure imposed. The water pressure level can be changed in the planted area to reduce or increase the pressure to help the system to get back to the default mode/condition.

14.6.5 Water Flow Sensor

Water flow measurement is the quantification of bulk fluid movement. Flow can be measured in a variety of ways. Positive displacement flow meters accumulate a fixed volume of fluid and then count the number of times the volume is filled to measure the flow. In the planted area, this reading is used to maintain the water flow in a normal mode. This measurement is used to improve the water level by regulating the controller, which can be done through IoT.

14.7 Sprinkling-Based Farming

Sprinkled-based farming (Figure 14.11) is also the advanced farming and irrigation system used by farmers in recent years. Sprinkling irrigation has some major advantages over the drip irrigation system (Figure 14.12). In this system, the water is irrigated to the farms through sprinklers, which is similar to natural rainfall (Patel et al., 2020). Also, it will create a moisture environment, which is required for plant growth during the summer season. However, in comparison, the utilization of water is slightly higher for sprinkling-based farming than that of drip irrigation. In addition to the moisture environment, the supply of essential nutrition could also augment the growth process of the plants. Here, renewable energy can be coupled with sprinkling-based farming to replace the fossil power source with a renewable power source for pumping and irrigation. Hence there will be a substantial reduction in greenhouse gas emissions and annual economic uplift in the livelihood of the farmers by reducing the cost of kerosene, diesel, and petrol.

14.7.1 IoT-Based Sprinkling Farming

Effective water management is a vital aspect and needs to be addressed in the agriculture sector because around 60% of water spent on agriculture is wasted due to various

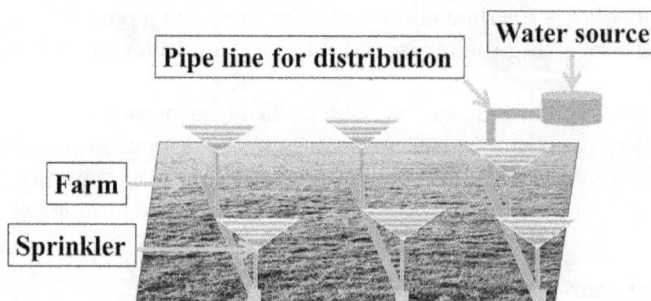

FIGURE 14.11
Sprinkle-based farming.

FIGURE 14.12
Drip irrigation (Yelagiri Hills, Tamil Nadu, India).

reasons such as contamination, overwatering, runoffs, and other related issues. Moreover, the crops were damaged due to either overwatering or under-watering. To overcome this deferred scenario, the technology deployed should be improved and optimized with the deployment of IoT in the agricultural field with advanced sensors to observe and monitor the field.

The fundamental requirements for irrigation of crops should be estimated along with the soil moisture content level. Based on the need, the system should effectively utilize water resources. Sprinkling-based farming added the advanced features of IoT technology to effectively monitor the moisture level of the field and to control the sprinklers with smart irrigation controllers for watering.

For the deployment of IoT devices in agriculture, a user interface is to be modeled to read/observe the value of the sensors used in the field through the service platform. This platform records information such as temperature, humidity, pH balance, nutrient levels, LED lights, and water levels (Kern et al., 2017). The service platform facilitates a web interface that accesses the data using mobile and saves the gathered sensor information using the connected IoT devices (Figure 14.13).

Raspberry PI Zero is used to gather information from the sensor to control the water pumping or dosing. Sequential query language (SQL) queries are used to transmit the saved information to the data server for processing. Also, this Raspberry PI Zero is connected with various sensors to record the temperature, humidity, water level, LED light, dosing pump, and submersible pump. IoT devices connected with the sensors gather the associated information and triggers appropriate action through relays. For example, if the nutrition level is lower than a certain threshold value, the connected dosing pump starts to work till the expected threshold level is reached. Similarly, it works in the same manner for all the rest of the cases.

To ensure objective irrigation, controllers are installed in the system to minimize water usage. Based on the type of crop field, either a channel system or sprinkler system or drip system will be used for efficient farming. In sandy soils, a sprinkler system is used even when the available water quantity is small. An irrigation controller consists of a mesh network that works at 868 MHz ISM frequency band (Figure 14.14) is designed to consume low energy with transmitting data capacity up to 15 km. This model is designed to consume less energy with a wide bandwidth for acquiring data through IoT devices (Cambra et al., 2017).

To process the data collected through various sensors and to make quick and smart decisions, a service-oriented architecture is applied using an intelligent farming system

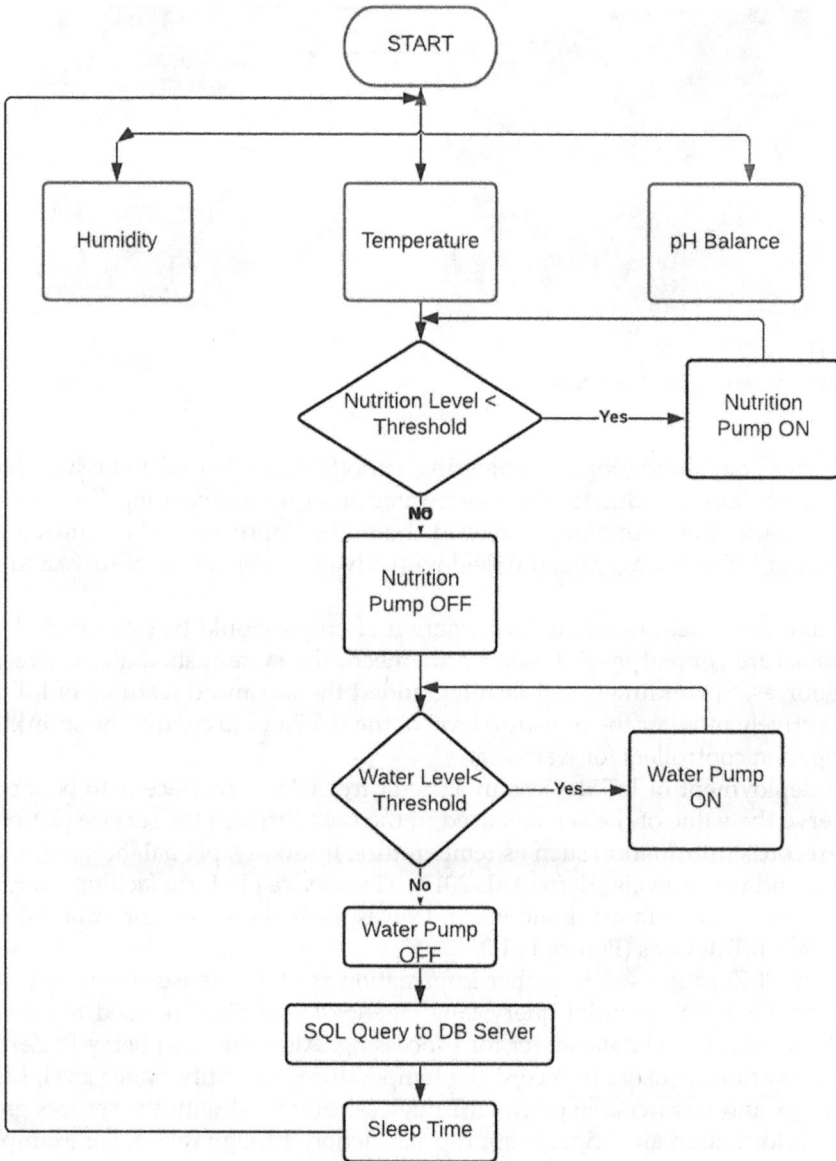

FIGURE 14.13
Flow diagram for IoT device.

(Figure 14.15), that eliminates the relationship between the service provider and farmers. Such intelligent farming comprises a service cloud, a service consumer, and a server. Service cloud deals with all sensor data, service consumers determine relevant services by exploring the cloud.

The server has major components such as application and rule container. The application component helps to register the sensor data in the system. The rule container uses a rete pattern matching algorithm. The objective of the algorithm is to reduce or remove redundancy and allows deleting the memory when facts are withdrawn.

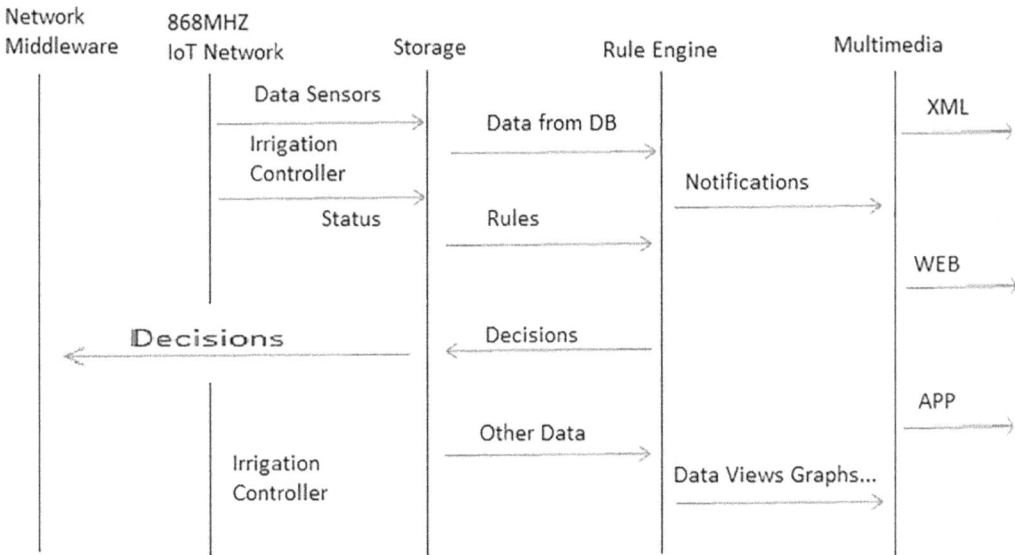

FIGURE 14.14
Data annotation process path.

14.8 Conclusions

The following conclusions were drawn from the detailed discussion on the various types of modern agricultural techniques, which can be integrated with renewable energy and IOT:

- Renewable energy could be an effective method in polyhouse cultivation for controlling the thermal environment and humidity with the help of solar photovoltaic (PV). Through IoT, an alert can be generated if the climate condition exceeds the particular threshold value. In addition, the various sensors for controlling the temperature, humidity, light monitoring, CO_2 level, soil temperature monitoring, and drip irrigation can be employed through IoT in polyhouse farming.

- The integration of IoT devices with shade net farming will monitor and control the soil moisture level. This ensures the required moisture level inside the system by enabling and disabling the hot airflow with the help of renewable energy resources.

- Renewable energy sources like solar PV or solar thermal can be integrated to control the temperature inside the polyhouse farming. The sensors like soil moisture sensor and lysimeter sensor can be used for the indirect measurement of volumetric water content in soil and the amount of actual evapotranspiration from the system.

- The nutrient of liquid or moisture inert materials is important for hydroponic farming. This can be achieved by maintaining the nutrient and oxygen level inside the farming house. For this, the oxygen extractor/separator (from air) filter, which operates using the renewable power source (solar/wind) can be preferred. The conductivity sensors and pH sensors are preferred to control the water reservoir.

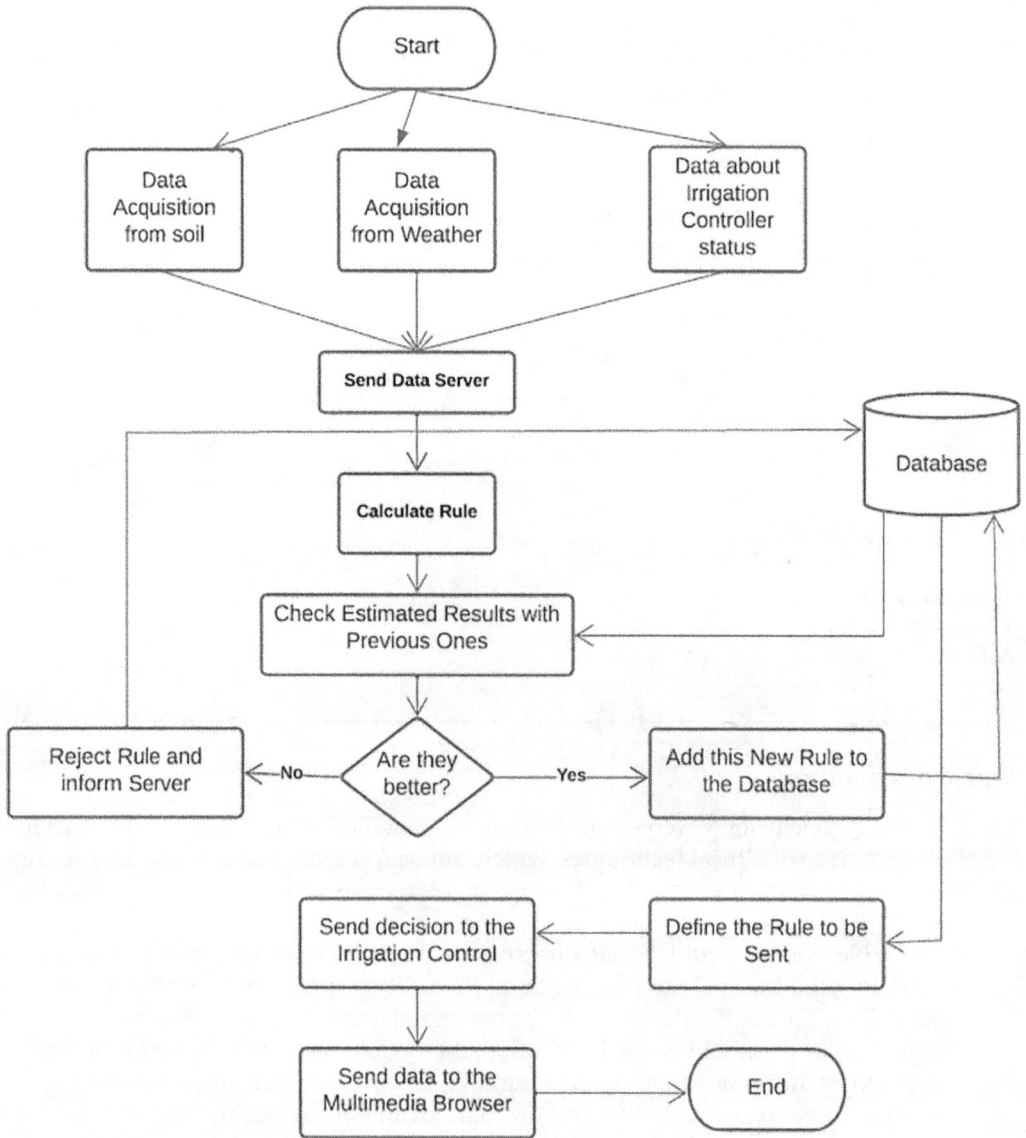

FIGURE 14.15
Flow chart for precision agriculture.

- Water management is the main advantage of drip irrigation. To control this with intelligent farming, the service cloud, service consumer, and a server is designed. Service cloud mainly deals with all sensor data and service consumers determine the relevant service by exploring the cloud, which ensures the minimal usage of water for irrigation.

It is thus concluded that each farming technique has unique functioning characteristics, and hence, the specific renewable energy and IoT can be preferred for the effectiveness of the farming and to improve the plant life, growth, and yield.

References

Al-Shrouf, Ali. "Hydroponics, Aeroponic and Aquaponic as Compared with Conventional Farming." American Scientific Research Journal for Engineering, Technology, and Sciences (ASRJETS) 27, no. 1 (2017): 247–255.

Arjun, Kekane Maruti. "Indian Agriculture – Status, Importance and Role in Indian Economy." International Journal of Agriculture and Food Science Technology 4, no. 4 (2013): 343–346.

Cambra, C., S. Sendra, J. Lloret, and L. Garcia. "An IoT Service-Oriented System for Agriculture Monitoring," 2017 IEEE International Conference on Communications (ICC), 2017, pp. 1–6.

Chadalavada, Hemadri. "Solar Powered Semi-Automated Multipurpose Agriculture Machine." Materials Today: Proceedings (2021), Paris, France. doi: 10.1109/ICC.2017.7996640.

González-Briones, Alfonso, Pablo Chamoso, Sara Rodríguez, Hyun Yoe, and Juan M. Corchado. "Reuse of Waste Energy from Power Plants in Greenhouses through MAS-Based Architecture." Wireless Communications and Mobile Computing 2018 (2018). doi: 10.1155/2018/6170718.

Gorjian, Shiva, Hossein Ebadi, Max Trommsdorff, H. Sharon, Matthias Demant, and Stephan Schindele. "The Advent of Modern Solar-Powered Electric Agricultural Machinery: A Solution for Sustainable Farm Operations." Journal of Cleaner Production 292 (2021): 126030.

Kasirajan, Subrahmaniyan, and Mathieu Ngouajio. "Polyethylene and Biodegradable Mulches for Agricultural Applications: A Review." Agronomy for Sustainable Development 32, no. 2 (2012): 501–529.

Kern, Stephen C., Joong-Lyul Lee, "Automated Aeroponics System Using IoT for Smart Farming," 8th International Scientific Forum, ISF 2017, USA, pp. 104–110, doi: 10.19044/esj.2017.c1p10.

La Notte, Luca, Lorena Giordano, Emanuele Calabrò, Roberto Bedini, Giuseppe Colla, Giovanni Puglisi, and Andrea Reale. "Hybrid and Organic Photovoltaics for Greenhouse Applications." Applied Energy 278 (2020): 115582.

Lakhiar, Imran Ali, Jianmin Gao, Tabinda Naz Syed, Farman Ali Chandio, and Noman Ali Buttar. "Modern Plant Cultivation Technologies in Agriculture under Controlled Environment: A Review on Aeroponics." Journal of Plant Interactions 13, no. 1 (2018): 338–352.

Lenka, Santosh Kumar. "Cultivation of Crops under Shade-Net Greenhouse." In Protected Cultivation and Smart Agriculture pp. 167–174. New Delhi Publishers, New Delhi, 2020.

Maraveas, Chrysanthos. "Environmental Sustainability of Greenhouse Covering Materials." Sustainability 11, no. 21 (2019): 6129.

Maucieri, Carmelo, Carlo Nicoletto, Erik Van Os, Dieter Anseeuw, Robin Van Havermaet, and Ranka Junge. "Hydroponic Technologies." Aquaponics Food Production Systems (2019): 77–110.

Nyarko, Daniel Ayisi, and József Kozári. "Information and Communication Technologies (ICTs) Usage among Agricultural Extension Officers and Its Impact on Extension Delivery in Ghana." Journal of the Saudi Society of Agricultural Sciences 3 (2021) 164–172.

Pantuwan, G, S. Fukai, M. Cooper, S. Rajatasereekul, and J. C. O'Toole. "Yield Response of Rice (Oryza sativa L.) Genotypes to Different Types of Drought under Rainfed Lowlands: Part 1. Grain Yield and Yield Components." Field Crops Research 73, no. 2–3 (2002): 153–168.

Panwar, N. L., S. C. Kaushik, and Surendra Kothari. "Solar Greenhouse an Option for Renewable and Sustainable Farming." Renewable and Sustainable Energy Reviews 15, no. 8 (2011): 3934–3945.

Pascaris, Alexis S., Chelsea Schelly, Laurie Burnham, and Joshua M. Pearce. "Integrating solar Energy with Agriculture: Industry Perspectives on the Market, Community, and Socio-Political Dimensions of Agrivoltaics." Energy Research & Social Science 75 (2021): 102023.

Patel, Nirali Hemant, and Chintan Rajnikant Prajapati. "Agricultural Sprinkler for Irrigation System." International Journal of Engineering and Technical Research 9, no. 5 (2020): 162–166.

Sadi, Meisam, Ahmad Arabkoohsar, and Asim Kumar Joshi. "Techno-Economic Optimization and Improvement of Combined Solar-Powered Cooling System for Storage of Agricultural Products." Sustainable Energy Technologies and Assessments 45 (2021): 101057.

Samreen, Tayyeba, Hamid Ullah Shah, Saleem Ullah, and Muhammad Javid. "Zinc Effect on Growth Rate, Chlorophyll, Protein and Mineral Contents of Hydroponically Grown Mungbeans Plant (*Vigna radiata*)." Arabian Journal of Chemistry 10 (2017): S1802–S1807.

Saran, Parmeshwar L., Susheel Singh, Vanrajsinh H. Solanki, Kuldeepsingh A. Kalariya, Ram P. Meena, and Riddhi B. Patel. "Impact of Shade-Net Intensities on Root Yield and Quality of Asparagus racemosus: A Viable Option as an Intercrop." Industrial Crops and Products 141 (2019): 111740.

Sarbu, Ioan, and Calin Sebarchievici. "A Comprehensive Review of Thermal Energy Storage." Sustainability 10, no. 1 (2018): 191.

Tarjuelo, J. Montero, J. Montero, F. T. Honrubia, J. J. Ortiz, and J. F. Ortega. "Analysis of Uniformity of Sprinkle Irrigation in a Semi-Arid Area." Agricultural Water Management 40, no. 2–3 (1999): 315–331.

Tiwari, Jagesh K., D. E. V. I. Sapna, Tanuja Buckseth, A. L. I. Nilofer, Rajesh K. Singh, Rasna Zinta, and Swarup K. Chakrabarti. "Precision Phenotyping of Contrasting Potato (*Solanum tuberosum* L.) Varieties in a Novel Aeroponics System for Improving Nitrogen Use Efficiency: In Search of Key Traits and Genes." Journal of Integrative Agriculture 19, no. 1 (2020): 51–61.

Trejo-Téllez, Libia I., and Fernando C. Gómez-Merino. "Nutrient Solutions for Hydroponic Systems." Hydroponics – A Standard Methodology for Plant Biological Researches 23 (2012): 1–22.

Von Zabeltitz, Christian. "Greenhouse Structures." In Integrated Greenhouse Systems for Mild Climates, pp. 59–135. Springer, Berlin, Heidelberg, 2011.

Xiong, Jing, Yongqiang Tian, Jingguo Wang, Wei Liu, and Qing Chen. "Comparison of Coconut Coir, Rockwool, and Peat Cultivations for Tomato Production: Nutrient Balance, Plant Growth and Fruit Quality." Frontiers in Plant Science 8 (2017): 1327.

15

SPLARE

A Smart Plant Healthcare System

S. Lodha, H. Malani, and N. Prasanth
School of Computer Science and Engineering
Vellore, India

K. Saravanan
Anna University Regional Campus
Tirunelveli, India

CONTENTS

15.1 Introduction 243
    15.1.1 Objectives 244
15.2 Related Works 244
15.3 Methodology 246
    15.3.1 System Architecture 246
    15.3.2 Technical Specifications 246
    15.3.3 Virtual Simulation 249
15.4 Hardware Implementation Procedures 250
    15.4.1 Power Supply and Testing 253
    15.4.2 Connecting to Cloud-Based Service 256
15.5 Results and Discussion 256
    15.5.1 Analysis 258
15.6 Conclusion 260
References 260

15.1 Introduction

With the increasing use of the Internet of Things (IoT) in day-to-day life, objects around us are becoming "smart," such that they can transmit data and automate tasks without significant human intervention [1]. With applications ranging from something as small and simple as a smart clock to a smart home to even a large-scale variance in the form of smart cities, IoT is the future of this generation. As a part of one of the IoT-based solutions, the idea is to develop a smart plant healthcare system named "SPLARE," which is, as it sounds, an automated system to not just water plants in an efficient manner but also keep a check of their health (including factors such as pH of soil and soil nutrient content). The aim is not inclined toward a large-scale smart irrigation system for agriculture purposes (which is already vastly used today) but toward a common man's domestic household. Growing plants at home for a wide range of purposes is a very common and popular phenomenon these days, and people admire such personal mini gardens, which form our target audience.

The intention is to tackle a variety of real-life problems. First of all, addressing an urgent issue of water scarcity. With existing irrigation systems such as drip irrigation and sprinkler irrigation, water wastage is reduced to a great extent in agricultural fields. However, with people's growing care and interest toward their own mini personal gardens or keeping plant pots, the agenda of water wastage kicks in on this level too. Imagine if thousands of such households are not able to efficiently control the amount of water needed for their plants? In case more than required water is used, along with affecting plants due to excessive moisture, some of this amount is wasted too. On the other hand, less than what is required can make the soil dry and consequently affect the plants and their growth [2]. With an efficient smart system, the agendas of soil moisture monitoring and creating an automated water supply are beautifully handled. Other than that, there shall also be a check on the surrounding temperature as well as the soil nutrient content of the plant. If either of these has any alarming or insufficient levels, an automated notification can be sent to the user stressing the current status and suggesting required actions (if needed) to restore a healthy habitat. By implementing this feature, a "smart watering system" can be extended to an overall smart healthcare system.

15.1.1 Objectives

The various objectives of this study include:

- To simplify and automate the plant watering system. The system will determine soil moisture content, temperature and humidity, amount of nitrogen, phosphorous, and potassium in the soil, pH of the soil, and the atmospheric pressure for each plant.
- A set of customized thresholds will be set for each plant for all the parameters mentioned above. Based on these thresholds, along with readings from respective sensors, the solenoid valve will be either turned on or off.
- Weather conditions can also shorten the life of the garden, as some crops/plants might perish due to a lack of rainfall, extreme heat, and humidity, among other things. We will also include the overall care properties and precautions of well-known plants, which will be displayed as a recommendation in response to certain parametric values, as a result enabling users to take better care of their plants.
- Along with other features, the system can be integrated with any cloud-based service like ThingSpeak, Firebase, or Blynk for easier tracking and user convenience.

15.2 Related Works

New systems based on sensors, software, and communication protocols for the automation of certain tasks have been developed using IoT concepts and knowledge [3]. One such exploration involved implementing an automatic plant irrigation system to control the measure of water utilizing Arduino Uno R3 and soil moisture sensors and controlling the system consistently via the observing station [4]. Another unique system from a study to monitor and operate a watering system employed an Arduino mega 2560 that has been modified with GSM technology, allowing the Arduino setup to receive/send SMS to/from the mobile phones of farmers/homeowners based on the soil's demand for water or the user's instructions [5]. However, these models took no measure to check soil fertility, nutrient levels, and environmental conditions. If we think about it, nowadays, water quality and pH play a major role in determining the soil's health – hence, indirectly, the plants' health. By not detecting these features such as the NPK value and temperature of the surrounding environment, direct watering based on just the soil moisture cannot just harm the plants but also affect soil dynamics. Moreover, extensive involvement of manual work in sending messages boils down to the problem of relying on the controller's remembrance, lethargy, and awareness for timely watering and maintenance of the plant.

The goal of another study was to create an autonomous watering system including hydroponic fertilizers, with the goal of developing, building, and testing a system that can automatically water plants, drain hydroponic nutrients, and fog the plant environment, as well as monitor the temperature in the plant house [6]. In this research, the model sprays the nutrients according to the needs of the plant. Again, here, there's a complete dependency on the "spray," a glitch in which can make the soil acidic/basic. The spray quantity, range, and management also add to the overheads. In our new model, we use a pH sensor to detect the pH value of the soil and all the details are sent using a third-party application. Hence, we get the information about the pH of the soil and all related information and if there is anything suspicious happening with our system, we can fix it.

One work was aimed at smart irrigation through intelligent control and decision-making based on precise real-time field data. It also included monitoring of the operations using an Android application [7]. There was indeed a very close application also explored here; dropping of moisture level would cause the Arduino board to activate a water pump. Sensing the water level in the water tank and sending this information to the microcontroller to fill the tank incorporated the benefit of reducing manual work as well as facilitate better watering. However, the problems that creep in here are mainly weather-related. Not accounting for the surrounding pressure, temperature and humidity can stultify the sensors' readings, often leading to the overflowing tank, water wastage, over or under watering the plants.

The primary goal of another system was to use color sensors to identify the levels of nitrogen (N), phosphorus (P), and potassium (K) in soil and adding appropriate quantities whenever required [8]. Similar implementation was depicted using optic sensors and Arduino Uno by passing a particular intensity of a particular wavelength through a soil sample and recording this intensity and shift in wavelength at the receiver end to determine the concentration of ions in the soil sample [9]. Although these implementations reduce the burden of measuring the amount of nutrients in the soil at a higher cost, there is still potential to give way to the use of unnecessary fertilizers in the soil and reducing the system's accuracy. This is because of sole reliance on NPK sensors, as well as not accounting for surrounding conditions. Moreover, failure to respond to the readings in an automated manner or linking the details to any cloud-based service makes the setup extremely inefficient and non-responsive. A soil moisture sensor was used to determine whether the watering pump should be turned on or off, controlled by Arduino, which would send a high or low signal response to the relay module [10, 11].

The threshold was set to 35% after certain computations, which was, however, only based on the readings from the soil moisture sensor and did not consider any other factors or sensors (as discussed earlier in this study). No application of any cloud-based system also made it difficult to track the estimated moisture level via any service. After considering other environmental factors and nutrients measurements, the appropriate threshold has been updated in this study.

A variety of sensors to evaluate soil and environmental variables was employed and compared to a threshold to determine whether the plant needs to be watered using a decision model, also considering many development circumstances for forecasting a plant's health and its maximum growth [12]. Including automatic start and stop of an electric motor based on a water threshold value, a flame detector with a buzzer to sound a warning in the event of a fire, etc., the computational complexity of the find-S algorithm was reasonably low, with a time complexity of $O(n)$. The same is applicable for our study with enabled cloud-based environment and services.

15.3 Methodology

15.3.1 System Architecture

Figure 15.1 shows the block diagram of the proposed system. It shows a basic data flow of the system and indicates how the various components are interconnected. NodeMCU facilitates as a central hub. All other sensors, actuators, and components are directly connected to it. Details of the components are discussed under Section 15.3.2.

Figure 15.2 shows the actual logic that's backing SPLARE. Once the system is initiated, data is collected and processed from all the concerned sensors. Based on this processed data, a preset threshold condition is checked and the solenoid valve is either turned on or off as required. Simultaneously, this processed data is uploaded to a third-party cloud-based service every few minutes so that end users can view it via a proper GUI.

15.3.2 Technical Specifications

Given below are the Specifications of the required components.

NodeMCU: NodeMCU is an open-source Wi-Fi development board on LUA. Its default version embeds an ESP8266, but other variants like ESP12 are also available. "MCU" in NodeMCU stands for Microcontroller Unit.

ESP8266: The ESP8266 is a low-cost Wi-Fi microchip that has full TCP/IP stack and microcontroller capabilities. The ESP8266 module enables any kind of microcontroller to connect to a single band Wi-Fi with a frequency of 2.4GHz, using IEEE 802.11 bgn.

Soil Moisture Sensor: A soil moisture sensor is used to measure the water content in the soil. Soil moisture sensors have two probes, which are responsible for the flow of current through the soil. Knowing the values of potential difference and amount of current flowing, the resistance of the soil can be calculated. This in turn, helps to measure the moisture level of the material (or soil in this case).

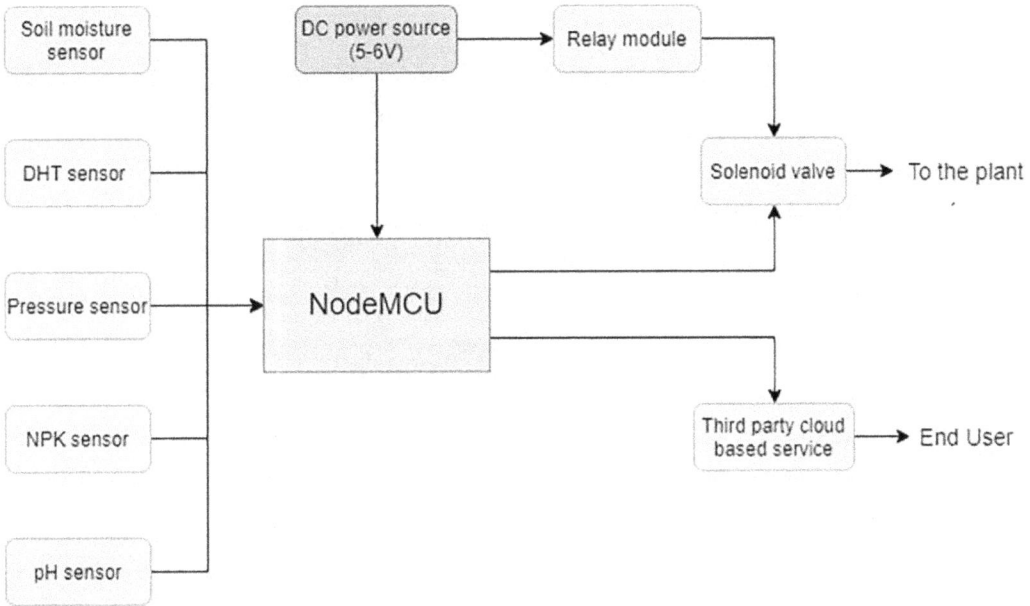

FIGURE 15.1
Block diagram of the proposed system.

Temperature and Humidity Sensor: The principle by which temperature and humidity sensors work is through changes in electrical currents due to changes in humidity and temperature in the air. A capacitive humidity sensor measures relative humidity. This is achieved by placing a thin strip of any metal oxide between two electrodes. The metal oxide's electrical capacity is altered with the atmosphere's relative humidity.

NPK Sensor: Used for detection of Nitrogen, Phosphorus, and Potassium nutrients of soil by using optical transducers. This optical transducer is implemented as a detection sensor consisting of three LEDs as a light source. The photodiode is used as a light detector. The wavelength of each of the LEDs is chosen to fit the absorption band of each compound.

pH Sensor: The working principle of this sensor depends upon the amount of exchanged ions from a sample solution to the inner solution, which is a pH 7 buffer.

Pressure Sensor: This sensor works by converting the pressure of the atmosphere to analog electrical sensors. The sensor is initially calibrated to a base value. Once a change in pressure is detected, an electric signal is generated. The magnitude of this signal helps in measuring the amount of change in the pressure.

Relay Module: Relays are basically electric switches that use electromagnetism to convert small electrical stimuli or pulses into larger electric currents. These conversions occur when electrical inputs activate an electromagnet inside the relay module to either form or break an existing electric circuit.

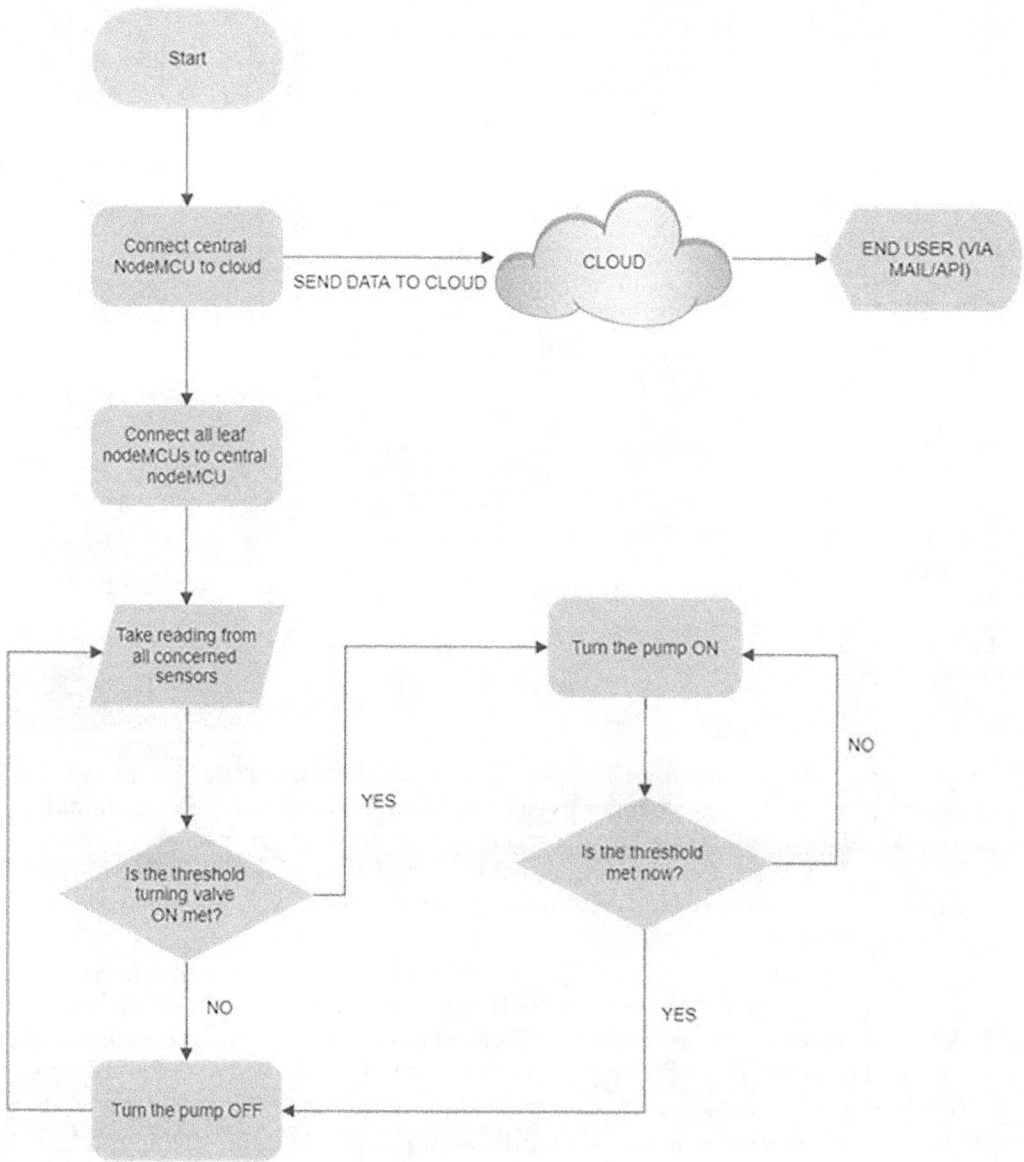

FIGURE 15.2
SPLARE working model.

Solenoid Valve: A solenoid valve is an electrically controlled valve. It is used to regulate the flow of water by applying a voltage over the coil. Due to the force on coil created by EMI, this coil is lifted up and the valve opens so that the liquid can flow through. Solenoid valves are also sometimes called electromagnetic valves.

Water Pump: The use of the water pump is to supply water to higher grounds in case the water source is at a comparatively lower level, as gravity would not help in such cases.

15.3.3 Virtual Simulation

Major concerns in the existing solutions were analyzed and worked upon to avoid the same. Before actually implementing hardware circuits, these were tested in a simulator (Proteus 8) and the overall design is shown in Figure 15.3. This was done as, even if there is a minor fault in the circuit, it can damage the microcontroller and other sensors, thus shorting them out. Initially, individual sensor circuits were built and tested. This includes the DHT (digital humidity and temperature) sensor, NPK (nitrogen, phosphorous, and potassium) sensor, pH sensor, soil moisture sensor, and pressure sensor (barometer based). Once all individual circuits worked as expected, a final circuit was built. This circuit consisted of all the sensors and other required components such as a solenoid valve and relay module. All the circuits which are a part of the entire setup have been developed and included below for reference.

There were certain changes that had to be made in the simulation circuit due to the unavailability of a few sensors and components in the software simulator used:

- Arduino Uno was used for simulation as NodeMCU is not supported by Proteus 8.
- Instead of actual NPK, pH, and soil moisture sensors, digital potentiometers were used for simulation. This was done due to the unavailability of support for these sensors in Proteus 8.

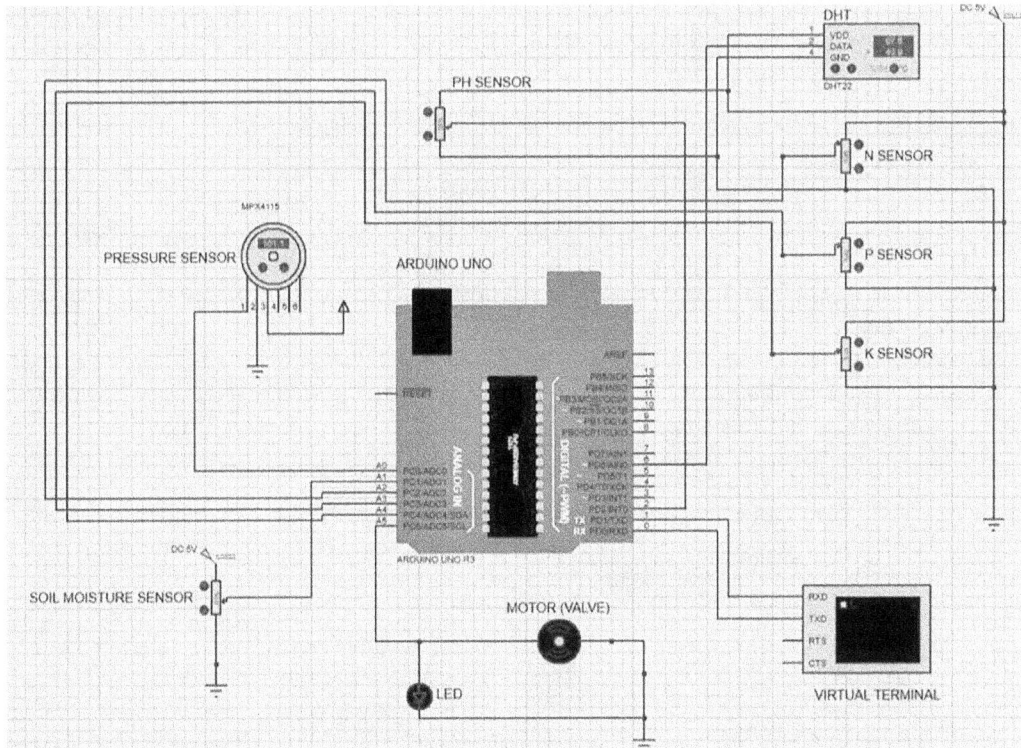

FIGURE 15.3
Proposed system circuit design.

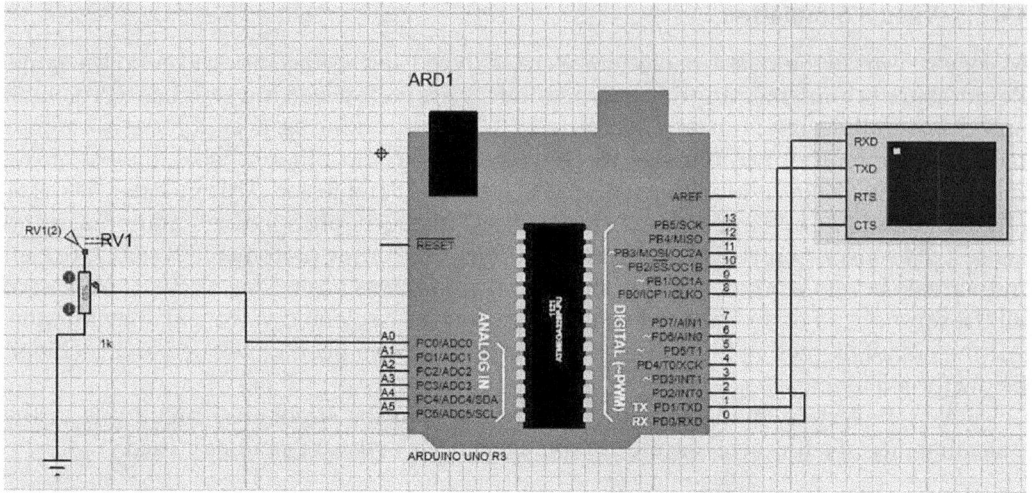

FIGURE 15.4
Soil moisture sensor virtual circuit.

- Instead of a solenoid valve, a DC motor was used in the simulation. This was done as Proteus 8 has no support for solenoid valves.

It was ensured that these changes had similar results as actual sensors and components.

Figure 15.4 presents a clear view of the connections and flow, i.e., virtual circuit of soil moisture sensor implemented using Proteus 8 software.

15.4 Hardware Implementation Procedures

The entire virtual simulation has to be replicated for the hardware implementation. As a sample, a smaller module with just soil moisture sensor and solenoid valve will be discussed in this section. This will help in making the rest of the implementation clear and easy to execute.

After collecting the equipment required, the circuit module has to be built (see Table 15.1). Soil moisture sensor is analog in nature. Hence, the data pin should be first connected to a potentiometer module and then to analog input of the microcontroller (A0 pin on NodeMCU). The NodeMCU (see Figure 15.5) has just one analog in pin, which might

TABLE 15.1

Connections to/from NodeMCU

Pin on NodeMCU	Connection
A0 (Pin: A0)	Input from soil moisture sensor
3V3	To Vin Pin on soil moisture sensor unit
GND	To GND Pin on soil moisture sensor unit
D1 (Pin: GPIO 5)	Control relay module
3V3	To Vin Pin on relay module
GND	To GND Pin on relay module

FIGURE 15.5
NodeMCU as microcontroller.

already be occupied by the soil moisture sensor (see Table 15.3). Hence, the alternative is to get a multiplexer that could serve as an extension to the analog in pin (for connection with other sensors).

Once the sensor is in place, the next step is to connect the solenoid valve to the NodeMCU, which will serve as a water outlet for plants. But as the node is capable of handling just 3.3V in output pins, a relay module is used to operate the solenoid valve, which requires an external power source of about 6V (see Table 15.2). After all the connections are set up,

TABLE 15.2

Connections to/from Relay Module

Pin on Relay Module	Connection
Vin (input side)	3V3 on NodeMCU
GND (input side)	GND on NodeMCU
Data Pin (input side)	D1 (Pin: GPIO 5) on NodeMCU
Neutral (output side)	+ve terminal of solenoid valve[a]
NO (normally open Pin, output side)	−ve terminal of solenoid valve[a]

[a] The terminals of the solenoid valve are not polarity sensitive. Hence these can be exchanged

TABLE 15.3

Connections to/from Soil Moisture Sensor Unit

Pin on Soil Moisture Sensor	Connection
Vin (potentiometer output side)	3V3 on NodeMCU
GND (potentiometer output side)	GND on NodeMCU
Neutral (potentiometer input side)	Any terminal of soil moisture sensor[b]
Positive (potentiometer input side)	Any terminal of soil moisture sensor[b]

[a] Make sure the connection is via analog input pin only (A0) and not digital input pin (D0).
[b] The terminals of the solenoid valve are not polarity sensitive. Hence these can be exchanged

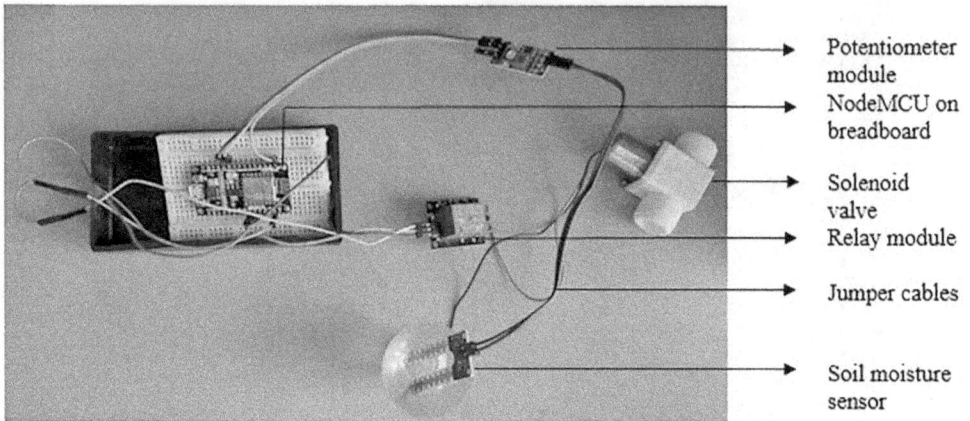

FIGURE 15.6
Soil moisture sensor–based hardware circuit.

verify the Arduino sketch and upload it to NodeMCU. The default baud rate is serialized at 9600. This has to be changed to 115200 to avoid garbage values and attain complete functionality of the system.

All the setup described was for one single module (see Figure 15.6). Figure 15.6 also contains all the components used in the setup and are labeled for reference. Eight such modules were implemented for eight different plants (see Figure 15.7). The proper working of the entire setup would guarantee that the system will work well for a mini-garden or a number of potted plants in an individual's house. The individual modules work

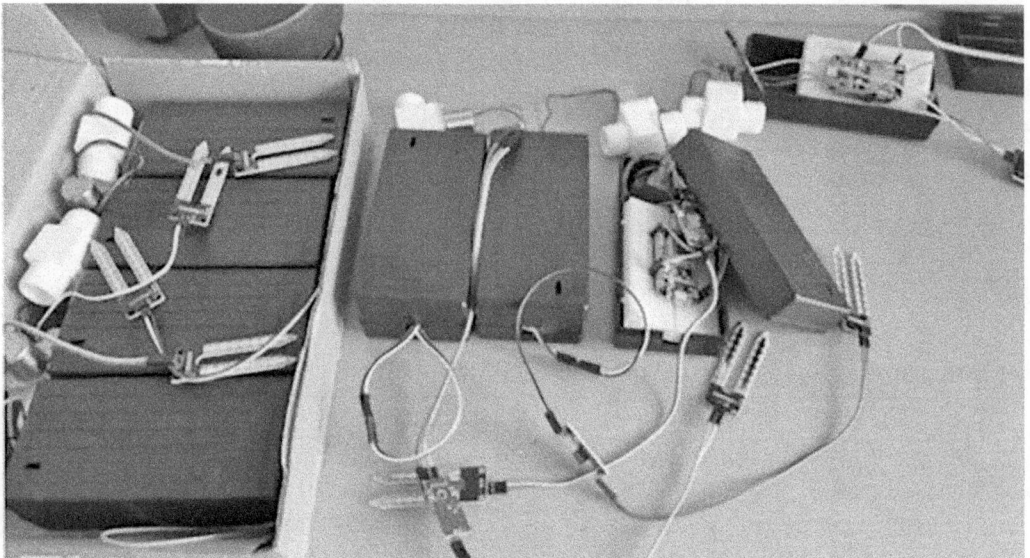

FIGURE 15.7
Replicated setup for eight different plants.

independently, hence, not affecting the working of other modules. Similarly, any damage in one particular module will not affect the others. The threshold was also dynamically set as per plant requirement.

Following are the tables that show all the connections in this system on NodeMCU, relay module, and soil moisture sensor unit.

Images of the working hardware for this smaller module are shown next in this section. These components were used to build a working model for detecting soil moisture and auto-watering different plants based on that. For the implementation of the entire hardware module, connections and other components shall be added to this smaller module itself.

As already mentioned, the soil moisture sensor is connected to the NodeMCU on the breadboard via the potentiometer. Similarly, the solenoid valve is connected via the relay module. This valve has two openings:

- First opening (from which water will enter) shall be connected to a water source with a pipe (preferably to a tap at a level higher than that of the plants if water pump or motor is not used, so that gravity assists the flow of water). However, in the case of a system containing multiple plants (like the one in this study), it should be connected to a central or branched pipe system, arranged using fittings/ adapters and various lengths of the pipe, which in turn is connected directly to the water source.
- The second (water outlet) is to water the plants using another piece of pipe.

By using individual setups for each plant, they will be watered based on their characteristic requirements. The opening or closing of one valve will not affect that of another.

15.4.1 Power Supply and Testing

One major challenge would be the power source. As USB power sources and batteries can't be used everywhere, an alternative was adopted for individual powering. The same has been explained here:

- AC to DC step-down converter was used to convert 240V AC to 5V DC.
- A voltage stabilizer and regulator were further attached to maintain max output potential difference at 5V and cap the current at 2A.

In this case, it would be possible to create a steady-state power source (i.e., directly from AC Mains supply), instead of relying on batteries and using cables. This would simplify the circuit as well as solve the issue of power running short.

NodeMCUs are highly electrostatic sensitive devices. Hence, just a sprinkle of water could damage it beyond repair. To overcome this problem, each module was sealed in a plastic enclosure to prevent water and moisture damage (see Figure. 15.8).

Once this was done in the above setup and there was a constant power supply, the only thing left to be done was to attach individual modules to plants and turn them on and see plants get perfect care without human intervention. Figures 15.9–15.11 show the testing of the setup. For this testing purpose, water in a container has been used.

The relay module has LED indicators (just 1 in this case). The default setting for relay is low logical output, indicating that the valve is closed and hence LED is not turned

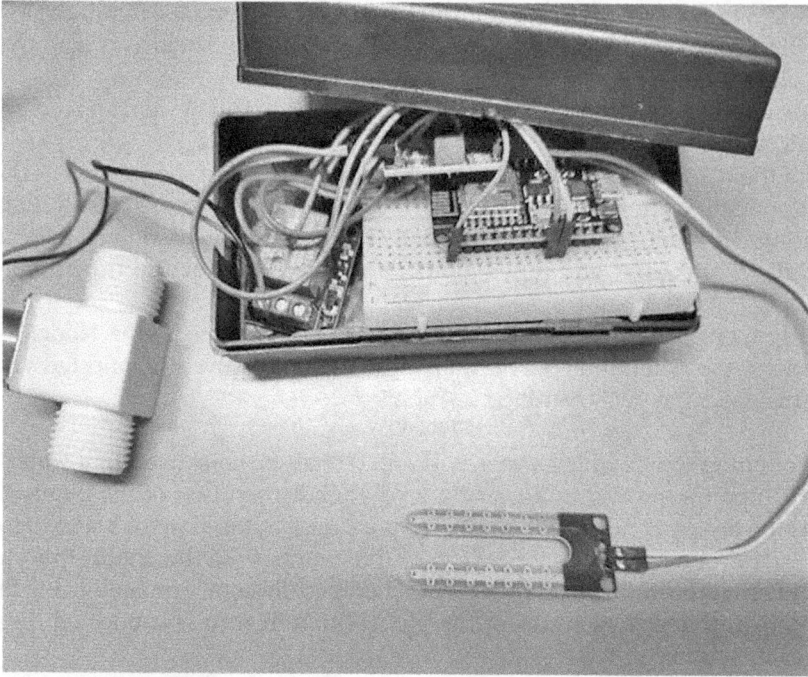

FIGURE 15.8
Setup inside each plastic encasing.

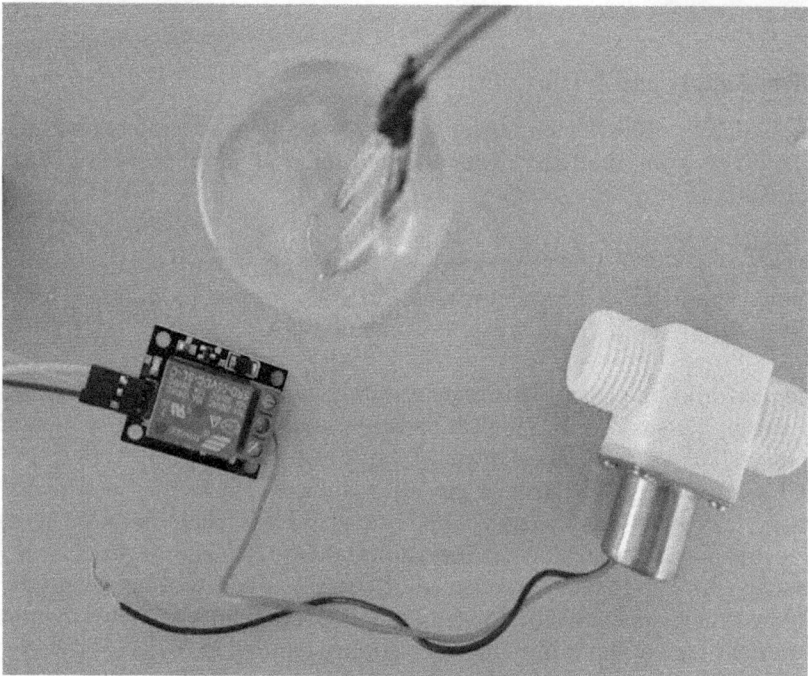

FIGURE 15.9
Valve closed (LED off).

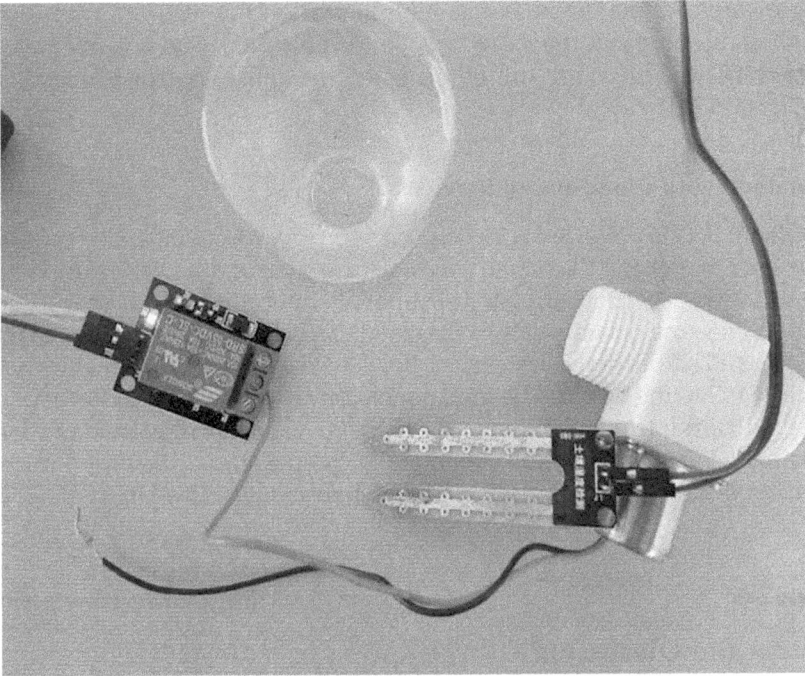

FIGURE 15.10
Valve open (LED on).

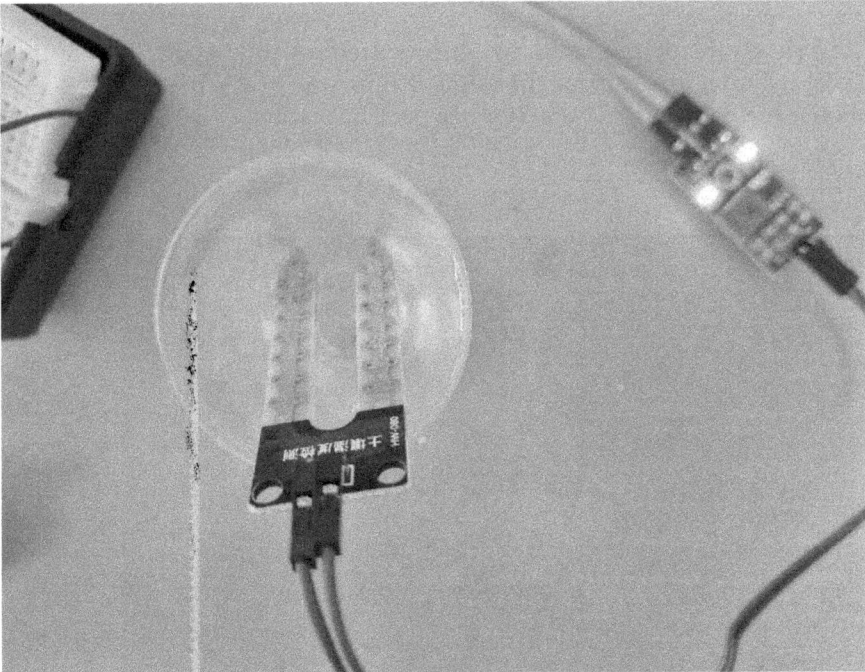

FIGURE 15.11
Working state of soil moisture sensor.

on (see Figure 15.9). When the relay has high logical output, this LED is turned on (see Figure 15.10), indicating that the valve is open (or powered). The potentiometer also has LED indicators. These LEDs are turned on when the analog sensor attached is working correctly (Figure 15.11).

15.4.2 Connecting to Cloud-Based Service

When the NodeMCU is uploaded with code and starts working independently, there is no way for the user to monitor the actual soil moisture level and if everything is working as expected. To tackle this problem, any third-party application which enables connecting IoT devices and writing data directly to the cloud can be used. This ensures that the actual moisture level and status of the system can be viewed by the user from anywhere around the world. The following can also be implemented by forming a local cluster and only the cluster head will send data to the cloud. This is more resource critical and efficient than all nodes sending data at the same time, but equally complex too.

One such application – ThingSpeak, was used in this scope (see Figure 15.15).

15.5 Results and Discussion

The screenshots for the simulations conducted and corresponding outputs obtained are given below. Figure 15.12 shows the output of the proposed system design (see Figure 15.3). Table 15.4 shows the output of each sensor in the proposed system design (results directly obtained from software simulation). This includes pH sensor, soil moisture sensor, NPK sensor, pressure sensor, and DHT sensor.

Table 15.4 shows the output of all the sensors used for simulation. Each of these readings is recorded after a particular time unit (1 time unit = 10 s here, hence total time of observation = 10×6 seconds = 1 min) for one particular sample.

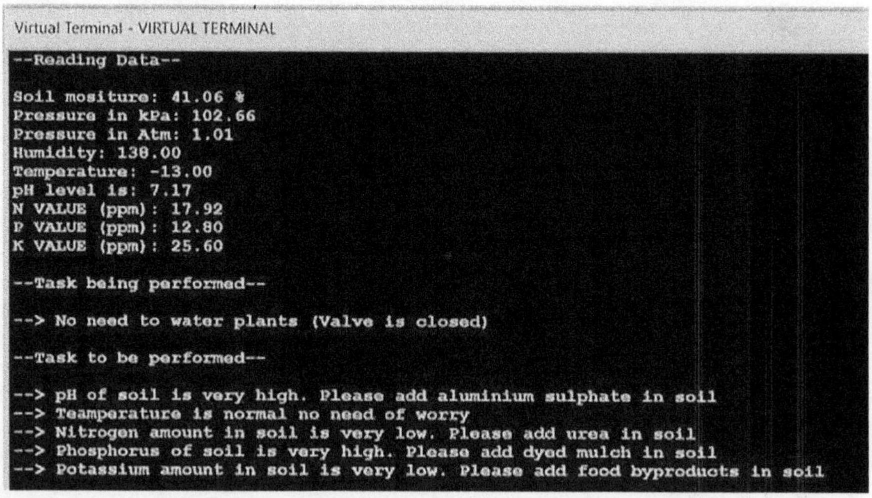

```
Virtual Terminal - VIRTUAL TERMINAL
--Reading Data--

Soil mositure: 41.06 %
Pressure in kPa: 102.66
Pressure in Atm: 1.01
Humidity: 138.00
Temperature: -13.00
pH level is: 7.17
N VALUE (ppm): 17.92
P VALUE (ppm): 12.80
K VALUE (ppm): 25.60

--Task being performed--

--> No need to water plants (Valve is closed)

--Task to be performed--

--> pH of soil is very high. Please add aluminium sulphate in soil
--> Teamperature is normal no need of worry
--> Nitrogen amount in soil is very low. Please add urea in soil
--> Phosphorus of soil is very high. Please add dyed mulch in soil
--> Potassium amount in soil is very low. Please add food byproducts in soil
```

FIGURE 15.12
Virtual terminal output of the final circuit.

TABLE 15.4

Output of All the Sensors Used for Simulation

Time Unit	pH Sensor Output	Soil Moisture Sensor Output	NPK Sensor Output (N, P, K)	Pressure Sensor Output (KPa, Atm)	DHT Sensor Output (humidity %, temperature)
1	7.00	594.00	676, 502, 604	102.55, 1.01	57.0, 24.00
2	7.00	594.00	676, 481, 604	102.66, 1.01	60.0, 25.20
3	7.00	604.00	674, 472, 626	102.88, 1.02	62.0, 26.30
4	7.00	666.00	646, 355, 475	102.66, 1.01	57.0, 24.50
5	7.00	686.00	676, 471, 600	102.88, 1.02	60.0, 20.00
6	7.00	669.00	670, 460, 600	102.50, 1.01	65.0, 22.00

As already established that soil moisture sensor is analog in nature, the output varies for each sensor and has to be calibrated as per requirement. In this case, the range for soil moisture sensors varies from 0 to 1024. The same applies to NPK sensors as well and they have to be calibrated before use. Pressure sensors, pH sensors, and DHT sensors usually do not require any further calibration and can be directly used.

In Figures 15.13 and 15.14, the virtual terminal of the actually implemented circuit can be seen (output of the hardware implementation explained above). In Figure 15.13, the status of the valve is closed as soil moisture level is > 40%, whereas, in Figure 15.14, it is open as soil moisture level is <40%.

Finally, Figure 15.15 shows the dashboard of the third-party cloud-based service (ThingSpeak here). This API is set up for four different plants (to show a sample). Each of them uses a separate channel. The readings from the soil moisture sensor and the status

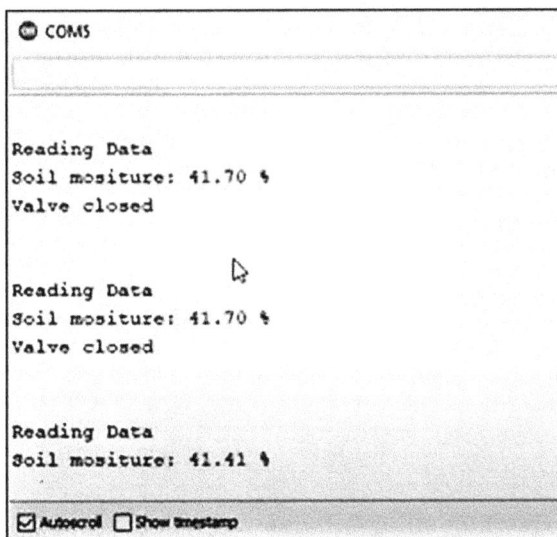

FIGURE 15.13
Serial monitor (valve closed).

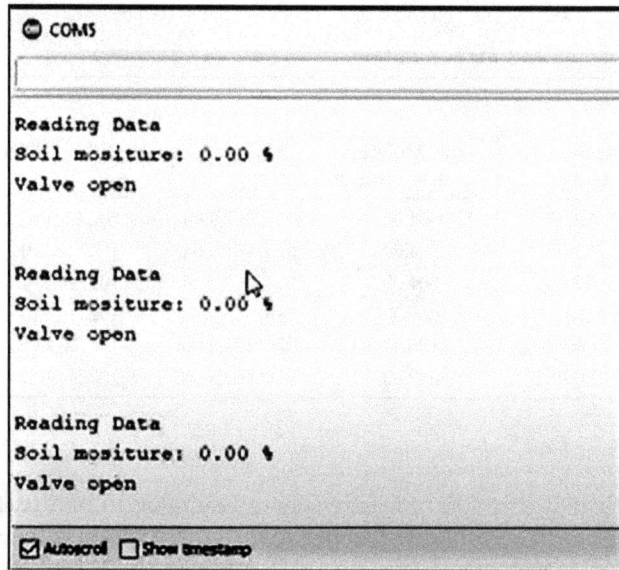

FIGURE 15.14
Serial monitor output (valve open).

of the solenoid valve will be updated here once every few minutes (can be customized) and can be viewed from any device with internet connectivity and authorization.

15.5.1 Analysis

After getting all the required levels from respective sensors, we can come up with the following inference and analysis.

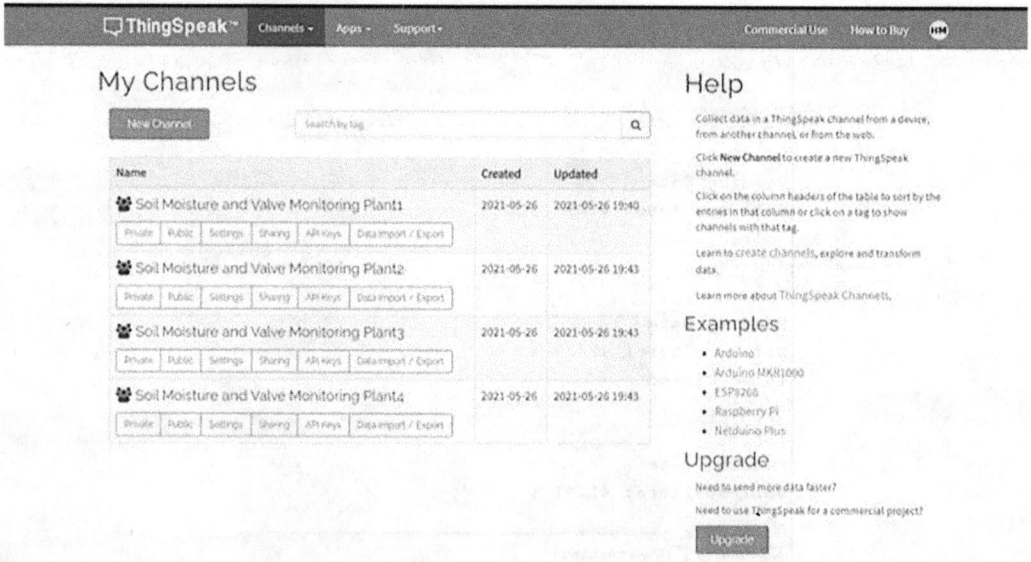

FIGURE 15.15
Dashboard of ThingSpeak.

TABLE 15.5

Results Obtained from Plant One

Soil Moisture Reading	Pressure in kPa	Pressure in Atm	Humidity %	Temperature	pH Level	N Value (ppm)	P Value (ppm)	K Value (ppm)	Valve Status
39.06%	102.66	1.01	63.8	26	7.17	17.92	12.8	25.6	On

Table 15.5 depicts a set of values obtained from the first plant.

- The soil moisture sensor reads 39.06%. So as per the thresholds set (which is 40% in this case), the solenoid valve should be turned on (as was observed).
- Humidity and temperature also play a slight role in determining whether the valve should be turned on or not. For example, if the temperature reading is too high or it crosses a certain threshold, then the valve should be turned on (considering a faster rate of evaporation).

Considering these two points, we can conclude that a combination of temperature reading and soil moisture reading should be used to determine the status of the valve.

Now, the other parameters will be used in determining the general health of the plant and what action needs to be taken.

- In this case, the pH of the soil was very high. Hence, to neutralize it, a suggestion that can be provided is: add aluminum phosphate to the soil.
- The N valve (i.e., the nitrogen content of the soil) was found out to be very low. Hence, a suggestion that can be provided is: add urea to the soil.
- The P valve (i.e., the phosphorus content of the soil) was found out to be very high. Hence, a suggestion that can be provided is: add dyed mulch to the soil.
- The K valve (i.e., the potassium content of the soil) was found out to be very low. Hence, a suggestion that can be provided is: add food byproducts to the soil.

These suggestions can directly be sent to the user via an API or a cloud-based platform that supports IoT (like ThingSpeak or Blynk), thus notifying the user if an action needs to be taken for a particular plant.

Consider another set of values for analysis, recorded from a different plant at the same time (as the first case) and the same is listed in Table 15.6.

Here, the current soil moisture level is at 46%, and hence the valve status is off (which was expected as the threshold was set to 40%). Other parameter readings such as pH valve, NPK values, and humidity and temperature were found out to be similar to the first set of readings. Hence, we can expect a similar kind of inference for the same. Also, the actions required and suggestions would be similar to that of the initial set of readings.

TABLE 15.6

Results Obtained from Plant Two

Soil Moisture Reading	Pressure in KPa	Pressure in atm	Humidity %	Temperature	pH Level	N Value (ppm)	P Value (ppm)	K Value (ppm)	Valve Status
46%	102.69	1.03	63.5	27	7.2	18.12	13.1	26.05	Off

15.6 Conclusion

Watering the plant and delivering information about the surrounding environment and soil nutrients could be accomplished by an autonomous watering system coupled with IoT platforms. A microcontroller (NodeMCU) is used to evaluate and convert readings from sensors into human-readable form. The entire system is programmed in such a way that the process of taking input of few parameters and deciding whether to start the valve or not is completely automated. With the use of NPK sensors, all essential soil radicals can be measured. Hence, the user has accurate real-time data of which fertilizer is currently required by the plant or if any other action is needed to maintain a healthy growing environment. This also solves the problem of adding more than required fertilizers which can hinder plant growth. In a similar way, the data from the pH sensor can be used to add an appropriate neutralizing agent if the pH of the soil is either too high or too low.

With slight modification, the current idea can also be exploited to determine whether the current soil type is suitable for a particular plant or not. To further enhance the system, small units consisting of fertilizers can be attached. With the help of this addition, the user needn't even worry about checking the nutrient content of soil and adding required fertilizers as the system will be able to do it as and when required.

References

1. S. P. Raja, T. Sampradeepraj, N. N. Prasanth, "Performance Evaluation on Internet of Things Protocols to Identify a Suitable One for Millions of Tiny Internet Nodes," International Journal of Intelligent Internet of Things Computing, Vol 1 No 1, pp. 222, 2019. Doi: 10.1504/IJIITC.2019.104733.
2. V. K. Akram M. Challenger, "A Smart Home Agriculture System Based on Internet of Things," 2021 10th Mediterranean Conference on Embedded Computing (MECO), 2021, pp. 1–4, Doi: 10.1109/MECO52532.2021.9460276.
3. Mubashir Ali, Nosheen Kanwal, Aamir Hussain, Fouzia Samiullah, Aqsa Iftikhar, Mehreen Qamar, "IoT Based Smart Garden Monitoring System Using NodeMCU Microcontroller," International Journal of Advanced and Applied Sciences, Vol 7, pp. 117–124, 2020.
4. Mohanad Ali Meteab Al-Obaidi, Muna Abdul Hussain Radhi, Rasha Shaker Ibrahim, Tole Sutikno, "Technique Smart Control Soil Moisture System to Watering Plant Based on IoT with Arduino Uno," Journal of Bulletin of Electrical Engineering and Informatics, Vol. 9, pp. 2038–2044, April 2020.
5. Hajar M. Yasin, Subhi R. M. Zeebaree, Ibrahim M. I. Zebari, "Arduino Based Automatic Irrigation System: Monitoring and SMS Controlling," Journal of 4th Scientific International Conference – Najaf – IRAQ (4th SICN-2019), Vol 4, pp. 109–114, 2019.
6. M. Mediawan, M. Yusro, J. Bintoro, "Automatic Watering System in Plant House – Using Arduino," Journal of 3rd Annual Applied Science and Engineering Conference (AASEC 2018), Vol 1, pp. 1–8, 2018.
7. G. Nandha Kumar, G. Nishanth, E. S. Praveen Kumar, B. Archana, "Aurdino Based Automatic Plant Watering System with Internet of Things," Journal of International Journal of Advanced Research in Electrical, Electronics and Instrumentation Engineering, Vol 6, pp. 2012–2019, March 2017.
8. Akriti Jain, Abizer Saify, Vandana Kate, "Prediction of Nutrients (N, P, K) in Soil Using Color Sensor (TCS3200)," Journal of International Journal of Innovative Technology and Exploring Engineering (IJITEE), Vol 9, pp. 1768–1771, January 2020.

9. R. Sindhuja, B. Krithiga, "Soil Nutrient Identification Using Arduino," Journal of Asian Journal of Applied Science and Technology (AJAST), Vol 1, pp. 40–42, May 2017.
10. Jacquline M.S. Waworundeng, Novian Chandra Suseno, Roberth Ricky Y Manaha, "Automatic Watering System for Plants with IoT Monitoring and Notification," Journal of Cogito Smart Journal, Vol 4, pp. 316–326, December 2018.
11. J. Dileep, V. Dhruthi, H. Nanditha, V. Deepthi, "Automatic Plant Watering System Using Arduino," International Research Journal of Engineering and Technology (IRJET), Vol 7, pp. 4142–4147, July 2020.
12. Arathi Reghukumar, Vaidehi Vijayakumar, "Smart Plant Watering System with Cloud Analysis and Plant Health Prediction," 2nd International Conference on Recent Trends in Advanced Computing ICRTAC – DISRUP – TIV INNOVATION, Vol 165, pp. 126–135, 2019.

Index

A

Accuracy 48–52
Activation layer 46
Adam optimizer 50
Advanced Message Queuing Protocol (AMQP) 201–202, 207
Aeroponics 25
Aeroponics farming 224, 234
Agricultural ontology 149
Agricultural science 134
Agricultural system 148
Agriculture 1, 2, 7–13, 75–81, 86, 87, 89–103, 109–116, 121–124, 126, 131, 224–226, 233, 239
Agro-Intelli algorithm 125–127, 129, 131
Agro-IoT 188–189, 191, 193, 195, 197–200, 207
Aircraft 76, 78
Algorithm 92–100
Aquaponics 27
Arduino board 3–6, 9, 12, 164, 180, 249
Artificial Intelligence (AI) 44, 112, 168
Auditory command 68
Automatic crop monitoring 149
Autonomous watering 245

B

Bag-of-words 128
Band 83
Behavior monitoring 57, 62
Big data 44, 79, 86, 90, 148, 149, 154
BLDC motors 171, 172, 174, 177
Bluetooth (BT) 5, 6, 7, 191, 195, 197–198, 207
Body temperature sensors 59
Brand 81
Building Structure Based on Plants 32

C

Category of models 28
Cellular structure 82
Chlorophyll 77
Circuit design 250, 252
Classification 79, 83–85, 92, 95, 96, 98–101, 103
Climate 86
Climate change(s) 102, 228
Climate prediction 150–152

Cloud 122, 124–126, 128, 129, 131, 137, 141, 149, 256
Cloud computing layer 57, 64, 90, 91
Cluster/Clustering 92, 96, 98, 99, 103, 256
Communication issues 211
Communication layer 57, 63
Communication protocols 197, 207
Confusion matrix 51
Constrained Application Protocol (CoAP) 200–202, 207
Continual flow system 28
Convolutional Neural Networks (CNNs) 44, 46
Crop 92–94, 96, 97, 99–103, 224, 225, 233, 237
Crop cultivation 136, 137, 141, 144
Crop damage assessment 164
Crop monitoring 166
Crop prediction 149, 151, 157
Crop yield 122, 124
Crop yield prediction 136, 137, 142, 144

D

Data 110, 112–116
Data collection 80, 83
Data Distribution Service (DDS) 199–200, 202, 207
Data management issues 217
Data mining 134, 137
Data security 124, 217
DC motor 250
Decision model 135
Decision tree 135, 141, 144
Deep learning 45
Detection 93, 94, 97–101
Device 76, 78, 79, 112, 113, 116, 117
Dimensionality vector 126, 128
Diseases 89, 93, 94, 96–98, 99, 101–103
Disruption of farming through the IoT 37
Distributed ledger 63
Drip systems 31

E

Ebb and flow system 30
Edge computing 63
Energy 112–114, 116

Energy consumption 218
Ensemble classification 137, 141, 143
Ensemble learning (EL) 97
Environment 224, 226, 228, 232, 235, 239
Environment monitoring 61
Error rate 50
ESP8266 246
Estimation 77, 80, 83, 85

F

Failure issues 215
Farm at hand 123
Farm OS 123
Farmer 134, 137, 140–142
FarmRexx 123
Fault anticipation systems 217
Feature map 46
Fertility 76
Fertilization 113, 115
Fixed solution system 28
Flat fan 175, 176, 178
Flatten layer 47
Flight controller 166, 171, 172, 173
Flooding fan 175, 176, 178
Flow rate 81
Frequency shift keying 11
Full cone 175, 176, 178
Future of Urban Farming 35
Fuzzification 149, 151, 152
Fuzzy Associate Memory (FAM) 151
Fuzzy-based crop prediction 149, 151

G

Gabor features 205–207
Gabor filter 204
Gas sensors 62
Gateway 92, 94
General Packet Radio Frequency (GPRS) 6–9, 11
Geo fencing 68
Geographical Information System (GIS) 75–77,
    113, 116
Global Positioning System (GPS) 8, 75, 76, 79, 81,
    113, 116
Global System for Mobile communication
    (GSM) 5, 7–9, 11, 13
Gray Level Co-occurrence Matrix (GLCM)
    204–206
Green Blue Red (GBR) 77
Greenhouse gas (GHG) 94, 97, 102, 103
Gross domestic product 164

H

Habitats 83, 84
Hardware 124
Heterogeneous resources 148, 154
Hollow cone 175, 176, 178
HTML 154
Human–wildlife conflict 168
Humidity 259
Humidity control 226
Humidity rate 126
Humidity sensor 149, 151, 154
Hydroponics 25
Hydroponics farming 224, 233
Hyperspectral 116, 119

I

Image 90, 92, 97, 99, 100, 101, 102
Information 92, 97, 101, 102, 113, 116, 117
Information gain 141
Information security 87
Infrastructure 124
Integration 230, 239
Intelligent rules 137, 140, 144
Intelligent virtual assistant 65
Intelli-Group Algorithm 128, 129
Internet of Things 2, 3, 7, 11, 44, 86, 87, 98, 109,
    110, 113, 115, 117, 124–126, 128, 131, 148,
    164, 188–190, 194, 197, 199–200, 202–203,
    206–207, 229, 230, 232–237, 239
IoT for smart agriculture systems 212
IPS 165, 176, 179, 184
Irrigation 164, 166, 167, 179, 182, 243–245
ISM band 190

K

Kaggle 49, 52

L

Learning rate 49, 50
Light Dependent Resistor (LDR) 4
Line of Sight (LOS) 10
Livestock 55
Location 113
Location tracking 65
Long Range (LoRa) 6, 9–11, 13, 212
LoRaWAN 194–195, 197–198, 207
Low-power wide-area network (LPWAN) 9
LP-WAN 213

M

Machine learning (ML) 89–102, 122
Management 92–98, 100–102
Map 75, 76, 80, 81, 83–86
Market factor 126
Market factor weight 128
Matthew Correlation Coefficient (MCC) 157, 160, 161
Message Queuing Telemetry Transport (MQTT) 197, 199–200, 202, 206–207
Method of selection 29
Mi-Trace 123
Micro services 64
Microwave radiation 232
Minimum shift Keying (MSK) 11
Mixed skyscraper buildings 25
Mobile computing 149
Monitoring 75, 78, 80, 81, 85, 86, 122, 123, 124
Multilayer feed-forward neural network 46
Multispectral 77, 83, 115, 116, 118

N

Natural language processing 154
NB-IoT 213
NDVI 167, 180
Near InfraRed (NIR) 77
Networking 124
Neural networks 93, 94, 101
Node 112
NodeMCU 246, 250, 251
Non-linearity 46
Normalized Difference Vegetation Index (NDVI) 77, 83, 84
NPK Sensors 137, 142
Nutrient concentration 234
Nutrient Film Technique Systems 32

O

Object-based 83
Objects 91, 95
Ontology 148, 149, 154–156, 160
Optimization 47

P

pH sensor 137, 142
Phase change materials 227
Photovoltaic 232
Pixel-based 83

Plant 259
    humidity 247, 259
    pH 245, 247, 259
    temperature 247, 259
    valve 247, 259
Plant growth 231
Plant structure 124, 127, 130
Plants 89, 97–99, 101, 102
Poly bag farming 224, 231, 232
Polyhouse farming 224, 226–228, 239
Precision 44, 50
Precision agriculture (PA) 75, 79, 80, 109, 110, 117, 133, 137
Precision farming 149
Precision livestock farming (PLF) 56
Predictions 92, 102
Predictive fault-tolerant systems 215
Propellers 171, 172, 174

Q

Quality 80, 89, 92, 94, 96, 98, 101, 103
Quantity 81, 83
Quickbird images 79

R

Radio frequency identification (RFID) 65, 195–198, 207
Raspberry Pi 4, 5, 12, 33
Received signal strength indicator (RSSI) 7
Receiver 171, 173, 174, 180
Recommendation system 136, 142
Rectified Linear Unit (ReLU) 47
Region-based Convolutional Neural Networks (RCNN) 101
Regression 92–95, 97, 100, 103
Relay 247, 251
Reliability 124, 218
Remote sensing (RS) 75, 115, 116
Renewable energy 223, 226, 229, 230, 231, 234, 239
Resnet 101
Rumination sensor 61

S

Scalability 217
Security issues 218
Security protocols 203
Seed pod planting 166
Segmentation 84

Semantic association 128
Semantic extraction 148
Semantic knowledgebase 149
Semantic retrieval 155
Semantic weight 128
Sensing layer 57
Sensitivity 50
Sensor 78, 79, 81, 86, 87, 109, 110, 112–115, 188–197,
    199, 206–207, 224–228, 231, 232, 234, 237,
    239, 245, 246, 257, 258
Service layer 57, 64
Shade net farming 224, 228–231, 239
Shallow learning 45
Short-wave Infrared (SWIR) 77
SigFox 6, 214
Sigmoid 47
Skyscraper buildings 24
Smart 89, 91, 93, 94, 96, 100–102
Smart agriculture system 211
Smart farming 43, 45, 47, 48, 50, 52, 122, 123
Smart feeding 69
SmartFarmNet 123
Software-as-a-Service (SaaS) 45
Soil 89, 91–93, 96, 97, 100, 101
Soil analysis for field planning 165
Soil fertility rate 127
Soil moisture 133, 136–140, 142
Soil moisture sensor 246, 250, 251, 259
Soil monitoring 225
Soil nutrient 136, 137, 141, 142, 144
Solenoid valve 248
Spatial data 76–79, 81, 83, 86
Spectral 77–79, 82, 83, 116
SPLARE 243, 244
    architecture 247
    circuit design 249
    working model 248
Sprinkling-based farming 224, 236
Stacked shipping containers 25
Standards 188–191, 193, 195, 197, 207
Sustainable technology 75
SVM 133, 135, 136, 141, 144

T

Tanh 47
TCP 201, 203, 207
Temperature 114, 118, 134, 135, 137, 138, 141, 142,
    144
Temperature sensors 149, 155
Testing 253
Tokenize the data 126
Training data 125

Transmission 76, 79, 112, 116
Transmission layer 113
Transmission power control 10
Trimble 123

U

Ultrasonic sensor 68
Unmanned Aerial Vehicles (UAVs) 87, 115
UDP 201–203, 207

V

Value iteration (VI) 83
Variable area nozzle sprayer 163, 175
Variable area nozzles 165
Vegetation 75, 77, 81–83, 94, 85, 115
Vehicles 115
Vertical Farming 24

W

Waste management 148
Water 91–94, 96, 97, 100–102
Water culture system 29
Water management 148
Water pump 248
Water supply 244, 245
Wavelength 82
Weather 78–80, 113, 117
Weather weight 127
Web crawling 154
Weed 97–102
Wi-Fi 193, 195, 197–198, 207
Wi-Fi HaLow 193
Wi-Fi standards 193
WiMAX 193–195, 197, 198, 207
Wireless sensor actuating network 11
Wireless sensor network (WSN) 5, 7, 9, 11, 65,
    92, 148
World Health Organization 169
Worldwide Interoperability for Microwave
    Access (WiMAX) 10

Y

Yield 75, 77, 79–83, 85–87, 110, 114, 115

Z

Zero-centered function 47
ZigBee 11, 117, 191–192, 195, 197–198, 207
Zone 82

For Product Safety Concerns and Information please contact our EU
representative GPSR@taylorandfrancis.com
Taylor & Francis Verlag GmbH, Kaufingerstraße 24, 80331 München, Germany

www.ingramcontent.com/pod-product-compliance
Lightning Source LLC
Chambersburg PA
CBHW082110220326
41598CB00066BA/5940

9 781032 028309